Feminist Care Ethics Confronts Mainstream Philosophy

Feminist Care Ethics Confronts Mainstream Philosophy

Editors

Maurice Hamington
Maggie FitzGerald

MDPI • Basel • Beijing • Wuhan • Barcelona • Belgrade • Manchester • Tokyo • Cluj • Tianjin

Editors
Maurice Hamington
Department of Philosophy
Portland State University
Portland
United States

Maggie FitzGerald
Department of Political
Studies
University of Saskatchewan
Saskatoon
Canada

Editorial Office
MDPI
St. Alban-Anlage 66
4052 Basel, Switzerland

This is a reprint of articles from the Special Issue published online in the open access journal *Philosophies* (ISSN 2409-9287) (available at: www.mdpi.com/journal/philosophies/special_issues/ Feminist_Care_Ethics).

For citation purposes, cite each article independently as indicated on the article page online and as indicated below:

LastName, A.A.; LastName, B.B.; LastName, C.C. Article Title. *Journal Name* **Year**, *Volume Number*, Page Range.

ISBN 978-3-0365-5330-6 (Hbk)
ISBN 978-3-0365-5329-0 (PDF)

Contents

 *philosophies*

Editorial

Feminist Care Ethics Confronts Mainstream Philosophy

Maurice Hamington [1] and **Maggie FitzGerald** [2],*

1   Department of Philosophy, Portland State University, Portland, OR 97201, USA
2   Department of Political Studies, University of Saskatchewan, Saskatoon, SK S7N 5A5, Canada
*   Correspondence: maggie.fitzgerald@usask.ca

Citation: Hamington, M.; FitzGerald, M. Feminist Care Ethics Confronts Mainstream Philosophy. *Philosophies* **2022**, 7, 91. https://doi.org/10.3390/philosophies7050091

Received: 9 May 2022
Accepted: 18 August 2022
Published: 23 August 2022

Publisher's Note: MDPI stays neutral with regard to jurisdictional claims in published maps and institutional affiliations.

1. Care and the History of Philosophy

The central role of care in human history is beyond questioning. We are vulnerable and needy creatures who require the goodwill and compassionate actions of others to survive. Yet, in the lengthy history of the "queen of the sciences", philosophy, care has only received sporadic attention. Today, mainly to the credit of feminist philosophers, care ethics has matured and received substantial consideration. This Special Issue of *Philosophies* places care theory in dialogue with modern philosophers and philosophical traditions to further that maturation. This volume aims to activate additional exchanges of ideas between care theorists and mainstream philosophical traditions.

The inconsistency of valuing care in philosophy does not mean that there are no historical threads to be found; they simply lack a continuous tradition. Through the voice of Socrates, Plato describes care, or "proper attention", as occurring when something is improved [1] (128b). In works such as *The Fables of Hyginus*, Ancient Romans advanced a Myth of Care around the Goddess Cura, whose name means care. The Goddess both creates humanity and must care for it. Literary scholar Halver Hanisch finds one of the points of the Fables of Hyginus being "the ontological role of care in human life" [2]. Soren Kierkegaard addresses care in terms of "concern" to discuss existential reality without the categorical and abstract methods found in the philosophical tradition [3]. David Hume's emphasis on emotion, particularly sympathy over rationalism, has caused some to identify him as a proto-care philosopher [4]. Similarly, in the American pragmatist tradition, figures such as John Dewey and Jane Addams emphasize sympathetic understanding, which resonates with care theory [5]. Western philosophy is not alone in having a care tradition. Several scholars have found resonance between care and Confucianism [6]. Furthermore, many Indigenous philosophies have manifested strong ties to care [7]. Many of these non-Western traditions far predate Western philosophy and there is much to be humbly learned from them.

2. Contemporary Feminist Care Ethics

Modern care ethics traces back to the field of developmental moral psychology. Before the publication of Carol Gilligan's *In a Different Voice* [8], the dominant understanding of moral development in the field of psychology was Lawrence Kohlberg's [9] Kantian-influenced model. Gilligan questions Kohlberg's framework and its masculinist biases towards individualism, abstraction, and rationalism. Noting that morality and moral thinking are conceived of in a very particular way within Kohlberg's model, Gilligan listens for—and hears—alternative conceptions of morality in the answers of her women participants. From these different voices, Gilligan identifies an approach to ethical deliberation that differs from the theoretical–juridical model [10], an approach which she calls an ethic of care. Rationalist approaches to justice conceive of morality as individualist. An ethic of care begins with the notion that relationships are primary and foregrounds the moral salience of connection, activities of care, and the fulfillment of responsibilities "based on a bond of attachment, rather than a contract of agreement" [8] (p. 57). Care theory thus eschews the idea that morals are principles to be accessed individually through rational thought. Instead, morality consists of the activities and practices of care and relationship maintenance that sustain webs of connections and minimize harm as best as possible.

Since Gilligan's trailblazing work, the ethics of care has ballooned into a fully interdisciplinary research agenda [11] with both theoretical and empirical [12] work on the ethics of care covering a variety of topics, issues, and avenues of inquiry. A related body of work on care practices and care labour in various contexts (see, for example, [13–16]) has also grown alongside the ethics of care literature. Care ethics, as a field, has sought to grapple with, understand, and uncover the diverse and contextually specific epistemological resources that emerge from the central role of care in human moral forms of life.

Of particular concern for this Special Issue, the thoroughgoing critique of rationalism—while important in creating space for a multiplicity of moral voices to be seen and heard as moral voices—also positioned (or seemed to position) care ethics in an adversarial relationship with a variety of dominant theories of justice and mainstream philosophical thought. The relational ontology underpinning a care ethics approach appears at odds with liberal philosophical standpoints, which presume an individualist subject as independent, atomistic, and self-sufficient (see, for example, [17,18]). Care ethics also locates the resources for making, interrogating, and validating knowledge claims in language, symbolisms, and practices that relationally tie us together. Accordingly, knowledge (particularly ethical knowledge) is situated, contextual, and emerges from specific social locations (see, for example, [19–21]). This epistemic position contrasts with theories that valorize objective and abstracted thinking and knowledge claims.

The contributors in this volume illuminate various methods to move beyond the boundary that has mitigated care ethics and care theory engagement with mainstream philosophical discourse. Some scholars demonstrate how the ontological and epistemological premises of care ethics, in fact, aligns with other philosophical theories, including the work of G.W.F. Hegel, Slavoj Žižek, Simone de Beauvoir, Edith Stein, Ludwig Wittgenstein, and Hannah Arendt. In establishing such shared ground, these authors also point to fruitful avenues for sustained dialogue that can mutually expand care and mainstream philosophy. Others take a more pragmatic approach, bracketing such meta-theoretical tensions and fruitfully demonstrating how care can expand and be expanded by the thought of notable figures like Bruno Latour, Frantz Fanon, John Locke, Jacques Rancière, and Jacques Derrida. Taken together, and in these different ways, these articles catalyze further intellectual interest and attention to how care enriches philosophy across various subjects.

3. Articles in This Issue

In *'Care Ethics, Bruno Latour, and the Anthropocene'*, Michael Flower and Maurice Hamington engage the relational morality, ontology, and epistemology of feminist care ethics with Bruno Latour's actor-network theory [22]. Specifically, these authors focus on fostering dialogue between Latour's recent publications, which focus on the new climate regime of the Anthropocene, and the emerging literature on posthuman approaches to care. In so doing, Flower and Hamington demonstrate that Latourian analysis, in conjunction with a care ethical framework, reinforces the notion that centering and valuing relationality across humans and non-human matter is essential to confronting the Anthropocene, which is perhaps the dire moral challenge of our time.

Thomas Randall's *'A Care Ethical Engagement with John Locke on Toleration'* draws upon care theory and John Locke's corpus on toleration to put forth a novel theory of toleration as care [23]. Arguing that Locke's thought and care ethics can converge by foregrounding the significance of an ethos of trustworthiness and civility, Randall further illustrates that such a convergence has the potential to enable toleration, particularly in contemporary pluralistic societies. This care-ethical toleration, Randall concludes, holds the potential to provide meaningful solutions to moral disagreement within such pluralistic societies—solutions that move us beyond the capacity of the liberal state.

Catherine Chaberty and Christine Noel Lemaitre's article, *'Thinking About the Institutionalization of Care with Hannah Arendt: A Nonsense Filiation?'*, examines the controversial partnering of Hannah Arendt's writings and an ethic of care [24]. Noting that feminist interpretations of Hannah Arendt are particularly contested, especially given her vital distinction between the private and public spheres, these authors maintain that Arendt's concepts are relevant to feminist and political ethics of care. To support this claim, Chaberty and Lemaitre skillfully show that the most recent analyses developed on the politics of care, especially those concerned with institutionalizing care, are shaped by Arendt's concepts such as power, amor mundi and by her conception of politics as a relationship.

In *'Care Ethics and Paternalism: A Beauvoirian Approach'*, Deniz Durmuş draws upon Simone de Beauvoir's existentialist ethics to address one of the most pressing critiques of care ethics today: the problematic Western-centric assumptions and registers underpinning much care ethics scholarship, particularly that work which universalizes care as practiced in the Global North [25]. More exactly, Durmuş argues that given the imperialist and colonial legacies embedded into the unequal distribution of care work across the globe, a Western-centric approach to care ethics may also carry the danger of paternalism. To help counter such Western-centric and paternalistic tendencies, Durmuş presents resources from Beauvoir's ethics, specifically the tenet of treating the Other as freedom, and lays crucial theoretical groundwork for continued engagement between care ethics theory and existentialist ethics.

Sacha Ghandeharian's article, *'Žižek's Hegel, Feminist Theory, and Care Ethics'*, presents conceptual bridges between the philosophy of G.W.F Hegel and feminist ethics of care [26]. More precisely, Ghandeharian engages with Slavoj Žižek's contemporary reading of Hegel in concert with existing feminist interpretations of Hegel's thought to demonstrate how such interpretations of Hegel's perspective on the nature of subjectivity foreground vulnerability as a condition of existence. Indeed, these readings of Hegel, Ghandeharian demonstrates, highlight the radical contingency of human subjectivity and the relationship between human subjectivity and the external world, and thus render Hegelian thought compatible with the feminist ethics of care's emphasis on the particularity, fluidity, and interdependency of human relationships. These conceptual bridges, Ghandeharian concludes, provide fruitful lenses for further analyzing the political and ethical significance of ontological vulnerability.

Petr Urban's contribution, *'Care Ethics and the Feminist Personalism of Edith Stein'*, asserts that the personalist ethics of Edith Stein and her feminist thought are intrinsically interrelated [27]. Further, Urban contends that Stein's ethical thought positions her well as an alley to care scholarship. His article offers an in-depth discussion of the overlaps and differences between Stein's ethical insights and the core ideas of care ethics, focusing on how both approaches relocate practices, values, and attitudes related to relationality, emotionality, and care from the periphery to the center of ethical reflection. Urban thereby lays crucial theoretical groundwork for sustained engagement between these two literatures, with the goal of mutually expanding both bodies of work in productive ways.

Sophie Bourgault proposes a conversation between Jacques Rancière and feminist care ethicists in her article, *'Jacques Rancière and Care Ethics: Four Lessons in (Feminist) Emancipation'* [28]. Demonstrating that both bodies of work share many common premises, including their indictments of Western hierarchies and binaries, their shared invitation to "blur boundaries" and embrace a politics of "impropriety", and their views on the significance of storytelling/narratives and the ordinary, Bourgault draws upon these shared premises and mobilizes Rancière's work to offer crucial insights for care ethicists related to feminist emancipation. These insights, Bourgault concludes, foreground the importance of attending to desire/hope in research, the inevitability of conflict in social transformation, and the need to think together the transformation of care work/practices and dominant social norms.

Maggie FitzGerald stages a dialogue between the political theory of Frantz Fanon and the ethics of care to rethink the relationship between violence and care in her contribution,

'*Violence and Care: Fanon and the Ethics of Care on Harm, Trauma, and Repair*' [29]. Noting that, at first glance, the ethics of care does not seem to sit well with violence (and thus Fanon's political theory more generally, given his assertion that violence is a necessary response to the colonial project), FitzGerald mobilizes a relational conceptualization of violence that allows for the possibility that specific violences may be justifiable from a care ethics perspective. Ultimately, this productive reading of Fanon's political theory and the ethics of care encourages both postcolonial philosophers and care ethicists alike to examine critically the relation between violence and care and how we cannot a priori draw lines between the two.

Anya Daly's contribution, '*Ontology and Attention: Addressing the Challenge of the Amoralist through Merleau-Ponty's Phenomenology and Care Ethics*', addresses the persistent philosophical problem posed by the amoralist—one who eschews moral values—by drawing on complementary resources within phenomenology and care ethics [30]. Asking, "How is it that the amoralist can reject ethical injunctions that serve the general good and be unpersuaded by ethical intuitions that for most would require neither explanation nor justification? And more generally, what is the basis for ethical motivation?", Daly draws on the work of Maurice Merleau-Ponty, and especially his analyses of perceptual attention, to articulate the nature and quality of perceptual attention that underpin our (in) capacities for care. In fostering this dialogue, Daly draws our attention to the compelling nature of care.

Tiina Vaittinen's article, '*An Ethics of Needs: Deconstructing Care Ethics with Derrida and Spivak*', asserts that the body in need of care is the subaltern of the neoliberal epistemic order: it is muted and unheard, partially so even in and from care ethics perspectives [31]. Building on Jacque Derrida's philosophy of deconstruction, Gayatri Chakravorty Spivak's notions of subalternity and epistemic violence, and care theories that emphasize corporeality, Vaittinen develops an ethics of needs that orients us toward the difficult task of attempting to read the world that care needs—that is, needy bodies—write with the relations they enact. This ethics, she ultimately concludes, opens an aporia for an ethical politics of life of needs.

In her article on care theory and liberal constructivism, titled '*Caring for Whom? Racial Practices of Care and Liberal Constructivism*', Asha Bhandary uses three autobiographical examples to draw our attention to a particular social construction that reinforces intersectional inequalities among women [32]. This social construction is related to white women's ready access to narrative positioning as victims, which, as Bhandary illustrates, creates additional burdens for people of color and, more specifically, for women of color. Bhandary defends neo-Rawlsian constructivism that, when combined with an analysis of care practices in the real world, can help us evaluate the fairness of any given arrangement—and particularly those arrangements that are shaped by social construction of white women as victims—from the perspective of a representative person who could occupy any position.

Finally, Sandra Laugier's contribution to this issue, '*Wittgenstein and Care Ethics as a Plea for Realism*', brings together the appeal to the ordinary in the ethics of care and the ordinary language philosophy as represented by Ludwig Wittgenstein, J.L. Austin, and Stanley Cavell, as read through a feminist perspective [33]. In so doing, Laugier emphasizes that the ethics of care fundamentally asserts a plea for "realism", meaning the necessity of seeing (or attending to) what lies close at hand. Such reflections on care, and this plea for sustained attention on the mundane, immediate, and concrete, brings ethics back to everyday practice. Care ethics, Laugier concludes, therefore shares a common political-ethical goal with Wittgenstein's work: much as Wittgenstein sought to bring language back from the metaphysical level to its everyday use, care ethics relocates ethics in the ordinary.

Of course, this Special Issue only offers a few overlaps between philosophy and care theory. We hope the articles here spark more intellectual dialogues between care theory and mainstream philosophical traditions in the future.

Author Contributions: Both authors contributed equally to the writing of this editorial, as well as to the editorial work involved in producing this Special Issue. All authors have read and agreed to the published version of the manuscript.

Funding: This research received no external funding.

Acknowledgments: The editors would like to extend heartfelt appreciation to Christina McRorie who skillfully helped us and several contributors with preparing the articles for publication.

Conflicts of Interest: The authors declare no conflict of interest.

References

1. Plato. Alcibiades 1. (Translated by Horan, D.). Available online: https://www.platonicfoundation.org/platos-alcibiades-1/ (accessed on 22 August 2022).
2. Hanisch, H. How Care Holds Humanity: The Myth of Cura and Theories of Care. *Med. Humanit.* **2022**, *48*, e1–e9. [CrossRef] [PubMed]
3. Reich, W.T. History of The Notion of Care. In *Encyclopedia of Bioethics*, 2nd ed.; Simon & Schuster Macmillan: New York, NY, USA, 1995; pp. 319–331.
4. Baier, A.C. Hume, The Women's Moral Theorist. In *Women and Moral Theory*; Kittay, E.F., Meyers, D.T., Eds.; Rowman and Littlefield: Lanham, MD, USA, 1989; pp. 37–55.
5. Leffers, R.M. Pragmatists Jane Addams and John Dewey Inform the Ethic of Care. *Hypatia* **1993**, *8*, 64–77. [CrossRef]
6. Staudt, M.S. Caring Reciprocity as a Relational and Political Ideal in Confucianism and Care Ethics. In *Care Ethics and Political Theory*; Engster, D., Hamington, M., Eds.; Oxford University Press: New York, NY, USA, 2015; pp. 187–207.
7. Hall, D.; du Toit, L.; Louw, D. Feminist Ethics of Care and Ubuntu. *Obstet. Gynecol. Forum* **2013**, *23*, 29–33.
8. Gilligan, C. *In a Different Voice*; Harvard University Press: Cambridge, MA, USA, 1982.
9. Kohlberg, L. *Essays on Moral Development, Vol. I*; Harper and Row Publishers: San Francisco, CA, USA, 1981.
10. Walker, M.U. *Moral Understandings: A Feminist Study in Ethics*, 2nd ed.; Routledge: New York, NY, USA, 2007.
11. Leget, C.; van Nistelrooij, I.; Visse, M. Beyond Demarcation: Care Ethics as an Interdisciplinary Field of Inquiry. *Nurs. Ethics* **2019**, *26*, 17–25. [CrossRef]
12. Nortvedt, P.; Vosman, F. An Ethics of Care: New Perspectives, Both Theoretically and Empirically? *Nurs. Ethics* **2014**, *21*, 753–754. [CrossRef]
13. Barnes, M.; Brannelly, T.; Ward, L.; Ward, N. *Ethics of Care: Critical Advances in International Perspective*; Policy Press: Bristol, UK, 2015.
14. Jordan, A. Masculinizing Care? Gender, Ethics of Care, and Fathers' Rights Groups. *Men Masc.* **2020**, *23*, 20–41. [CrossRef]
15. Kelly, C. Care and Violence through the Lens of Personal Support Workers. *Int. J. Care Caring* **2017**, *1*, 97–113. [CrossRef]
16. Mullin, A. Gratitude and Caring Labour. *Ethics Soc. Welf.* **2011**, *5*, 110–122. [CrossRef]
17. Kittay, E.F. When Caring is Just and Justice is Caring: Justice and Mental Retardation. In *The Subject of Care: Feminist Perspectives on Dependency*; Kittay, E.F., Feder, E.K., Eds.; Rowman & Littlefield Publishers, Inc.: London, UK, 2002; pp. 257–276.
18. Kittay, E.F.; Jennings, B.; Wasunna, A.A. Dependency, Difference and the Global Ethic of Longterm Care. *J. Political Philos.* **2005**, *13*, 443–469. [CrossRef]
19. Hankivsky, O. *Social Policy and the Ethic of Care*; UBC Press: Vancouver, BC, Canada, 2004.
20. Sevenhuijsen, S. *Citizenship and the Ethics of Care: Feminist Considerations on Justice, Morality, and Politics*; Routledge: London, UK, 1998.
21. Tronto, J. *Moral Boundaries: A Political Argument for an Ethic of Care*; Routledge: New York, NY, USA, 1993.
22. Flower, M.; Hamington, M. Care Ethics, Bruno Latour, and the Anthropocene. *Philosophies* **2022**, *7*, 31. [CrossRef]
23. Randall, T. A Care Ethical Engagement with John Locke on Toleration. *Philosophies* **2022**, *7*, 49. [CrossRef]
24. Chaberty, C.; Lemaitre, C.N. Thinking about the Institutionalization of Care with Hannah Arendt: A Nonsense Filiation? *Philosophies* **2022**, *7*, 51. [CrossRef]
25. Durmuş, D. Care Ethics and Paternalism: A Beauvoirian Approach. *Philosophies* **2022**, *7*, 53. [CrossRef]
26. Ghandeharian, S. Žižek's Hegel, Feminist Theory, and Care Ethics. *Philosophies* **2022**, *7*, 59. [CrossRef]
27. Urban, P. Care Ethics and the Feminist Personalism of Edith Stein. *Philosophies* **2022**, *7*, 60. [CrossRef]
28. Bourgault, S. Jacques Rancière and Care Ethics: Four Lessons in (Feminist) Emancipation. *Philosophies* **2022**, *7*, 62. [CrossRef]
29. FitzGerald, M. Violence and Care: Fanon and the Ethics of Care on Harm, Trauma, and Repair. *Philosophies* **2022**, *7*, 64. [CrossRef]
30. Daly, A. Ontology and Attention: Addressing the Challenge of the Amoralist through Merleau-Ponty's Phenomenology and Care Ethics. *Philosophies* **2022**, *7*, 67. [CrossRef]
31. Vaittinen, T. An Ethics of Needs: Deconstructing Neoliberal Biopolitics and Care Ethics with Derrida and Spivak. *Philosophies* **2022**, *7*, 73. [CrossRef]
32. Bhandary, A.L. Caring for Whom? Racial Practices of Care and Liberal Constructivism. *Philosophies* **2022**, *7*, 78. [CrossRef]
33. Laugier, S. Wittgenstein and Care Ethics as a Plea for Realism. *Philosophies* **2022**, *7*, 86. [CrossRef]

philosophies

MDPI

Article

Care Ethics, Bruno Latour, and the Anthropocene

Michael Flower [1] and Maurice Hamington [2,*]

[1]   University Studies, Portland State University, Portland, OR 97201, USA; hcmf@pdx.edu
[2]   Department of Philosophy, Portland State University, Portland, OR 97201, USA
*   Correspondence: maurice4@pdx.edu

Abstract: Bruno Latour is one of the founding figures in social network theory and a broadly influential systems thinker. Although his work has always been relational, little scholarship has engaged the relational morality, ontology, and epistemology of feminist care ethics with Latour's actor–network theory. This article is intended as a translation and a prompt to spur further interactions. Latour's recent publications, in particular, have focused on the new climate regime of the Anthropocene. Care theorists are just beginning to address posthuman approaches to care. The argument here is that Latourian analysis is helpful for such explorations, given that caring for the earth and its inhabitants is the dire moral challenge of our time. The aim here is not to characterize Latour as a care theorist but rather as a provocative scholar who has much to say that is significant to care thinking. We begin with a brief introduction to Latour's scholarship and lexicon, followed by a discussion of care theorist Puig de la Bellacasa's work on Latour. We then explore recent work on care and the environment consistent with a Latourian approach. The conclusion reinforces the notion that valuing relationality across humans and non-human matter is essential to confronting the Anthropocene.

Keywords: actor-network theory; Anthropocene; Bruno Latour; care ethics; hesitation; Gaia; hiatus; modes of existence; relationality; translation

Citation: Flower, M.; Hamington, M. Care Ethics, Bruno Latour, and the Anthropocene. *Philosophies* **2022**, *7*, 31. https://doi.org/10.3390/philosophies7020031

Academic Editor: Marcin J. Schroeder

Received: 7 February 2022
Accepted: 9 March 2022
Published: 14 March 2022

Publisher's Note: MDPI stays neutral with regard to jurisdictional claims in published maps and institutional affiliations.

1. Introduction

The 8 October 2021 issue of the journal *Science* offered a sobering yet not unfamiliar warning: "Intergenerational Inequities in Exposure to Climate Extremes: Young Generations are Severely Threatened by Climate Change" [1]. Co-authored by 37 scientists, the article applied historical data to environmental modeling to suggest that continued global temperature rise will result in future generations being exposed to many more extreme environmental events, including heatwaves, crop failures, river floods, and droughts than generations past. The authors indicate, "Aggregated across all the event categories, lifetime exposure to extremes is unprecedented at all warming levels and cohorts" [1] (p. 159). Furthermore, the scientists speculate that beyond the events they modeled, mass environmental migration and declines in life expectancy will occur. For anyone paying attention to the scientific community on climate change issues, these conclusions are not surprising, although the bleak reminder remains discouraging. Embedded in the article is a critical ethical note representing a dispassionate cry for readers to care about the climate data and to take action. After stating that younger generations are expected to face more extreme environmental events, the authors suggest, "this raises important issues of solidarity and fairness across generations" [1] (p. 158). Restating the obvious, if the body and its ability to perceive is our vehicle for having a world, the world is the vehicle for having our bodies and a world to perceive. We cannot disentangle human morality from environmental morality. Feminist care theorists have faced vexing and wicked social problems, but none looms as large as climate degradation. Furthermore, care scholars have generally not addressed non-human care as extensively as interpersonal care. Dire global necessity requires a broader construct of care.

In *Down to Earth: Politics in the New Climatic Regime*, French philosopher, anthropologist, and sociologist Bruno Latour frames the Anthropocene as a fundamental crisis of modernity, in part because of modernism's abstract assumptions and emphasis on human detachment from objects of the material world. Rather than simply an issue for the left or progressives, Latour indicates that "we can understand nothing about the politics of the last 50 years if we do not put the question of climate change and its denial front and center" [2] (p. 2). In particular, leaders of nation-states and multinational corporations have treated the Earth as an object for manipulation and extraction, as mere means to human ends, rather than an essential constituent of our shared being. Latour notes that the powerful and super-rich have realized that their goals of unending growth are incompatible with the limits of our environment. Thus, the new "fairytales of unending economic growth," to borrow a term from activist Greta Thunberg [3], including their schemes for escaping the Earth and finding new worlds to exploit. Latour points to the irony in this privileged dream of geo escapism. He describes how this turn of events gives new meaning to the term "postcolonialism" in that the settlers stole the land but are about to squander and lose it anew [2] (pp. 7–8). *Down to Earth* continues Latour's ongoing indictment of modernity in its valorization of human exceptionalism. He concludes that until we reframe our thinking to address the Earth and its elements—rocks, trees, oceans, etc.—as actantial participants in shared solutions, the Anthropocene cannot be adequately addressed. Latour calls for an inclusive, reflective process that respects interdependency to

> avoid the trap of thinking that it would be possible to live in sympathy, in harmony, with so-called "natural" agents. We are not seeking agreement among all these overlapping agents, but we are learning to be dependent on them. No reduction, no harmony. The list of actors simply grows longer; the actors' interests are encroaching on one another; all our powers of investigation are needed if we are to begin to find our place among these other actors [2] (p. 87).

The concern for an expansive notion of relatedness is nothing new for Latour, who has built an intellectual legacy in highlighting the associational and heterogeneous as part of the nature of networks. Similarly, relationality is the currency of care theorists, and, unlike modernism, care theory has emphasized the personal and particular; that is, how do we care for unfamiliar others? The Anthropocene forces us to consider how we care for the non-human material world—very unknown others.

A few care scholars such as Vivienne Bozalek and Maria Puig de la Bellacasa have pushed care thinking into considering the realm of posthumanism in a manner consistent with Latour. Keep in mind that posthumanism is not antagonistic to the insights of the humanities or caring relations but problematizes human-exclusive thinking and human exceptionalism. This reframing or decentering of the human is essential in understanding and analyzing the Anthropocene, a term proposed as a successor era to previous geologic periods (Holocene, Pleistocene, etc.). Although the start date of the Anthropocene is disputed, many scientists favor the beginning as the 1950s, when human involvement in climate change began to grow exponentially. Posthumanism suggests a shifting of values to counter dominant worldviews of human primacy that resulted in a devaluing and degrading of the non-human world. Philosopher Francesca Ferrando describes, "As the Anthropocene marks the extent of the impact of human activities on a planetary level, the posthuman focuses on de-centering the human from the primary focus of the discourse" [4] (p. 32). This article addresses how Bruno Latour's work might assist care theorists in developing a posthuman framework, not by overlaying another philosophical paradigm but rather because Latour's thinking is fully relational.

Puig de la Bellacasa is one scholar who has endeavored to translate aspects of Latour's work for care theorists in a limited manner. Accordingly, in this project, we extend Puig de la Bellacasa's analysis to further place Latour's work in conversation with contemporary feminist care theory. Latour's body of scholarship is vast, eclectic, and complex, so any short treatment is somewhat of a caricature; however, we endeavor to draw some particular insights to prompt further discussion. The aim here is not to characterize Latour as a care theorist but rather as a provocative scholar who has much to say that is significant to care thinking. We begin with a brief introduction to Latour's scholarship and lexicon, followed by a discussion of Puig de la Bellacasa's work on caring for the "more than human worlds of technoscience and naturecultures." Her book, *Matters of Care: Speculative Ethics in More Than Human Worlds*, is one of the few lengthy treatments of Latour from a care theory framework. Moreover, it is also one of the few works addressing non-human-centered care explicitly.

To foreshadow the liminal spaces that Latour operates in, a brief review of his notion of translation (discussed further below) is useful as both a Latourian hermeneutic and a metaphor for this Special Issue of *Philosophies*. Translation is an essential concept for Latour in making networks functional. The contributors to this issue are endeavoring to translate the work of significant philosophers to those interested in care theory and, reciprocally, care ethics for those steeped in mainstream philosophy. Latour describes his actor–network theory (ANT), a framework of dynamic relational existence including humans, non-human beings, and matter, as a "sociology of translations" [5] (p. 106). For Latour, translation is not merely transportation, a conduit of passing from one point to another. Instead, he describes translation as relational and impactful: "a relation that does not transport causality but induces two mediators into coexisting" [5] (p. 108). This characterization is far from a transactional account of interaction so common in neoliberal modernity. Similarly, caring can be viewed as a rich form of translation. Effective caring is not a simple process of need satisfaction through an exchange. Good quality or effective care involves humble inquiry, inclusive connection, and responsive action. Humble inquiry is an active effort to gain knowledge of a particular other; inclusive connection is the empathetic and emotional attachment that we form, which motivates and participates in our engagement; and responsive action consists of the practices required to meet the needs of the other. The actors are not just mechanically reciprocating as in a transaction. Their mutual engagement is transformative to all involved, engendering more than additive impacts. Care can be described as deep translation work in the spirit of Latour. Effective caring is a translation of engrossment and inhabiting of the other, such that connections are built. The response not only foments the flourishing of the other, but the one caring is no longer the same being as prior to the encounter.

2. An Introduction to Latour

From his first book co-authored by Steve Woolgar, *Laboratory Life: The Construction of Scientific Facts* (1979), to *The Politics of Nature: How to Bring the Sciences into Democracy* (2004), to *An Inquiry into Modes of Existence* (2013)—his magnum opus—to his recent books on the ecological mutation of the Anthropocene, *Facing Gaia* (2017), *Down to Earth* (2018), and *After the Lockdown: A Metamorphosis* (2021), Latour has challenged how we produce knowledge, how we account for experience and the manner of life amidst an intimately entangled world of fellow humans, plants, animals, microbes, instruments, texts, habits, political disputes, geological forces, disintegrating ecosystems, and much, much else that comprises the heterogeneous tapestry of our world. He has developed and deployed several provocative concepts and practices from his studies in neuroendocrinology at the Salk Institute, the work of Louis Pasteur, the nature of religious speech, the French administrative law court (the Conseil d'État), and most recently, the earth scientists whose work defines the critical zone within which life on Earth subsists [6,7]. Out of this rich array, we draw upon a handful of Latour's concepts that we wager are useful in understanding how care ethics can best respond to the challenges of climate change: the operation of translation and the role of the hiatus in the composition of continuity from the discontinuity of experience, the tonality of modes of existence, modal crossings where modes engage one another, and the necessity of hesitation which allows for scrupulous reflection in the face of ethical dilemma.

In the unpublished English translation of *Cogitamus. Six lettres sur les humanités scientifiques*, a primer for students, Latour early on makes the Archimedean claim: "Give me the concepts of translation and composition, and I will move the earth." What does

Latour mean by these terms? We begin with Latour's notion of translation and composition. With regard to translation, Latour states, "I use translation to mean displacement, drift, invention, mediation, the creation of a link that did not exist before and that to some degree modifies two elements or agents" [8] (p. 32). By employing the term "composition," Latour is endeavoring to navigate the notion that "things have to be put together" while "retaining their heterogeneity" [9] (pp. 473–474). These terms are indicative of Latour's amodernist liminality that seeks to frame the world and its processes as it is rather than forcing it into discrete linguistic categories.

To see translation and composition in action, let us accompany Latour to a study site near Boa Vista, the capital of the Brazilian state of Roraima. This journey may seem far from the concerns of care ethics, but it is a trip worth taking [8]. In the early 1990s, Latour accompanied a small group of naturalists who were studying a boundary between the forest and savanna, seeking to establish whether the savanna was pushing into the forest or the forest into the savanna—a matter of importance at a time when the extent of tropical forests and their removal of carbon dioxide from the atmosphere was of concern, as it is now. What does Latour observe as he follows the naturalists? First, the botanist collected plants at the boundary, attaching metal tags to trees to create a Cartesian grid and recording those locations in her field notebook. This is a translation. The land is mapped, translated from a tangle of plants rooted in the soil onto a grid of markers; the collected plants are referenced to the site map. The botanist has performed a rudimentary translation; a piece of the savanna/forest boundary is now a paper record. Next, the plants are taken to her laboratory, where they are keyed out, pressed, and stored in a coded cabinet. The study site stays in place while its coded and translated representations travel step-by-step; all the while, the botanist's carefully articulated practices work to establish continuity.

Other members of the naturalist team are pedologists; they study soil. They, too, create a map to mark the places from which they take soil samples, boring down into the Earth and removing meter-long earthen cores. Each of the cores is cut into small pieces along its length, and the pieces are carefully placed into a pedocomparator, a wooden box containing empty cardboard cubes, in an order according to their position along with the core. Each soil core has become a pedolibrary; the clods of earth are dropped from the scientist's hand across a gap of a few centimeters and land in the appropriate cardboard cubes. In an instant, a translation has occurred. The earthen clods are now signs in a boxed Cartesian grid, each cube within the grid being individually designated. A clod of *matter* has been translated into a *form* that locates the clod along the core's length; it has been differently materialized. There is more: the two pedologists, Armand and René, are not done with the soil clods. Small pinches of soil are rolled around in the hands of these expert pedologists who engage in a back-and-forth conversation as their embodied, tactile, material experience of soils is used to categorize the texture of the earthen clods: "Sandy clay or clayey-sand?" "No, I would say clayey, sandy, no sandy-clay." "Wait, mold it a bit more, give it some more time" "Okay, yes, let's say between sandy-clay and clayey-sand" [10] (p. 63). Hence, a translation of the earthen clods into verbal and written descriptions. Additionally, the soil samples are described using a tool, the Munsell color codebook, to reference soil color to soil type. Thus, the pedocomparator "form" (an ordered array of the boxed clods) has become the "matter" upon which other "forms"—the tactile "feel" and the color coding—can be applied. Much later, the collected samples are taken to a laboratory for chemical analysis and translated into still other paper forms—those of the soil chemist.

We are witnessing a series of *matter–form–matter–form* translations—small leaps taken across gaps as a series of transformations. We see a *network* being composed. Each transformation, each translation, is a *different* composition. Different tools, instruments, and standard protocols are utilized at each step. Here, we should note that those tools and protocols, now stable and taken for granted, are each possessed of earlier histories of development and use, which, too, consisted of a series of networked translations and compositions. No matter where you look, there are networks of translation, of stepwise recomposition, of scientific *reference*. Moreover, what of the research findings? The field study results are published, and the core samples, now represented in data tables and a schematic of the forest–savanna boundary's soil composition, travel in journals to libraries and the desks of fellow scientists. The findings are presented at conferences; the results appear in biology classes. The article, encoded as a PDF, can travel across the globe in seconds. The Boa Vista forest–savanna border is mobile. The findings may enter into other compositions—of tropical forest management, the design of timber harvesting practices, the comparative study of different tropical soils—thus elaborating further networks where different scientific, political, and economic stakes are at issue. Should a controversy about the results arise in those subsequent uses, one can move back along the trajectory of translations to the study site, where the land is once again beneath one's feet. In short, we can see a webwork of compositions preceding the Boa Vista study and following it; wherever we look, we see discontinuous leaps, but not disarray. Agreed-to practices are available to establish and stabilize continuities of stepwise translations. Thus do we carry on *in medias res*—networks of translation behind us, an uncertain future of networked transformations in front of us. The work of translation never ceases for long.

This process may seem all well and good, but what of situations that are more complex than a field study involving only scientists using relatively simple tools and stable and uncontroversial protocols? What do composition and translation look like when, as in the case of the growing turmoil resulting from dramatic climate change, decisions about the actions involve science, law, politics, religion, morality, and economic forces? How heterogeneous are the composites that take shape amid dispute, amid institutional forms of practice that differ (law is not science, politics is not economics)? Where and how, amid the tangles, does care come into play? What more do we need to know of Latour's approach when dealing with networks that are dramatically more perplexing than those of the Boa Vista study?

In 2012, a largely invisible project in the works for over 20 years made its public appearance with the publication of Bruno Latour's *Enquête sur les modes d'existence: Une anthropologie des Modernes*, the English translation being published a year later [11]. In addition to the book, an interactive website was created to aid in portraying and understanding the modes of existence.[1] In this book, Latour identifies fifteen modes of existence. Each mode operates according to a rationality or tonality of its own. We, as well as non-human actors, make our world and our way through it by deploying these modes. Each mode of existence makes sense differently, each being characterized by the nature of the gaps (the discontinuities) that must be bridged, the mode-specific trajectories of subsistence that are employed, the felicity/infelicity conditions involved, the beings that are engendered, and the alterations that are aimed for [11] (pp. 488–489). So, for example, science advances its claims in a many-fold manner quite distinct from law's form of rationality; the mode of political existence is unlike that of religion or morality. As with the case of soil science at work in Boa Vista, each mode operates in a stepwise fashion; all modes extend their operations and the fashioning of beings across a series of discontinuities: "if no alteration, then no being" [12] (p. 39). All compose the continuity of their trajectories of subsistence across gaps that Latour calls *hiatuses* (as we illustrate in our discussion of Figure 1 below). None of the modes advances its continued existence on its own; rather, "We try out the Other" [12] (p. 319). The extension of a mode is a plural affair—and substantially multimodal. For in Latour's notion of being-as-other (his alternative to being-as-being), "it is always via the other that being is extracted" [12] (p. 327). Thus, modes routinely cross through one another and, as a consequence, quite different modal rationalities come into play, producing differences of compositional practice that must be resolved. As Latour states, "Each mode will define itself through its own way of differing and obtaining being by way of the other" [12] (p. 316). As we will see, it is for this reason of ongoing negotiation and reconfiguration of modes that translational crossings are often perplexing and equivocal.

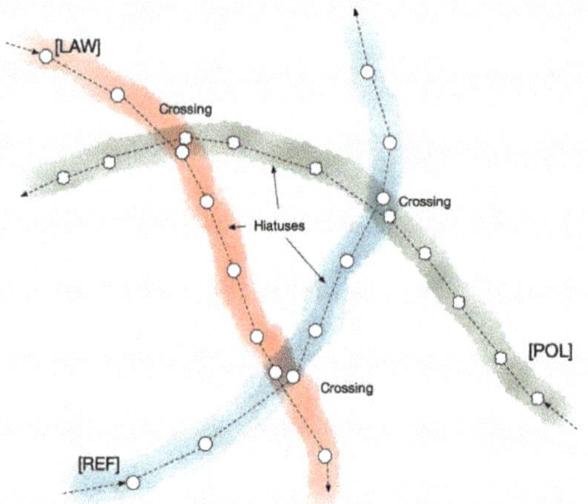

Figure 1. The crossing of different modes of existence: establishing continuity at hiatuses.

Before turning to our discussion of the crossings and entanglements of modes represented in Figures 1–3,[2] it is essential to describe, if only briefly, the fifteen modes of existence that Latour has gathered into five groups of three related modes. The first group—reproduction [REP], metamorphosis [MET], and habit [HAB]—concerns forms of *being as alterity* that multiply forms of persistence, multiply transformations, or rush forth into existence with dispatch, respectively [11] (p. 285). A second group—technology [TEC], fiction [FIC], and reference [REF] (the mode of science we saw in the Boa Vista study)—concerns *quasi-objects*,[3] while the third group [11] (p. 372)—politics [POL], [LAW], and religion [REL]—addresses *quasi-subjects*.[4] With these two modal groups, we see the conventional binary pair of subject and object replaced by entangled actors. Latour says succinctly: "What is an object? The set of *quasi* subjects that are attached to it. What is a subject? The set of *quasi* objects that are attached to it" [11] (p. 428). We are "sort of" subjects, interestingly and integrally adorned with a variety of objects. Objects are accompanied by a retinue of subjects. We are compositions; entities exist as filigreed, multimodal collectives.

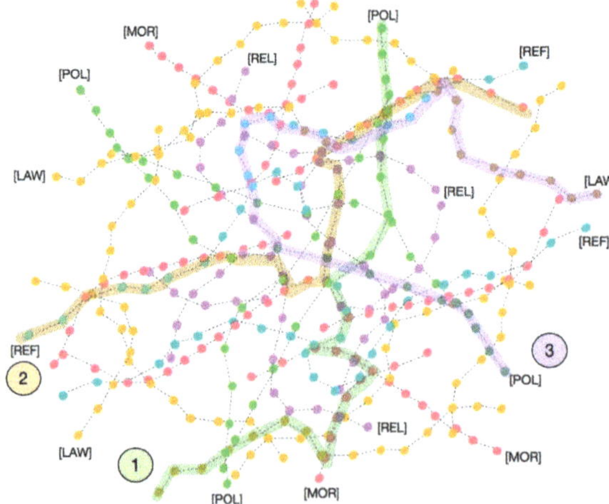

Figure 2. Depicting the rhizomatic entangling of modes of existence—a snapshot of the multiple translations of the pluriverse of possibility.

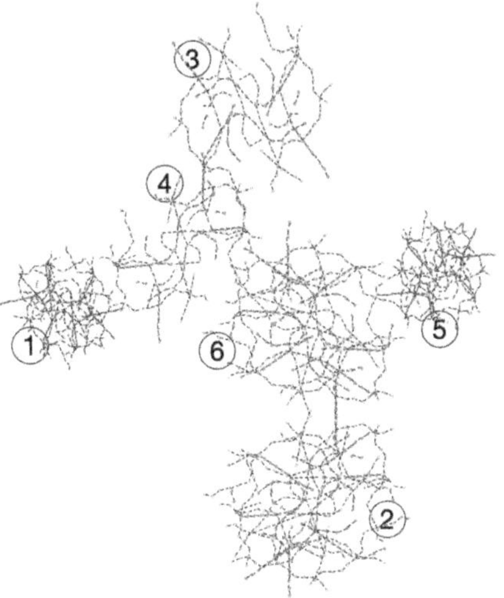

Figure 3. The entanglement of tangles.

The fourth modal group, perhaps the most difficult to understand (and indeed the one that suffers most from the necessarily minimal characterizations we have offered here)—attachment [ATT], organization [ORG], and morality [MOR]—mingles quasi-objects and quasi-subjects in such a way as to move us Moderns to a position of agnosticism about "The Economy," thereby freed up to compose a novel and scrupulous underpinning for economic valuation and exchange.

Finally, the fifth group—network [NET], preposition [PRE], and double click [DC]. The first two of these might be thought of as the primary gearworks of translation and composition. Before moving on, we will put [DC] aside, even as we recognize how troubling (and sometimes dangerous) it can be as the mode of taken-for-granted certainty; the stuff of bold-face terms in textbooks, of unquestioned claims, of the too-easily accepted bit of knowledge readily accessible with clicks of the mouse. For our purposes [and Latour's), [NET] and [PRE], and their crossing [NET·PRE], are crucially important; they authorize the whole of Latour's *Inquiry* and are central to our consideration of translation and composition and our attempt to enrich the conception of situations in which we seek to understand care in a climate-changing world of human and non-human actors. Let us allow Latour to lead the way as he crosses [NET] and [PRE]:

> We shall thus say of any situation that it can be grasped first of all in the [NET] mode—we shall unfold its network of associations as far as necessary—and then in the [PRE] mode—we shall try to qualify the type of connections that allow its extension. The first makes it possible to capture the multiplicity of associations, the second the plurality of the modes identified during the course of the Moderns' complicated history. In order to exist, a being must not only pass *by way of another* [NET] but also *in another manner* [PRE], by exploring other ways, as it were, of ALTERING itself [11] (p. 62).

Consider Figure 1, where three different modes are represented: [POL], [LAW], and [REF]. The hazy colors represent the tonal envelopes established by the three couples—[PRE·POL], [PRE·REF], and [PRE·LAW]—within which modes can fashion the next translation. [PRE] essentially says to a mode seeking to subsist, "Hey, there is the need for a proper envelope. What sort of envelope is this to be? You and I have to elaborate our continued subsistence in the proper direction and tone." To which the mode [LAW] in this case) answers, "I am [LAW]. The hiatus I must leap across risks the dispersal of cases and actions; the trajectory I am to extend requires that I link cases and actions via legal means; I must felicitously reconnect levels of enunciation and avoid the infelicity of breaking them. My job is to institute beings that are safety-bearers. And the alterations I am charged with? I am to ensure the continuity of actions and actors".

Giving substance to the general example in Figure 1, it is not difficult to imagine cases of environmental law (the red trajectory)—be it about pipeline construction, the clearing of forests to establish a soybean plantation, or the ill effects of exposure to glyphosate—that will at some point involve political considerations (the crossing of the green trajectory where [POL) indicates political matters seen to be at stake and about which there may be controversy] as well as scientific findings (the blue trajectory where [REF) indicates the type of scientific leaps of reference like those we saw in the Boa Vista example). As the case proceeds, there is a juncture (the red/green crossing, [LAW·POL]) where matters of political rationality come into compositional play; later, a finding of science may be taken up (the red/blue intersection, [LAW·REF]). Although the law consists not only of law (i.e., it must pass in translation "in another manner"), the critical point is this: the *composition*, the *making* of law at the crossings occurs in accordance with *its* terms of rationality, even as those terms wrestle with, say, the terms of scientific or religious modes of existence. Though it is now in part comprised of quasi-objects and quasi-subjects of different modes, its envelope must be extended as that of [LAW]. At each step of the legal proceeding, different and heterogeneous compositions—of legal precedent, the effect of glyphosate on human cells, the ecological impacts of vast acreage being given over to palm tree plantations, and struggles for climate justice—must be crafted in a legal fashion, each step requiring uncertain decisions as to what collection of considerations, what translation, will best extend the life and power of the case. One must decide which composition amongst all the contingent possibilities will amount to a good engendering of means and ends, a [LAW·MOR] crossing.

Such decisions are complex; they are trying. One approaches a juncture, and another translation is required. The modal crossing at the juncture entails a difficult hiatus; there is hesitation, a weighing of uncertainties as heterogenous possibilities present themselves, as one considers the room for maneuver before new associations are made. Consider a legal case involving a damaged ecosystem. There are many scientific studies to draw upon. Which will best advance our case when put into evidence? To what degree do we have to contend with the mix of local political sensibilities at trial? How should we fashion our next legal move? As we hesitate, which of the assembled case elements might we put aside, putting forth a somewhat different collection, an alternative translation, of political and scientific elements? What, in one's estimation, is the right way to proceed? What moral scruples are relevant? Among the several possibilities, what trajectory of anticipated future translations across hiatuses ought one seek to extend?

Consider the additional complexity represented in Figure 2, a webwork of Figure 1 writ large. Here many and different trajectories of the several modes have made their way, tenuous leap after tenuous leap. Three courses of mode-crossing action are highlighted; they have twisted and turned as their multimodal translations sketch different paths of continuity through a discontinuous pluriverse of possibility. There are no domains here; there are no well-demarcated zones of science, politics, law, religion, and economy. The crossings are many, as are the occasions for scrupulous hesitation. This is a world of translation and composition in which one must pay constant and diverse attention to the uncertainties being faced.

Yet this is but a part of even more extensive entanglements. Figure 3 ups the ante; six somewhat distinct composites are depicted as an entanglement of tangles. Again, there are no clear boundaries; the courses of action that have produced the entanglements are many and varied. Imagine cases of environmental law being pressed differently in different countries; different scientific studies of ocean acidification at different study sites; different political responses to news of court judgments and scientific findings; the diverse array of concerns expressed by members of COP 26 (the United Nations' 26th Climate Change Conference held in 2021) delegations inside the meeting halls as compared to those of demonstrators in the streets outside; and much more—always much more.

In the face of the burgeoning Anthropocene, consider for just a moment what Latour, following Isabelle Stengers, has devoted most of his attention to over the past decade: the intrusion, the irruption of Gaia [13]—the rumbling, disorienting emergence into view of the Earth as an actor, as a mighty distributed force requiring us to deploy the modes of existence in new ways. What *is* this Gaia that now requires us to confront human/nonhuman collectives in our political, economic, moral, technical, scientific, artistic, and legal struggles for well-being, justice, and security? First, Latour endeavors to rescue the notion of "Gaia" from a new form of religiosity that references a goddess entity by considering a more holistic sense of collective agency: "the Gaia idea does not involve adding a soul to the terrestrial globe, or intentionality to living things, but it does recognize the prodigious ingenuity in the way living things fashion their own worlds" [14]. More recently, Lenton, Dutreuil, and Latour [15] argue that to trace Gaia's extent and consequences, to identify the manner of that ingenious fashioning, we must identify all living beings on Earth, trace the material interactions between these living beings and what is outside their membranes, and establish which of those connections are relevant to Earthly habitability—Gaia being the entity isolated by the resulting network of such connections [15] (p. 253). In such a view, the currently witnessed events of the Anthropocene might then be understood as early indicators of the emergence of a new state of Gaia set into motion by global warming. We are thus facing an Earth whose composition will involve changing arrays of living beings, different chemical and material interactions, differently articulated conditions for habitability—a changed Gaia, a differently connected biotic/abiotic webwork within which we terrestrials are differently entangled. The Earth that is one is a new one. In line with the Latourian view that the Earth is one, Patrice Maniglier has advanced the especially important claim that a key meaning of Gaia, perhaps its first meaning, is *continuity of entanglements* [16] (pp. 67–71)—precisely the point of our focus on the risky composition of jumps across hiatuses. Gaia is the high dimensional manifold that occasions the transformations created by the continuous scrambling of entanglements as glaciers melt, seas rise, desertification increases, severe weather events grow in number, lobbyists argue for continued fossil fuel subsidies, croplands wither, and climate refugees flee their homelands.

The pervasiveness of Gaia's irruption surely means that the number of novel crossings of our modes of existence will dramatically increase. At the crossings, there is equivocation; we are faced with the task, time after time, of composing continuity anew amidst conditions of heterogeneity and multiplicity. Imagine the trajectories of Figures 1–3 in motion (before we rendered the traces of such activity in the schematic snapshots) as they extend jump by jump, facing contingency after contingency. Multiple and varied trajectories are possible as science *meets* law *meets* morality *meets* politics *meets* the organization [ORG], attachment [ATT], and moral [MOR] modes that comprise the entanglement we call the economy. We move from equivocation to equivocation. Maniglier describes our current situation succinctly; we face "The Earth, this Great Equivocation" [16] (p. 123). No wonder we must hesitate at the edge of hiatuses, choosing our attachments, allies, and the terms of our leaps of composition with care.

A great many of the complex entanglements we have engaged with in the past must now be changed; we must disentangle from them to re-entangle anew, and differently so. We must become compositionists of a different sort [9]—and of necessity, the terms of our multi-fold translations must change. The occasions for hesitation will grow in number; the terms of the hesitation will be different. We must be more scrupulous than ever. We must take care differently, a shift we find in the work of Maria Puig de la Bellacasa.

3. Bellacasa on Latour (Matters of Care)

Through a series of articles, and in her 2017 *Matters of Care: Speculative Ethics in More Than Human Worlds*, philosopher and transdisciplinary scholar at the University of Warwick, Maria Puig de la Bellacasa pioneered applying feminist care theory to natureculture, bringing together humanist and post-humanist thinking. She acknowledges that care is a "human trouble" but finds it absurd to disentangle human and nonhuman care relations [17]

(p. 2). Bellacasa offers a rich and layered analysis of care worthy of much reflection. Still, for this project, we focus on the first chapter of *Matters of Care,* where she engages Latour's notion of matters of concern (and from which she derives the title of her book).

It was inevitable that Latour's thinking would find its way into care theory. Philosopher Graham Harman suggests that "Bruno Latour is starting to look like Michael Foucault's eventual replacement as the default citation in the humanities" [18] (p. 249). Indeed, like Foucault and other modern philosophers such as Judith Butler, Latour questions critique's very nature and efficacy. In "'Why Has Critique Run Out of Steam? From Matters of Fact to Matters of Concern," Latour is concerned that both society and philosophers have gone astray in their use of critical analysis. Donna Haraway agrees, describing Latour's article as "a major landmark in our collective understanding of the corrosive, self-certain, and self-contained traps of nothing-but-critique. Cultivating response-ability requires much more from us. It requires the risk of being for some worlds rather than others and helping to compose those worlds with others" [19] (p. 178). This risk of being for only some worlds hints at the particularism of care which begins in inquiry and understanding broadly construed. For Latour, the emphasis on propositional knowledge in the form of facts to be understood misses the point about how knowledge and meaning contextually exist in the world. The isolated and abstract fact cannot subsist without the gatherings or assemblages which bring them into existence. He problematizes the factual, propositional "thing": "A thing is, in one sense, an object out there and, in another sense, an *issue* very much *in* there, at any rate, a *gathering* the same word *thing* designates matters of fact and matters of concern" [20] (p. 233). A Husserlian phenomenologist may wish to peel away social meaning and noise to study the artifact or phenomenon. However, for Latour, the search for pre-phenomenal essence misses the point of seeking the whole: a gathering, a thing. The concern for matter is what inspires Bellacasa.

For Bellacasa, Latour's work, and precisely his notion of matters of concern, holds promise but lacks an explicit connection to care: "Concern brings us closer to care. However, there is a 'critical' edge to care that the politics of making things matter as gatherings of concern tends to neglect" [17] (p. 18). Bellacasa discusses the divergent connotations of the words "concern" and "care." The former, according to Bellacasa, suggests worry and thoughtfulness, while the latter adds a sense of attachment and commitment [17] (p. 42). Bellacasa credits Latour for connecting these forms of regard to the technoscientific world. For example, she cites Latour's concern for the maintenance of technology and infrastructure [21] as a type of care. This concern is not simply for an artifact, a matter of fact, independent of its contextual assemblage—its thingyness. This concern is a care for the technology's becomings or what matter of concern it will become. A subway is not merely a collection of parts but an ethico-political assemblage. Concern for the maintenance and efficiency of the subway is more significant than a concern for metal, rubber, and plastic. It represents concern for the well-being of riders, the Earth, and the future. These gatherings of ideas, forces, players, and arenas in which "things" and issues, not facts, come to be and persist because they are supported, cared for, worried over. Such concern also embraces values as possible becomings or worlds are chosen over others.

Citing Latour, Bellacasa refers to Science, Technology, and Society scholars as "liberators of things" [17] (p. 45). Reminiscent of some indigenous ontologies in honoring the Earth and its artifacts, Bellacasa quotes Latour in imbuing (or instead, finding) the agential quality in matter: "Humanists see the imposture of treating humans as objects ("objectification" in feminist theory)—but what they don't realize is that there is an imposture also to treat objects as objects" [17] (p. 45). Western thought is so ingrained with human exceptionalism that grasping any investing of matter with value is challenging. Latour offers the contrast between two German terms, *Realpolitik,* which is a "positive, materialist, no-nonsense interest only matter-of-fact way of dealing with naked power relations," [22] and *Dingpolitik,* which Latour suggests is much more realistic than *Realpolitik* because politics has always been concerning things. As Latour states, "We might be more connected to each other by our worries, our matters of concern, the issues we care for, than by any other set of values, opinions, attitudes or principles" [22]. Bellacasa employs the example of the SUV, an object of concern. Both environmentalists and certain car enthusiasts have strong feelings about SUVs. Bellacasa leverages Latour to find that strident positions on either side of the SUV issue cannot forget to care about those who hold positions about this object dear [17] (p. 48). Care for things is intertwined with care for people.

To reiterate, Latour is not a care theorist. Bellacasa modifies Latour's thinking to add feminist care ethical sensibilities. However, she understands and appreciates the relational potential and insight of his work in developing posthumanist care. Given the enormity of the current environmental threat, although care theory remains an influential innovation in moral philosophy, it must extend its relational reach to the other-than-human world.

4. Current Work on Care Ethics and the Environment

Care ethics is a relational approach to morality that differs from traditional approaches to morality by valuing context, emotions, and empathy. Although care has normative implications, it also has epistemological and ontological significance beyond adjudicating moral dilemmas. The most quoted definition of care ethics is that of Joan Tronto and Berenice Fisher, and it is indeed flexible enough to include the possibility of care for the environment:

> On the most general level, we suggest that caring be viewed as a species activity that includes everything that we do to maintain, continue, and repair our 'world' so that we can live in it as well as possible. That world includes our bodies, our selves, and our environment, all of which we seek to interweave in a complex, life-sustaining web [23] (p. 19).

Historically, care has been a morally ambiguous word given that much oppression has been wrought in the name of care, as in, for example, colonialism. Care can be applied in harmfully paternalistic ways, bringing coercion to repressed communities [24]. Nevertheless, social and political care theorists have emerged across a wide variety of disciplines that are endeavoring to offer a care ideal to many of modernity's vexing challenges. There is no more wicked problem than the environmental crisis.

There have been several promising efforts to apply feminist care ethics to the environmental crisis [25,26], but only a few have engaged the work of Latour. One of the challenges for feminist care studies is that so much of the originary work in the field focused on naming the previously unnamed morality of care and drew upon easily identified manifestations of care in familiar relationships such as between a parent and a child. Although the field has moved on to consider social, political, and institutional approaches to care, care maintains much of its interpersonal connotation. Indeed, ultimately care is experienced on a personal level, but that does not mean that care cannot extend across time and space in a manner that is not proximal. As Thomas Randall has observed, environmental ethics is a demanding subject for care ethics given the intergenerational nature of environmental concern [27]. Such temporal and spatial abstraction makes the object of care much less relationally tangible. For example, care theorists have often emphasized reciprocity [28]. Noddings claims, "There is, necessarily, a form of reciprocity in caring" [29] (p. 71). Furthermore, some care theorists have emphasized the embodied basis of care [30] grounded in the face-to-face knowledge and presence of the other [31]. Because future generations have no possibility of reciprocity or presence, caring seems complicated under these terms.

Randall finds the work of political care theorists such as Joan Tronto more promising for intergenerational care because political constructs of care lack the focus on reciprocity. However, Tronto assumes the social, moral force of interdependency in her work on care [23] (p. xv). Interdependency is also absent when addressing the moral challenge of future generations. Dependency is unidirectional when it comes to environmental heritage. Christopher Groves explicitly offers a nonreciprocal intergenerational theory of care grounded in the idea that we can care about future generations, but not in a controlling manner [32]. Instead, care is a means for living with and through uncertainty and precarity [33]. Although Randall finds value in Groves's approach, he offers an alternative method to thinking about intergenerational care akin to creating a present ethos of care. Two central notions for Randall's blueprint for intergenerational care are imagination and inheritance. Imagination is a necessary element of care, as manifested in empathy and considering responsive actions [29]. However, Randall employs the term "imaginal" as a hybrid between that which is imaginary and that which is constrained by reality [27] (p. 537). Imaginal activity opens up the possibility of a relational understanding with future generations beyond immediate proximal interplay. Randall suggests that we should create the best possible caring practices and spirit in the present with an eye toward leaving them as an inheritance to future generations. Accordingly, "the present generation forms imaginal relations with future generations by virtue of their being part of a transgenerational community. That is, persons use their imaginal content to envision how their community will be inherited by, and make better the lives of, future generations" [27] (p. 540). Imagination stitches together present and future in a relationship of inherited care.

This introduction to the current work on care ethics and the environment offers a glimpse into the nature of the recent discussion for which the work of Latour might be brought to care theory. One scholar who has endeavored to make an explicit connection between Latour and care ethics on the environment is Chinese literary scholar, Adeline Johns-Putra. She argues that "our construction of "sustainability" is driven by a notion of care—care for the nonhuman environment enfolded with a concern for our human descendants" [34] (p. 126). Johns-Putra applies a new materialist approach to the environmental crisis, and she finds the liminal relationality of care to provide a vital moral force. New materialism is a family of interdisciplinary theories representing a post-constructionist, ontological, and/or material turn in contemporary theory. The dynamic notion of emergence is a focus of new materialism. Furthermore, this nascent field has a solid feminist influence and is sometimes framed as feminist new materialism [35] (p. 132). Bruno Latour, Karen Barad, Rosi Braidotti, and Elizabeth Grosz are among those associated with new materialism. Dissatisfied with extreme linguistic and constructivist approaches, new materialists turn to the question of matter. Johns-Putra credits Latour's analysis of materiality and being as motivating new materialism. She describes new materialism as reorienting theory discussions in terms of verbs and action. Accordingly, ontology is reframed as agency, and the notion of being is associated with "becoming".

In this manner, Johns-Putra characterizes care as always becoming and agential ("a means by which agency occurs") [34] (p. 134). She draws upon Latour for a vision of relational care that is not dependent on human initiation but part of the relational context of existence. To make this construction, Johns-Putra utilizes Latour's term "actants" to describe a universe (human and non-human material) as a collection of intra-active units or, again in Latour's words, a network [5] (p. 10). She asserts, "Care is part of the discursive and material mesh from which objects emerge. Care—in the act of being named and purportedly exercised—emerges from and re-submerges into that mesh" [34] (p. 134). Admittedly, this is a far cry from the notion of care put forth by Carol Gilligan in 1982 as an alternative to traditional justice approaches to human morality. However, this new materialism, fueled by Latourian thinking, demonstrates the broader potential of care as a way to think about being and becoming rather than simply as a moral theory.

5. Conclusions: Valuing Care Relationships to Confront the Anthropocene

Latour's desire to reassemble the social through actor–network theory, an analytical framework for which he is considered a founding father, along with Michel Callon and John Law, has relational implications even if he does not place the same kind of emphases as care theorists. For Latour, this reassemblage has at least three characteristics: (1) non-humans are raised in status to actors rather than "hapless bearers of symbolic projection" [5] (p. 10); (2) society must be viewed as dynamic without fixed or static categorical understanding; and (3) social network thinking is more than postmodern deconstruction and seeks "new institutions, procedures, and concepts able to collect and reconnect the social" [5] (p. 11). ANT is already fundamentally relational; however, in the spirit of Bellacasa, a care inflection can help translate ANT more centrally into care theory. In this conclusion, we briefly examine how translation, hiatus, tonality, modal crossings, and hesitation can be framed as essential relational understandings in a caring reassemblage of the social. Furthermore, we explore several theorists taking on this work, employing a care lens in the face of the Anthropocene.

Returning to the notion of translation discussed earlier, scientists are collecting data on climate change and engaging in matter–form translation that not only makes the information intelligible to non-specialists but "induces two mediators into co-existing." Given the relational ontology of care theory, such matter–form translation is not just a passing-through or even a creation of propositional knowledge but rather imbued with affective elements: it is matter to be cared about, an aspect of the *dingpolitik*. The matter of the Anthropocene is not simply one fact among many that we have to hold, but it is a complex assemblage of ideas that we are compelled to engage with. Effective caring—humble inquiry, inclusive connection, and responsive action—is taking the work of translation seriously in the profound sense that Latour describes.

Thus, hiatus can be understood as a respect for the task ahead: traversing the gap or interruptions between modes of existence. Care must be taken in the translation, and thus, methodology is just as crucial as epistemology in knowledge production and participation. Failure to respect hiatus damages relational integrity, as each mode has its tonality. One can see why Latour's project can be interpreted in care theoretical terms because respecting, inquiring into, and understanding the depth of hiatus and tonality facilitates reassembling the social. It is a constructive project of effective caring that requires and builds trust—as was one of Latour's stated goals in developing ANT. Hesitation, then, is not a sign of weakness, as masculinist and neoliberal narratives might have one believe. On the contrary, hesitation reflects the enormity of the task ahead, the gap between modes of existence to be forded, an example of the examined life. Humility and regard for the other are manifested in hesitation. In his book on the religious mode of existence, *Rejoicing: On the Torments of Religious Speech*, Latour employs the terms of his lexicon to remind his readers, "There is no other world, but there are several ways of living in this one and several ways, too, of knowing it" [36] (p. 34). We must address the Anthropocene together and without supernatural intervention, but we have tremendous capabilities at our disposal:

> There is no control and no all-powerful creator, either—no more 'God' than man—but there is care, scruple, cautiousness, attention, contemplation, hesitation and revival. To understand each other, all we have is what comes from our hands, but that does not mean our hands have to be taken for the origin [36] (p. 144).

Latour often dances around the language of care without employing the feminist literature. Nevertheless, the resonance is there. Several scholars of late have also seen the connections and have taken Latour's work in the direction of care theory.

Australian geographer Emma R. Power integrates Tronto's notion of "caring-with" into Latourian concepts of social assemblage to understand how local networks of care operate. Tronto proposes "caring-with" as a communal and generational practice of caring about care. Pertinent to this article, Tronto's "caring-with" suggests posthuman qualities. She describes caring about a future that is "not only about oneself and one's family and friends, but also about those with whom one disagrees, as well as the natural world and one's place in it" [23] (p. xii). Power is less interested in "caring-with" as a category of caring but rather as a conceptualization of care that "brings together related work from assemblage theories, actor–network approaches, and theories of dwelling" [37] (p. 765). Accordingly, à la Latour, Power conducts field research and then engages in matter–form translation employing a care hermeneutic. She empirically studies the care experiences of single older women living in precarious housing in Sydney, Australia. Her interviews reveal how care capacity is assembled under adverse circumstances:

> Caring-with shows that the capacity to care is not a sum of the properties of entities embroiled in a relation of care, but rather emerges from the ways they

come together in relation—as bodies, markets, material structures, and so on generatively combine to shape the possibility of care. Caring capacity was constituted from within and without as women adapter their caring practices, accommodated to poor housing, and assembled new resources and networks in the hope of sustaining their capacity to care [37] (p. 774).

This work exemplifies what we are describing as a care-infused ANT. However, what of the Anthropocene? What does an analysis of caring practices have to do with environmental devastation? Network and assemblage thinking are efforts of leaving no agents out of the discussion. Modernist categories of value that do not honor the connection between the plight of people and the planet are set aside for a more inclusive approach.

Anna Krzywoszynska, a geographer from the U.K., brings together care and network theory to address how expanded attentiveness to soils is critical in the age of the Anthropocene. In particular, she is concerned with soil biota or the ecosystem that exists just below the ground. Krzywoszynska notes that the Food and Agriculture Organization of the United Nations has indicated that humans have degraded one-third of Earth's soil [38] (p. 662). She proposes a care network as an inclusive method of enacting attentiveness to soils, particularly farm soils. Attentiveness has been a central component of care theorizing from its origin, but as Krzywoszynska points out, it is not often applied to the non-human world. Similarly, she notes the lack of attention given to soils generally. Citing Latour, Krzywoszynska contends that soils are shaped within human interactions rather than mere matters of fact [38] (p. 663). She describes her approach as employing a care network model defined as "the assemblage of interconnected entities whose existence enables the well-being of the primary object of care" [38] (p. 664). Her field research took her to speak with U.K. farmers about soil practices and their dispositions toward the soil. Krzywosznska found farmers who were indeed attentive to the needs of the soil. Coinciding with the effective care cycle of humble inquiry, inclusive connection, and responsive action, those who truly "listened" to the soil found themselves in opposition to neoliberal agribusiness practices that favored short-term productivity above all other considerations. The results included decisions to change crops grown, alter crop rotation patterns, and in some cases, return some farmland to a more natural state (thus out of production). Like Power, Krzywosznska is concerned that traditional individualistic approaches to ethical thinking are inadequate to the systemic and relational frameworks needed in light of the Anthropocene: "In the face of ecological and resource crises of the Anthropocene, a pragmatic and anthropocentric form of eco-sociality is emerging, in which caring for human survival and well-being starts to implicate caring for the survival and well-being of non-human entities" [38] (p. 672). Krzywosznska and Power are just two recent examples of the potential for integrating Latourian insight with care theory. Bellacasa may have blazed the trail, but others are quickly following.

A final example of the expansive notion of Latourian-like care can be found in the recent work of Serbian-born French anthropologist Dusan Kazic. Like Latour, Kazic makes systemic observations about the inadequacy of dominant economic and political narratives while finding examples of new relationships between matter and people that suggest a different way of being in the world. Here is an extended quote from an article written during the COVID-19 pandemic:

> farmers have never been in a relationship of "production" with their plants, but of "co-domestication." Farmers domesticate plants just as plants domesticate farmers, and this has been going on since the dawn of time. In concrete terms, this means that neither farmer, nor carrot plant, nor tomato plant, nor courgette [zucchini] plant, nor chicken, cow, pig or sheep has ever "produced" a single carrot, tomato, courgette, chick, calf or lamb. If we eat and live on this Earth, it is thanks to our relationships with living beings, without which no-one could live. This is why we cannot say we are suffering with hunger more *during* the lockdown than *before it*. As we enter into a new world, then, it is entirely possible "to imagine preventative measures against the resumption of pre-crisis

production," to use the title of Bruno Latour's article—because we have never lived *in* production, but always in a world that is about more than merely the human [39].

In *Quand les plantes n'en font qu'à leur tête* (*When Plants Do As They Please*), Kazic tells the stories of poor farmers and their relationships to the land and the land's inhabitants. He begins the book with an anecdote about one farmer who has a field covered with slugs. However, rather than eradicate them, he collects and observes them. He engages them with humble inquiry and connects with the slugs. Through the time and effort of this relationship, the farmer discovers that the slugs' slime heals the cracks in his hands that have arisen from tending the soil. The farmer declares that the slugs have a right to be there, particularly since neighboring farms have forced them away from other lands. Kazic explicitly describes the farmer as in a caring relationship with the slugs [40] (pp. 11–13). Like Latour, Kazic leverages the micro-observations of the particular, not to argue against farming for the purposes of feeding humans, but rather to make the point that thinking in terms of "economic production" does not provide all the answers. Resonating with care theory, Kazic desires a recentering on relationships, particularly human and non-human relationships. In one of his stories, Kazic engages a farmer whose relationship with weeds calls for apologizing to them for what must be done. Reminiscent of indigenous practices, apologizing recognizes the awareness of violence endemic to farming. For Kazic, apologizing is part of human–plant cohabitation [40] (p. 332). Given the threat of the Anthropocene, new ways of thinking are in dire need. Care can help provide a revolutionary means of seeing the world. As Kazic insists, his approach and what he recommends to others is not to bring a new critical analysis of neoliberal narratives but instead to imagine and speculate what is possible by observing what is going on [40] (p. 363).

Although care theory has garnered widespread attention and application up to this juncture, the liminality of care thinking and its implications for expansive ontology, epistemology, ethics, and politics provide perhaps the best opportunity for the complex and dynamic analysis needed to care for people and things in the shadow of the Anthropocene. Latour may not have foreseen such developments, but his insights and methodology cultivate the possibility of a care network assemblage approach. As mentioned earlier, care theorists have only begun to publish on the environment, and the available literature is not vast. Latour's notions of translation, composition, hiatus, tonality, modal crossings, and hesitation are relational terms that invite care thinking. Furthermore, Latour's method of translation and composition are able to address the political while maintaining the personal in a manner consistent with recent care scholarship. As Latour has stated, we were never really modern, and the Anthropocene may force scholars to think about collective and systematic care on a scale that defies the comfort of single-issue problem-solving. A co-created ethos of care that permeates each translation and composition may be required, given the enormity of the existential threat to humanity. Accordingly, the crisis of care and empathy is not a discrete issue from the Anthropocene, institutional racism, neoliberal precarity, and social alienation. The age of deconstruction has witnessed a warranted lack of trust in meta-narratives, but perhaps an inclusive, humble, and democratic meta-narrative of care still holds potential for both insight and hope for a better future. It will be exciting to see how scholars continue this research trajectory.

Author Contributions: Conceptualization, M.F. and M.H.; resources, M.F. and M.H.; schematic preparation, M.F.; writing—original draft preparation, M.F. and M.H.; writing—review and editing, M.F. and M.H. All authors have read and agreed to the published version of the manuscript.

Funding: This research received no external funding.

Acknowledgments: We thank Patrice Maniglier and Dusan Kazic for generously allowing us to read pre-publication manuscripts of their books.

Conflicts of Interest: The authors declare no conflict of interest.

Notes

[1]   There are no limitations on the use of the site, but visitors are asked to register. The website (modesofexistence.org; accessed on 19 January 2022) is available in both English and French. The full text of An Inquiry into Modes of Existence is present on the website and includes a number of interactive features. An especially powerful feature of the site is the ability to interactively examine several examples of all possible two-fold modal "crossings" of the sort imagined in Figure 1 of the present text.

[2]   All the schematics were created by Michael Flower. They are not Bruno Latour's.

[3]   Understanding Objects as quasi-objects, as instances of dynamic entanglement, offers us an avenue for seeing things differently. Consider COVID-19. It is a dynamic actant; it is an instance of faire faire. It forces a "making [others] to do". COVID-19 is entangled with bodies that differ in their response to infection, but within which the virus replicates and mutates. The quasi-object COVID-19 does not travel alone. It is accompanied by testing protocols, epidemiologists, public health directives, political disputes, vaccines and disparities in vaccine availability, and more. When COVID-19 mutates, its successful variants (e.g., Delta, then the dramatically more transmissible form, Omicron) result in a change in its associates: testing protocols cannot keep up with the rapidity of transmission, epidemiological profiles change, health directives are rethought, antivax battle lines are redrawn, and more. The entanglement that is the quasi-object COVID-19, var. Omicron is a "making [others] to do" differently; it is dynamically different.

[4]   With this third group we see the rendering of persons as assemblages in a way that illustrates the inadequacy of the notion of Subject. "To sum up the originality of this THIRD GROUP in an overhasty sentence, let us say that, while following along the political Circle, humans become capable of opining and of articulating positions in a collective—they become free and autonomous citizens; by being attached to the forms of law, they become capable of continuity in time and space—they become assured, attributable selves responsible for their acts; by receiving the religious Word, they become capable of salvation and perdition—they are now PERSONS, recognized, loved, and sometimes saved".

References

1.   Thiery, W.; Lange, S.; Rogelj, J.; Schleussner, C.F.; Gudmundsson, L.; Seneviratne, S.I.; Andrijevic, M.; Frieler, K.; Emanuel, K.; Geiger, T.; et al. Intergenerational Inequities In Exposure to Climate Extremes: Young Generations are Severely Threatened by Climate Change. *Science* **2021**, *374*, 158–160. [CrossRef] [PubMed]

2.   Latour, B. *Down to Earth: Politics in the New Climatic Regime*; Polity Press: Cambridge, UK, 2018.

3.   Full text of Greta Thunberg's "How Dare You" speech at the UN Climate Action Summit. Available online: https://www.cntraveller.in/story/full-text-greta-thunberg-speech-how-dare-you-un-climate-action-summit/ (accessed on 5 January 2022).

4.   Ferrando, F. Posthumanism, Transhumanism, Antihumanism, Metahumanism, and New Materialisms: Differences and Relations. *Existenz* **2013**, *8*, 26–32.

5.   Latour, B. *Reassembling the Social: An Introduction to Action-Network-Theory*; Oxford University Press: New York, NY, USA, 2005.

6.   Latour, B. Some advantages of the notion of "Critical Zone" for Geopolitics. *Procedia Earth Planet. Sci.* **2014**, *10*, 3–6. [CrossRef]

7.   Latour, B. Seven Objections against Landing on Earth. In *Critical Zones: The Science and Politics of Landing on Earth*; Latour, B., Weibel, P., Eds.; The MIT Press: Cambridge, MA, USA, 2020; pp. 12–19.

8.   Latour, B. On Technical Mediation—Philosophy, Sociology, Genealogy. *Common Knowl.* **1994**, *3*, 29–64.

9.   Latour, B. An Attempt at a "Compositionist Manifesto. " *New Lit. Hist.* **2010**, *41*, 471–490.

10.   Latour, B. Circulating Reference: Sampling the Soil in the Amazon Forest. In *Pandora's Hope: Essays on the Reality of Science Studies*; Harvard University Press: Cambridge, MA, USA, 1999; pp. 24–79.

11.   Latour, B. *An Inquiry into Modes of Existence*; Harvard University Press: Cambridge, MA, USA, 2013.

12.   Latour, B. Reflections on Etienne Souriau's Les différents modes d'existence. In *The Speculative Turn: Continental Materialism and Realism*; Bryant, L., Srnicek, N., Harman, G., Eds.; re.press: Melbourne, Australia, 2011; pp. 304–333.

13.   Stengers, I. *Catastrophic Times: Resisting the Coming Barbarism*; Open Humanities Press: London, UK, 2015.

14.   Latour, B. Bruno Latour Tracks Down Gaia. Los Angeles Review of Books, 3 July 2018. Available online: https://lareviewofbooks.org/article/bruno-latour-tracks-down-gaia/ (accessed on 7 March 2022).

15.   Lenton, T.; Dutreuil, S.; Latour, B. Life on Earth is hard to spot. *Anthropol. Rev.* **2020**, *7*, 248–272. [CrossRef]

16.   Manglier, P. *Le Philosophe, la Terre et le Virus: Bruno Latour Expliqué par l'Actualité*; Les Liens Qui Libèrent: Paris, France, 2021.

17.   Puig de la Bellacasa, M. *Matters of Care: Speculative Ethics in More Than Human Worlds*; University of Minnesota Press: Minneapolis, MN, USA, 2017.

18.   Harman, G. Demodernizing the Humanities with Latour. *New Lit. Hist.* **2016**, *47*, 249–274. [CrossRef]

19.   Haraway, D. *Staying with the Trouble: Making Kin in the Chthulucene*; Duke University Press: Durham, NC, USA, 2016.

20.   Latour, B. Why Has Critique Run out of Steam? From Matters of Fact to Matters of Concern. *Crit. Inq.* **2004**, *30*, 225–248. [CrossRef]

21.   Latour, B. *Aramis, Or the Love of Technology*; Harvard University Press: Cambridge, MA, USA, 1996.

22.   B. Latour: From Realpolitik to Dingpolitik—or How to Make Things Public. Available online: https://www.pavilionmagazine.org/bruno-latour-from-realpolitik-to-dingpolitik-or-how-to-make-things-public/ (accessed on 5 January 2022).

23.   Tronto, J. *Care Democracy: Markets, Equality, and Justice*; New York University Press: New York, NY, USA, 2013.

24.   Nakano Glenn, E. *Forced to Carre: Coercion and Caregiving in America*; Harvard University Press: Cambridge, MA, USA, 2010.

25. Whyte, K.P.; Cuomo, C. Ethics of Caring in Environmental Ethics: Indigenous and Feminist Philosophies. In *The Oxford Handbook of Environmental Ethics*; Gardiner, S., Thompson, A., Eds.; Oxford University Press: New York, NY, USA, 2017; pp. 234–247.
26. Allison, E. Toward a Feminist Care Ethic for Climate Change. *J. Fem. Stud. Relig.* **2017**, *33*, 152–158. [CrossRef]
27. Randall, T. Care Ethics and Obligations to Future Generations. *Hypatia* **2019**, *3*, 527–545. [CrossRef]
28. Staudt, M.S. Caring reciprocity as a relational and political ideal in Confucianism and care ethics. In *Care Ethics and Political Theory*; Engster, D., Hamington, M., Eds.; Oxford University Press: New York, NY, USA, 2015; pp. 187–207.
29. Noddings, N. *Caring: A Feminine Approach to Ethics and Moral Education*; University of California Press: Berkeley, CA, USA, 1984.
30. Hamington, M. *Embodied Care: Jane Addams, Maurice Merleau-Ponty, and Feminist Ethics*; University of Illinois Press: Urbana, IL, USA, 2001.
31. Klaver, K.; Baart, A. Attentiveness in care: Towards a theoretical framework. *Nurs. Ethics* **2011**, *18*, 686–693. [CrossRef]
32. Groves, C. *Care, Uncertainty, and Intergenerational Ethics*; Palgrave Macmillan: London, UK, 2014.
33. Hamington, M.; Flower, M. (Eds.) *Care Ethics in the Age of Precarity*; University of Minnesota Press: Minneapolis, MN, USA, 2021.
34. Johns-Putra, A. Environmental Care Ethics: Notes toward a New Materialist Critique. *Symplokē* **2013**, *21*, 125–135. [CrossRef]
35. Dionne, E. Resisting Neoliberalism: A Feminist New Materialist Ethics of Care to Respond to Precarious World(s). In *Care Ethics in the Age of Precarity*; Hamington, M., Flower, M., Eds.; University of Minnesota Press: Minneapolis, MN, USA, 2021; pp. 229–259.
36. Latour, B. *Rejoicing, or the Torments of Religious Speech*; Polity Press: Cambridge, UK, 2002.
37. Power, E.R. Assembling the capacity to care: Caring-with precarious housing. *Trans. Inst. Br. Geogr.* **2019**, *44*, 763–777. [CrossRef]
38. Krzywoszynska, A. Caring for soil life in the Anthropocene: The role of attentiveness in more-than-human ethics. *Trans. Inst. Br. Geogr.* **2019**, *44*, 661–675. [CrossRef]
39. Kazic, D. COVID-19, My Ambivalent Ally. Trans. Tim Howles. AOC 15 September 2020. Available online: https://logisticsofreligionblog.wordpress.com/2020/09/16/exiting-from-the-economy-translation-of-a-piece-by-dusan-kazic/ (accessed on 11 January 2022).
40. Kazic, D. *Quand les Plantes n'en font qu'à leur Tête [When Plants Do as They Please]*; Renaud-Bray, FR: Montreal, QC, USA, 2022.

 *philosophies*

MDPI

Article

A Care Ethical Engagement with John Locke on Toleration

Thomas Randall

Department of Political Science, Western University, London, ON N6G 2V4, Canada;
thomaserandall@outlook.com

Abstract: Care theorists have yet to outline an account of how the concept of toleration should function in their normative framework. This lack of outline is a notable gap in the literature, particularly for demonstrating whether care ethics can appropriately address cases of moral disagreement within contemporary pluralistic societies; in other words, does care ethics have the conceptual resources to recognize the disapproval that is inherent in an act of toleration while simultaneously upholding the positive values of care without contradiction? By engaging care ethics with John Locke's (1632–1704) influential corpus on toleration, I answer the above question by building the bases for a novel theory of *toleration as care*. Specifically, I argue that care theorists can home in on an oft-overlooked aspect of Locke's later thought: that the possibility of a tolerant society is dependent on a societal ethos of trustworthiness and civility, to the point where Locke sets out positive ethical demands on both persons and the state to ensure this ethos can grow and be sustained. By leveraging and augmenting Locke's thought within the care ethical framework, I clarify how care ethics can provide meaningful solutions to moral disagreement within contemporary pluralistic societies in ways preferable to the capability of a liberal state.

Keywords: care ethics; John Locke; toleration; liberalism; feminism; trustworthiness; civility; Anna Galeotti; recognition; neutrality

Citation: Randall, T. A Care Ethical Engagement with John Locke on Toleration. *Philosophies* **2022**, *7*, 49. https://doi.org/10.3390/philosophies7030049

Academic Editor: Marcin J. Schroeder

Received: 24 February 2022
Accepted: 18 April 2022
Published: 26 April 2022

Publisher's Note: MDPI stays neutral with regard to jurisdictional claims in published maps and institutional affiliations.

1. Introduction

Care theorists have yet to outline an account of how the concept of toleration should function in their normative framework. This lack of outline is a notable gap in the literature, particularly for demonstrating whether care ethics can address moral disagreement within contemporary pluralistic societies. Care theorists certainly have the vision to do this— Virginia Held writes that "the value of caring that can be seen most clearly in such activities as mothering is just what must be extended, in less intense but not entirely different forms, to fellow members of society and the world" [1] (p. 89). Yet Held does not elucidate what should happen when this extension of care is immobilized due to moral disagreement. For instance, despite the United States' Supreme Court 2015 ruling in Obergefell v. Hodges that requires all states to issue licenses for and recognize same-sex marriage, various states still refuse to tolerate (let alone celebrate) same-sex marriage. Even five years after the Supreme Court ruling, the religious right continued to block attempts to remove same-sex marriage bans in Republican-controlled Legislatures in at least Indiana and Florida. How does care ethics leverage its conceptual resources to appropriately understand and respond to issues such as this? Are care theorists able to recognize the disapproval that is inherent in an act of toleration while simultaneously upholding the positive values of care without contradiction? In other words, what could a theory of *toleration as care* look like?

I address these questions by engaging care ethics with John Locke's (1632–1704) thought on toleration. The reason for focusing on Locke is because there is a rich literature that care theorists can draw on here to begin their foray into developing a theory of toleration—not only within Locke's own influential corpus but also in the voluminous secondary literature on Locke that has concurrently played a substantial role in shaping contemporary liberal theories of justice. Care ethics and Locke (and, by extension, liberal theories of justice) will diverge on some foundational conceptual schema, preventing care theorists from straightforwardly appropriating Locke's thought (as I will later detail in this paper). However, there are still significant opportunities for care theorists to leverage Locke's thought toward a theory of toleration as care. Specifically, care theorists can home in on an oft-overlooked aspect of Locke's later thought: that the possibility of a tolerant society is dependent on a societal ethos of trustworthiness and civility, to the point where Locke sets out positive ethical demands on both persons and the state to ensure this ethos can grow and be sustained.

This paper moves in three steps. First, it traces Locke's changing views on toleration over his lifetime, highlighting Locke's later focus on the criteria needed for toleration to emerge in society: in essence, the need for a societal ethos of trustworthiness and civility. Second, I connect common ground between care theorists and Locke on the importance of this ethos and the role of the state in enabling such an ethos. An interesting discovery at this point will be that contemporary liberal thinkers (operating under the guise of 'Lockean liberalism') inadvertently reject Locke's requirement for the state to play an active role in promoting the conditions for toleration to emerge. Third, then, I examine how care ethics can model a theory of toleration by drawing inspiration from Locke's criteria for toleration to emerge—and, as an upshot, reveal that Locke's criteria could be better realized within the care ethical framework. The resulting theory of toleration as care is defended as the preferable interpretation of toleration for care theorists to utilize moving forward. For toleration as care, the point of toleration is not for setting up a neutral space for peaceful coexistence (albeit in an uncomfortable *modus vivendi*). Toleration as care goes further, obligating the state and its residents to develop an ethos of trustworthiness and civility to build the space for sincere discourse and/or praxis that supports and maintains good caring relations.

2. Tracing Locke's Thought on Toleration

Toleration, as generally understood in contemporary liberal theories of justice, refers to "the principle of peaceful coexistence where there are conflicting, incompatible, and irreducible differences in ways of life, practices, habits, and characters" [2] (p. 22). While care theorists have remained largely silent on how toleration ought to function within their normative framework, Daniel Engster [3] has been one of the few care theorists to address issues of toleration. However, I believe Engster's approach does not provide a suitable outline for how care theorists should think about toleration—thus underscoring the need for a stronger theory of toleration as care. Engster's approach is derived from his minimal definition of care. This definition of care is minimal because it restricts 'care' to only encompass three aims: (1) help "individuals to satisfy their vital biological needs"; (2) aid "individuals to develop and sustain their basic or innate capabilities, including the abilities for sensation, movement, emotion, imagination, reason, speech, affiliation, and in most societies today, the ability to read, write, and perform basic math"; and, (3) "helping individuals to avoid harm and relieve unnecessary or unwanted suffering" [3] (pp. 25–28). These three aims of care are reinforced through three virtues of caring: attentiveness ("Do you need something?"); responsiveness ("What do you need?"); and respect ("the recognition that others are worthy of our attention and responsiveness") [3] (pp. 30–31). This minimal care definition informs Engster's argument on the proper function of toleration: "Practices that effectively support adequate care for all individuals are moral, practices that impede adequate care are immoral, and practices that neither support nor impede adequate care are indifferent" [3] (p. 98).

The problem with Engster's theory is that, when confronted with resolving non-trivial cases of toleration within a pluralistic society, the minimal care argument does not have anything of use to say. If the only aspect of a religious, cultural, and moral system that is considered by Engster is the minimal level of care provided to its members, then Engster's theory becomes blind to serious cases of multicultural clashes that move beyond minimal caring. Consider, for instance, Engster's comments on same-sex marriage.

Engster is aware that the symbolic public gesture of legalizing same-sex marriage would legitimize this practice as "normal" in the realm of the viable options open to society. However, it is not clear that the minimal definition of care can capture the importance of this symbolic recognition. Engster even explicitly states, "The argument for and against symbolic recognition fall outside the scope of care theory" [3] (p. 108). As Engster continues, his care theory "is neutral on other issues of multicultural justice that do not directly affect the ability of individuals to give or receive care" [3] (p. 107). Consequently, Engster's theory is silent on non-trivial cases of moral disagreement—cases that care ethics ought to at least recognize and have something meaningful to say about how they ought to be resolved. This is especially if care ethics aims to be a feminist ethic—that is, an ethic that can identify and critically evaluate instances of suppression and dominance.

I will argue that care theorists can generate a more robust theory of toleration, and that they can do so via an engaged study with prominent authors in the history of liberal ideas—specifically, Locke's thought on toleration. Given that the main goals of liberalism "are the protection and the fostering of individual freedom and the limitation of justifiable coercion on the part of the state," toleration constitutes an essential element of the liberal project [2] (p. 22). Typically, the evolution of this position is traced to Locke's writings on religious toleration. Above being a seminal 17th century philosopher in the development of empiricism and medical science, Locke's political works were an important precursor to liberalism [4] (p. xxxii). Typically, 'Lockean liberalism' is understood to have proclaimed "the irrationality of persecution, the neutrality of the state, and individual rights and equal protection" [2] (p. 23)—ideas that have become the norm in contemporary liberal democracies. For instance, Jeremy Waldron, on his influential reading of Locke's *Letter Concerning Toleration* (1689), writes that Locke's position on toleration "is a negative one […]. Nothing is entailed about the positive value of religious or moral diversity" nor the role of the state in ushering such diversity [5] (p. 76). We can trace this history of neutrality to contemporary influential liberal thought, exemplified, for instance, in John Rawls' idea of justifying principles of justice via overlapping consensus of differing values across society [6] (pp. 189–193).

Now, based on the current care ethical literature, it may seem initially odd to look to Locke for inspiration on building a theory of toleration. After all, care theorists have frequently targeted the above interpretation of 'Lockean liberalism' in their critical analysis of the liberal normative framework. For instance, Fiona Robinson critiques liberalism's *laissez-faire* stance on the contractual role of the state in only enforcing negative rights—all of which, Robison states, is based on a faulty Lockean "ontology of atomistic individualism that privileges the norm of self-sufficiency" [7] (p. 50)[1]. However, I believe that care theorists have not questioned whether this interpretation of 'Lockean liberalism' is actually an accurate reflection of Locke's thought. For even when some care theorists *have* argued in favour of caregiving operating within a liberal framework (such as Asha Bhandary's liberal account of dependency care [9]), this understanding of 'Lockean liberalism' (especially as expressed by Waldron) has been taken at face value without critical analysis.

The reason for my above suspicion stems from Nicholas Jolley [10], Teresa Bejan [11], and others, who have recently argued this 'Lockean liberalism' is a caricature of Locke's thought. Though Locke's *Letter* is "one of the foundational documents of the liberal tradition," Locke wrote several other important works that give broader context to his views on toleration [12] (p. 989). This not only includes the precursory *Tracts on Government* (1660–1662) [13,14] and *An Essay on Toleration* (1667) [15] that Locke did not publish, but three further letters Locke wrote in an ongoing exchange with critic Jonas Proast (throughout 1690–1692) [16,17] and *Some Thoughts Concerning Education and Of the Conduct of the Understanding* (1693) [18]. As suggested by the broad timeline over which these works were written, Locke had a "lifelong preoccupation" with the issue of toleration [10] (p. 4). Consequently, limiting our understanding of Locke's thought to the most obvious work— the *Letter*—does not give us the whole story. For even then, the *Letter* is a loose English translation of the original Latin *Epistola de Tolerantia*—a translation by William Popple and published without Locke's consent. If this were not enough, we do further disservice to Locke's thought by only concentrating on particular sections within the *Letter* to generalize a specific 'Lockean' argument—an exercise that John William Tate has accused Waldron of doing [12]. It is also notable that one of Locke's most well-known works on political thought, the *Second Treatise of Government* (1689) [19], does not even mention toleration—with Jolley suggesting that the *Second Treatise* is but a complimentary text to the *Epistola* and Locke's subsequent letter exchanges with Proast [10]. Exploring texts beyond the traditional Lockean canon, then, provides a better context for understanding Locke's complex views on the role of the state and the enablement of toleration in society.

This prelude has sought to problematize how we might traditionally approach Locke's thought on toleration and the veracity of influential secondary literature. In what follows for this section, I follow Jolley [10] and Bejan [11] in outlining a more accurate evolution of Locke's thought on toleration via his broader works and the context in which they were written. In turn, this tracing exercise will demonstrate where and why Locke's thought has the potential to converge with care ethics.

Locke's formative years saw the eruption of the English Civil War in 1642, a result of deep-seated conflicts between the Crown and Parliament over English governance and the manner of religious freedom between Christian denominations. By 1651, the Parliamentarians had beheaded Charles I and Oliver Cromwell began his Protectorate. The following year, Locke joined Christ Church, Oxford, to complete his B.A. It is perhaps no surprise, then, to find that Locke's *Tracts* (written during his time at Oxford) reveal how impacted he had been by this political turmoil: "I no sooner perceived myself in the world but I found myself in a storm, which hath lasted almost hitherto" [13] (p. 7). What we also find in the *Tracts*, though, is that Locke's early thought had considerable Hobbesian sympathies. Thomas Hobbes (1588–1679) had published *Leviathan* (1651) [20] at the end of the Civil War, a work representative of preventing future political upheavals via justification of legitimate political power centered in an absolute sovereign. Locke was among those students at Oxford during the 1650s who had read *Leviathan* and was particularly inspired by Hobbes' latitudinarian arguments about religion [21] (pp. 238–239).

In the *Tracts*, Locke followed Hobbes in arguing that a magistrate ought to hold absolute authority vis-à-vis which religious practices were fundamental (*fundamenta*) and which were matters of indifference (*adiaphora*). Here, Locke believed that the Civil War was illustrative of what happens when such magistrate authority does not exist: worsening discourse between religious denominations leads to civil anarchy and religious fanaticism. For Locke, "this war of words had exacerbated spiritual, social, and political divisions and corroded civil society from within" [11] (p. 119). In particular, Locke initially believed toleration to be infeasible due the English having a propensity for being "ready to conclude God dishonoured upon every small deviation from [their] way of his worship [and] apt to judge every other exercise of religion as an affront to theirs" [13] (p. 42). Throughout his life, Locke maintained that religious enthusiasm was a bane of the prospect of a tolerant society, devoting an entire section to it in his *Letter* some thirty years later [22,23] (pp. 177–180). Hobbes offered the young Locke an eirenic hope through the guise of an absolute sovereign to regulate societal interactions. Without it, Locke was deeply suspicious of toleration emerging—that is, for persons of different religious denominations to behave peaceably toward one another in civil *adiaphora*. Locke's *Tracts* outlined the means for bringing religious denominations together through the outward union of inwardly divided minds, enforced by a civil magistrate.

Yet, as the 1660s went on, Locke's early intolerance began to mitigate. By the time of his writing the *Essay* in 1667, Locke's thought had shifted from the *Tracts*. Contemporary Lockean scholars tend to point to biographical information to explain this change. In particular, Locke's first extended leave from the Anglican walls of Christ Church to Cleves in 1665 left Locke astounded with how different religious denominations (Lutherans, Calvinists, and even Catholics) worshipped peaceably and publicly: "I cannot observe any quarrels or animosities amongst them on account of religion [. . .] [They] quietly permit one another to choose their way to heaven" [24] (1.228). His move thereafter to a cosmopolitan London under his association with Lord Ashley (the future Earl of Shaftesbury), revealed to Locke that living with civil *adiaphora* was possible after all.

This realization triggered a crucial distinction Locke makes in his *Essay*, written as a memorandum for Shaftesbury and only privately circulated. Whereas in the *Tracts* Locke perceived incivility as almost necessarily accompanying religious disagreement, Locke now saw it possible for religious disagreement to exist in peaceable and civil circumstances. Rather than religious uniformity imposed by magistrate authority, toleration could be possible if persons had *trust* in one another to be peaceable in the face of religious disagreement—or what Locke referred to more strongly as the "Bond of Society" [25] (p. 132). If this bond should exist, in which persons trusted one another to remain civil in the face of religious disagreement, meaningful toleration could emerge from such conditions. Whereas Locke had initially followed Hobbes to focus on enforced outward civility, the *Essay* demonstrated Locke's growing interest in the "beliefs, attitudes, and dispositions of individuals" [11] (p. 126). This emphasis on societal trust and the sincerity of the outward expression of one's internal beliefs would mature over the next 20 years and take a prominent form in Locke's *Letter*.

Despite these signs of Locke's maturing thought, the *Essay* retained continuities with the *Tracts*. Locke still believed, for instance, that the magistrate should be allowed to govern what civil *adiaphora* looked like in the best interest of suppressing civil unrest—especially if any religious denomination's signs, symbols, and worshippers, could be judged "numerous enough to become dangerous to the state" [11] (p. 148). Locke did not explicate in the *Essay* what counted as being dangerous—though, considering Locke's introduction of societal trust, it is likely that 'dangerous' referred to that which caused civility between persons to break down.

In the 20 years between the *Essay* and Locke's writing of the *Letter*, Locke was exiled for his suspected role in the 1683 Rye House plot to assassinate Charles II and his brother, James, Duke of York. Yet, as Bejan notes, it is significant that Locke (like Hobbes before him), "wrote his most systematic treatment of toleration after an experience of exile—albeit in the commercial and cosmopolitan cities of the Netherlands" [11] (p. 127). In the *Letter*, Locke elaborated his distinction in the *Essay* between 'religious disagreement' and 'incivility', rather than conflating the two as he had done in the *Tracts*: "It is not the Diversity of Opinions (which cannot be avoided), but the Refusal of Tolerating those that are different Opinions [. . .] that has produced all the Bustles and Wars that have been in the Christian World" [22] (p. 60). Furthermore, in a step that appeared to finally shake off his Hobbesian sympathies, Locke now argued that the magistrate should not have absolute control in civil nor religious *adiaphora*—an important move that sought to conceptually separate civil and church communities into different domains. Against Hobbes, Locke wrote that "there is absolutely no such thing, under the Gospel, as a Christian commonwealth" [22] (p. 42).

With this conceptual separation of church and civil communities, Locke moved away from his previous notion that toleration emerged as part of some minimal state enforcing legislation. Instead, Locke focused on toleration as emerging from individual behavior, beliefs, attitudes, and interactions with others. In the *Letter*, Locke now wrote that the "narrow Measures of bare Justice" would not suffice [22] (p. 20). Rather, toleration emerges as a possibility within civil society through the recognition that Christian "Charity, Bounty, and Liberality [. . .] Equity [and] Friendship" must always mutually "be observed" by fellow citizens [22] (p. 20). In saying so, Locke emphasized the need for positive ethical demands to be placed on individuals to avoid the "rough Usage of Word or Action" and amplify "the softness of Civility" in disputes [22] (pp. 19, 23).

What is meant by the 'softness of civility' is contested, given that the phrase comes from Popple in his translation of the *Letter* (from the Latin *Epistola*) in 1689. However, some of Locke's later works (including those that aimed to defend the *Letter*) provide insight into what 'civility' could mean here. *Some Thoughts*, in particular, is explicit in Locke's definition of civility as part of one's beliefs and attitudes, rather than a Hobbesian expression of outward politeness for the sake of mere *modus vivendi*. In *Some Thoughts*, civility is referred to as "that general good will and regard for all people which makes anyone have a care not to show [them] any contempt, disrespect, or neglect" [18] (p. 43). Civility was the "disposition of the mind not to offend others" [18] (p. 107). Maintaining toleration was about how disagreement was handled and not disagreement *per se*. While Locke's elitism shines through in *Some Thoughts* (that civility concerned the social virtues expressed by a gentleman engaging in reasoned discourse in the appropriate manner), the overall point is a significant one: that resolving disagreement and maintaining toleration lay in teaching others (children especially) to "love and respect other people" [18] (p. 110).

Written while he assembled *Some Thoughts*, Locke's *Second Letter Concerning Toleration* and *A Third Letter for Toleration* [16,17] make the full link between charity and civility for the basis of toleration. Composed in response to Proast's criticism of the *Letter*, the second and third letters reveal Locke's expanding global view and the possibility for persons of differing faiths (and not merely Christian denominations) to live peaceably together, including with "a rational Turk or Infidel" or "a sober sensible Heathen" [17] (p. 82). The *Letter* itself was quite clear that discrimination was not grounded on one's religion: "neither Pagan, nor Mahumetan, nor Jew, ought to be excluded from the Civil Rights of the Commonwealth because of his religion" [22] (pp. 58–59). Locke's emphasis on trust (or the worthiness of a person's trust) that was first laid out in his *Essay* now became central to his argument for a tolerant society. In the second and third letters, Locke explored the conditions for such trust: namely, removing the binary of inward mental states and outward expressions. Instead, Locke defended the idea that one's outward civil expression ought to be a sincere representation of one's inner beliefs. Combining an outward expression of good faith and the sincerity of one's internal beliefs would continually create and reinforce mutual trust "through the practice of disagreement itself" [11] (p. 137). That is, moral disagreement between persons could be tolerated via trustworthiness built by civil honesty and sincere discourse through the very act of disagreement.

Of course, Locke retained limits on toleration and who could be tolerated. This included not tolerating those whose beliefs would threaten societal trust in others, such as atheists (who threatened secure promises, covenants, and oaths if they believed no God was there to punish deviance) and Catholics (whose loyalties, Locke accused, were always to the Pope ahead of their country or fellow patriots) [4] (p. xiii). Yet the magistrate was not just there to enforce negative rights for upholding a tolerant society; Locke's positive demands were also to encourage magistrates to change persons' manners and dispositions instead of performing mere punishment [11] (p. 140). The magistrate was not intended to be a neutral supervisor but to set fellow citizens a good example for civil conduct and evangelism. The later Locke even considered state-sponsored evangelism, "cooperating with the church to ensure that citizens divided in their religious opinions would remain safely united through virtuous living" [11] (p. 140).

A tolerant society, then, was dependent on this ethos of trustworthiness and civility, in which toleration could only function if fellow citizens were able to trust one another to be civil during religious disagreement. In this, the role of the magistrate was not just to enforce the negative rights of persons by preventing harm that might emerge during disagreement; it was also to lay the groundwork for influencing the attitudinal mindset of persons to be sincere and respectful in their interactions with one another. Such sincerity abetted trustworthiness, in turn abetting civility in matters of moral disagreement.

In the context of toleration, then, the Locke we find at the end of this section is a thinker concerned with the bonds of society, trustworthiness and civility, and the role of the state beyond mere enforcement of negative rights. In siding with Jolley, Bejan, and others, I hope to have provided here some nuance that lays the groundwork for how care ethics could converge with Locke's thought, decrying the 'Lockean liberalism' exhibited at this paper's outset.

3. Engaging Care Ethics with Locke on Toleration

By focusing on the three points of concern listed at the end of the previous section (the bonds of society, trustworthiness and civility, and the role of the state beyond mere enforcement of negative rights), this section aims to demonstrate where Locke's thought on toleration converges with care ethics in important ways. This is not to say that there is perfect convergence, whereby simply transposing Locke into the care ethical framework produces an out-of-the-box theory of toleration. What it is to say is that care ethics can provide the conceptual resources that Locke's criteria require, including the source of obligation for why individuals and the state ought to build and reinforce an ethos of trustworthiness and civility. In turn, this will provide the foundation for a theory of toleration as care.

The first point of concern is Locke's interest in the "Bond of Society" [25] (p. 132), which aimed at ensuring persons viewed others as having trustworthiness to be civil in the face of religious disagreement. On first reading, Locke's concern appears to align with a standard liberal interpretation of the self—that is, we find the self as a vulnerable being situated in relations to others, requiring the state to shield their liberties and opportunities to pursue their life plan without hinderance. As Immanuel Kant later put it, the need for political society emerges from our interdependency of "living side by side" [26] (p. 96), whereby all our actions are potentially other-regarding. If we understand Locke to simply mean that the strength of this societal bond lies in the ability of the state to enforce negative rights to prevent harmful actions, then this liberal view holds. However, I think it is more complicated than this. It is not a mere connection from simply recognizing our actions are other-regarding; it is to promote "love and respect" for "other people," particularly teaching children such behaviour [18] (p. 110). This moves beyond mere outward politeness for the sake of prudence and instead toward sincere concern for others and their wellbeing. Care theorists will find more in common with this view of Locke's bond of society.

Within the care ethical framework, the self is perceived as being both situated in *and* constituted by their relations [27] (p. 152)[2]. To be a person is to be a "temporally extended embodied subject whose identity is constituted in and through one's lived bodily engagement with the world and with others" [28] (p. 119). Though we are embodied in a bound and discrete body, the self is a node that emerges via its sociality with other persons, saturating our embodiment with meaning [3,29–31]. Relations do not just join us together; they weave into our very being. Now, I do not say here that Locke holds this relational view. There are genuine irreconcilable differences between care theorists' relational view of the self and Locke's (and the broader liberal framework's) individualistic view of the self [32]. Yet there is still plenty of overlap between Locke and the care ethical framework on the importance of positive duties between persons for the functioning of society in meaningful ways. There is already a broad literature that care theorists have developed that emphasizes the importance of societal connection, alongside teaching others the importance of empathy, love, and respect. For instance, Michael Slote has argued that developing our empathic capacities in schools solidifies in children the moral obligations that we have to others beyond our nearest and dearest. One method is to expose "children to literature, films, or television programs that make the troubles and tragedies of distant or otherwise unknown (groups of) people vivid to them" [33] (p. 29). Another method is to "provide for more international student exchanges than now exist" to develop an understanding of distant other's livelihoods [33] (p. 29). All aspects point to the possibility of toleration by way of understanding the *other* and our first interactions being ones of love and respect, rather than initial suspicion.

The value that Locke places on trustworthiness and civility comes to the fore here, as the second point of concern. Without trustworthiness and civility, the enterprise of toleration fails. It is significant, then, that trustworthiness and the components of civility form similar fundamental values for how care theorists assess the moral worth of relations—personal, societal, and beyond. Care theorists' fundamental normative claim centers on our relational interdependencies of care: while we all need care, this by itself is not enough—we need an *ethics* of care, one that continually presses for the moral evaluation of care provision and receipt to avoid and prevent asymmetrical relations of power becoming "dominating, exploitative, hostile, mistrustful, or negligent" [1] (p. 37). In turn, this normative claim guides us toward identifying what moral values are intertwined in the activity of good caring, and what responsibilities emerge (individually and collectively) through these values to ensure our needs are appropriately understood, respected, and allowed to flourish [1] (pp. 31–33). There are four principal caring values: attentiveness, mutual concern, responsiveness, and trustworthiness [34]. Relations are identified as holding moral worth when they exemplify these caring values in fulfilling successful caring practices. We can call these relations "good caring relations" [34] (p. 70).

Locke's reference to trustworthiness maps onto care theorists' own understanding of trustworthiness. For care theorists, trustworthiness is a value exemplified when persons in a relation uphold certain normative expectations and do not pursue deceitful or hostile actions toward each other [1,35]. As Annette Baier has argued, activities of trust-building are mutually reinforcing, creating a "climate of trust" in which relations become increasingly meaningful over time [35] (p. 177). There are two different accounts of trustworthiness that can emerge here, which Karen Jones labels "risk-assessment" and "will-based" [36] (p. 68). A "risk-assessment" account of trustworthiness underpins non-intimate relations and is especially found in Hobbes' thought. This view describes a minimal climate of trust in which persons expect others to act in a certain way because it is in these person's self-interest to act in that way. However, care theorists do not fully subscribe to this risk-assessment view. Instead, care theorists are more amenable to the "will-based" account of trustworthiness, which follows Baier's interpretation given above. This account asserts that trustworthiness emerges through a trustee motivated by goodwill towards another person—that the trustee actually cares for the trustor's wellbeing. Whereas the risk-assessment account does not assume that persons care for each other intimately, the will-based account does assume the possibility of doing so.

While the early Locke can be charged with holding a risk-assessment account of trustworthiness (in his following of Hobbes), the later Locke appears to follow a will-based account of trustworthiness that enables the possibility of toleration to emerge. Here, the conditions of trust depend upon the outward expression of good faith and the sincerity of one's internal beliefs. Enabling these conditions would continually create and reinforce mutual trust, concretizing the bond of society and forming the context for peaceable moral disagreement to exist. For care theorists, these conditions are then further exemplified by the value of mutual concern, which is expressed between related beings when there exists a shared, intertwined interest to make possible the cooperation required to develop and sustain association for the benefit of all involved [1].

The components of civility are also found in the values of care. We have seen that, for Locke, civility refers to the disposition of the mind not to offend others [18]. This definition is concerned with how to handle disagreement in appropriate ways such that doing so reinforces the bond of society. Civility, though, is subsumed into the values of attentiveness and responsiveness. Attentiveness, at its base, is the recognition of a need that requires attending to, and at most is a critical awareness about what psychological and social biases could be preventing the recognition of certain needs [30]. Without attentiveness, we cannot hope to understand the boundaries for where offense lies with others; critical awareness of knowing how one's actions and words will negatively impact others are fundamental to the growth of becoming a civil person. Moreover, responsiveness refers to the ability of someone to respond to another's needs and recognize how useful a response was to determine if one's action(s) were well-received [30]. Handling disagreement relies upon one's skill in navigating difficult conversations and recognizing when one's response has aided or faltered discourse. Locke's civility, then, forms close ties with the values of care and their requirement to create the space for toleration to exist.

The last point of concern references the role of the state in enabling these conditions for toleration to emerge. We saw in the previous section that Locke does place positive demands on a magistrate to encourage persons to change their manners and dispositions, setting a good example for civil conduct. We straightforwardly find similar recommendations within the care literature, though taken to a fuller extent. The state is understood within care ethics as providing the conditions for the successful cultivation of caring practices and values "in families, neighbourhoods, churches, the workplace, and voluntary associations of various sorts" [37] (p. 198). Though not a care theorist, Martha Nussbaum captures what is at stake for care ethics in the political context: "institutions teach citizens definite conceptions of basic goods, responsibility, and appropriate concern, which will inform any compassion they learn" [38] (p. 405). For instance, there is good evidence that supports a generalized fostering of trust among people through state funding of welfare programs to overcome economic inequality [39], alongside the promotion of pro-social behavior through featuring social emotional learning programs in young children's curricula [40]. The state in the caring society is not neutral and actively supports the promotion of developing and sustaining good caring relations. In turn, good caring relations reinforce "the wider network of relations within which issues of rights and justice, utility, and the virtues should be raised" [1] (p. 136). This continuous feedback effect is vital for solidifying strong civic associations and social cohesion.

Importantly, the above remarks should not be read as if care ethics is built on faulty images of peace: "the ethics of care is quite capable of examining the social structures of power within which the activities of caring take place" [1] (p. 151). After all, care ethics recognizes and denounces patriarchal systems that create power asymmetries. The point for the care theorist is to uphold the moral standards of care and emphasize the ways in which caring practices can overcome violence or prevent it before it occurs. Such care is a central part of any good life and society; it is "the work we will do that creates the relationships, families, and communities within which our lives are made pleasurable and connected to something larger than ourselves" [41] (p. 89). A significant part of what makes our lives go better or worse, then, depends on how the interdependent relations of care that we are embedded in, and their surrounding institutional context, are structured. As Virginia Held puts it, "prospects for human progress and flourishing hinge fundamentally on the care that those needing it receive" [1] (p. 10). A major component of how this happens is the need for societies to actively "cultivate trust between citizens and governments; to achieve whatever improvements of which societies are capable, the cooperation that trust makes possible is needed" [1] (p. 42). In this way, the care ethical framework is preferable to the liberal framework in its defense of a caring and involved state that enables good caring relations to flourish.

Once more, this is not to say that Locke would advocate for exactly the same things that care theorists do with regards to the role of the state. Locke's state is certainly more minimal in its involvement for setting good examples and promoting civil discourse[3]. Locke's primary objective is prudence as peaceability, in contrast to care theorists, who see prudence as a means to good caring relations. However, the point is that both care ethics and Locke see a role for the state in positively building the conditions for trustworthiness and civility to emerge. Without the state's involvement in such a way, a mere *modus vivendi* will not suffice for enabling toleration between citizens.

What this section intended was to demonstrate that care ethics holds the conceptual resources to meet the conditions Locke sets for the possibility of toleration. There is important overlap with societal connections, the values of care, and the state taking a positive role in enabling conditions for toleration (and not maintain a neutral stance). Of course, this does not yet create a theory of toleration within the care ethical framework, only that there is good plausibility for one.

4. Toleration as Care

Now we have seen that care ethics has the conceptual resources to meet Locke's criteria for toleration to function, the pressing question is: why focus on care ethics and not liberal theories of justice? Given Locke's influence on contemporary liberalism, it may seem odd to not consider how this paper's outline of Locke's thought can be integrated within a liberal framework. The answer, though, is that there has been a fundamental rejection in the contemporary liberal literature of the state taking positive actions in fostering an ethos of trustworthiness and civility between citizens. Instead, liberal theorists have tended toward 'Lockean liberalism', whereby the liberal state is said to only justify intervention if an individual's rights have been violated in some form; otherwise, the state is difference-blind to pursuits of the good [42] (p. 644).

This rejection has been played out in liberal theorists' critiques of Anna Galeotti's theory of toleration as recognition [2,43]. Galeotti inadvertently channels Locke's thought by attempting to justify that the liberal state should play more of a role in influencing the attitudinal mindsets of citizens. However, Galeotti has faced criticism that her theory is incompatible with a liberal state, precisely because the liberal state should not intervene to influence citizens mindsets—that is, the role of the liberal state should be to neutrally maintain a *modus vivendi*, intervening only when negative rights are threatened.

This section explores Galeotti's theory and its criticisms, for two reasons. First, Galeotti's theory takes steps that enhance the enterprise of toleration beyond Locke's thought, developing the requirements for a modern theory of toleration. Second, given the rejection of toleration as recognition by the liberal framework, I argue that Galeotti's theory can instead find a suitable normative home within the care ethical framework. With some remodeling and alignment of Galeotti's theory within care ethics, a theory of toleration as care can be formed. Of course, at a future point, it is plausible to think that liberalism could reinterpret itself and eventually include toleration as recognition within its framework. However, as I argue later in this section, care ethics has unique conceptual resources that can enhance toleration as recognition beyond what the liberal framework could offer.

Galeotti's theory begins by considering limitations to Locke's thought on toleration—namely, for considering examples of moral disagreement beyond religious disagreement. As Galeotti points out, the scope of non-trivial moral disagreement in pluralistic liberal democracies is far broader, which any comprehensive theory of toleration should consider; that is, to acknowledge those cases whereby "the issue is not only perceived by the general public as highly controversial but as one which also requires the intervention of the state to settle it" [2] (p. 3). For Galeotti, the source of cases of intolerance in the modern day comes from "pluralism, understood as the coexistence within the same society of a plurality of groups and cultures with unequal social standing" [2] (p. 86)[4].

Galeotti goes much further than Locke here in describing how moral disagreement emerges. For Galeotti, conflicts appear when a majority group contests an attempt by a minority group to publicly display their collective identity, creating a case that calls for political settlement by the state. The majority group is not understood in terms of membership number, but rather by the group's power (by virtue of their dominant social position) to "define the characteristics, physical traits, habits, practices, and beliefs of other groups as deviant compared to their own, which they assume, implicitly, to be normal" [2] (p. 90). Minority groups then, are "socially constructed artefacts of the beliefs and perceptions of the majority" [44] (p. 294). As such, minority groups only become publicly distinct due to certain perceived differences that a majority identifies that group with. The majority then construes these differences as ascriptive, regardless of whether these differences genuinely are [2] (p. 89). With the majority group able to identify and label what characteristics and behaviors count as normal and what counts as different or threatening to the public order (and therefore *prima facie* intolerable), minority groups with different (supposed ascriptive) characteristics are pressured either to: (1) conform to the majority group's norms, or (2) to only display their collective identity in private. This pressure has a serious upshot: the act of publicly excluding minority groups in this way "reinforces the feeling of humiliation and shame by keeping the different identities publicly invisible and socially marginal" [2] (p. 98). Consequently, due to these power asymmetries of group membership, minority groups become incapacitated to function as civic members, and involuntarily become perceived as a *de facto* second-class citizenry.

Galeotti thus acknowledges power asymmetries in society in a way that Locke does not. But the basis of the problem that Locke posed remains: the distrust of the *other*, which undermines civility and the possibility for peaceful coexistence. Indeed, Galeotti argues that contemporary liberal theorists that defend the difference-blind neutrality of the state cannot adequately discern this new source of non-trivial cases of toleration. As no negative rights are technically violated through the majority's social pressure, difference-blind toleration does not have the resources to intervene and resolve societal mistrust. And yet, as Galeotti notes quite devastatingly, overlooking these group power asymmetries undermines the very principle of neutrality: by not recognizing differences in its citizenry and acting to accommodate for minority groups, the liberal state implicitly endorses the prevailing majority group's norms—a state itself often represented "by cultural majorities" [2] (pp. 99–100).

Galeotti believes this problem can (at least begin to) be addressed through the state adopting a new conception of toleration: toleration as recognition. What this entails is the liberal state overcoming its difference-blindness and recognizing the differences of minority groups as legitimate options within a pluralist liberal society. What this theory calls for, then, is "redrawing the map of social standards" and developing the grounds for trust between different groups to emerge [43] (p. 102). The state pursuing toleration as recognition achieves this goal of legitimization through symbolic public gestures, to "signify the end of the public exclusion of certain social differences and certain identities" [2] (p. 105). Other than the literal freedom extended to minority groups through these gestures, they become symbolic because the "public visibility of differences that has resulted symbolically represents the legitimization of [the minority group's] presence in public. In its turn, the legitimization of their presence in public signifies their inclusion in the public sphere on the same footing as those whose practices and behaviour are 'normal'" [2] (p. 101). With the public presence of a minority group's practice declared acceptable, this can begin to remove mistrust and stigmas attached to a minority group's collective identity.

Galeotti argues that the theory of toleration as recognition can straightforwardly function within a liberal state. However, liberal critics of Galeotti disagree that this is the case. In particular, Sune Lægaard [44], Peter Jones [45], and Andrea Baumeister [46] have all argued that toleration as recognition is incompatible with the proper limits of the liberal state. The significant problem is whether, without any additional interventions by the state to alter a majority's mindset, we can be sure that toleration as recognition provides the first step to achieve full social inclusion for minority groups. As Baumeister comments, "Conceiving of 'toleration as recognition' in purely institutional terms arguably falls well short of Galeotti's aspirations. After all, it is difficult to see how, in the face of the majority's grudging toleration, minorities could begin to feel genuinely respected" [46] (p. 108). State gestures of recognition only provide symbolic inclusion into the public sphere. As such, this recognition is not equivalent to full social inclusion. If toleration as recognition does end up requiring additional state intervention for minority social inclusion, does this theory remain compatible with the neutralist liberal framework?

It is Lægaard who explicitly expresses these above concerns as a genuine difficulty for Galeotti: "The question regarding the ideal of full inclusion into full citizenship is first of all whether it is plausible as an interpretation of neutralist liberalism, as claimed by Galeotti" [44] (p. 299). Here, Lægaard draws on Jonathan Seglow's distinction between "wide" and "narrow" recognition [47] (pp. 83–84). Wide recognition concerns the "attitudes, perceptions and resulting actions of other people, i.e., dependent on social factors"; narrow recognition is the equal treatment of the majority and minority groups through the state, such as by legislation [44] (p. 304). Though toleration as recognition appears to fall under the narrow recognition label, Lægaard argues that this does not match Galeotti's full aim: for minorities to fully access their citizenship (a feat that can only really be achieved under the wide recognition label). A tension thus arises in Galeotti's thought: though Galeotti aims for wide recognition for minority groups, toleration as recognition can only be limited to the narrow range when operating within a liberal framework. For if one assumes that an extension of state policies "aimed at changing social attitudes toward minorities will be at least *prima facie* problematic on any recognizable form of liberalism, a commitment to wide recognition as a requirement of justice may therefore be problematic" [44] (p. 305).

Another concern with toleration as recognition that Baumeister raises is that Galeotti does not pay sufficient attention to the social and political processes shaping (dynamic) cultural practices that appear to require state recognition. Galeotti appears to have neglected "the impact of power relations within minority communities," when deciding what the genuine meanings of a minority practice are [46] (p. 104). Toleration as recognition risks overlooking the fluidity of a collective identity and could inadvertently misrecognize an entire cultural practice through essentializing a perceived "authentic" trait of a minority, which the minority then rejects. For instance, are veils worn by Muslim women a symbol of patriarchal oppression, free choice of religious clothing, or otherwise? Without identifying how minorities themselves perceive certain practices, and the very power dynamics within the minority groups that shape how even they perceive the practice, this could have serious consequences. Misrecognition may cause the reverse intended effect of toleration as recognition, creating a backlash not only from the majority but various factions within the minority group as well.

Moreover, Galeotti's focus on recognition may also appear to simply reinforce power dynamics between majority and minority groups. Kelly Oliver (though not referencing Galeotti) writes that the politics of recognition "makes oppressed peoples beholden to their oppressors for recognition, even if that recognition afford them political rights and improved social standing" [48] (p. 477). For even as we saw Galeotti state earlier in this section, those in political power are typically the cultural majority that ultimately retains the authority to grant symbolic recognition. It is not clear how toleration of recognition could respond to Oliver's claim, either.

There is doubt, then, over the efficacy of Galeotti's theory of toleration as recognition. Galeotti requires an obligation for the attitudinal mindset change of a state's residents in accordance with symbolic recognitions performed by the state. Without both individuals and the state building and reinforcing an ethos of trustworthiness and civility, a tolerant society—short of a mere *modus vivendi*—remains unachievable. Galeotti implicitly relies on Locke's criteria as a result. Yet, with liberal critics rejecting the demand Galeotti places on the liberal state, this ironically seems to cast a shadow on Locke's thought in the process. Moreover, following Baumeister and Oliver, several improvements are required to critically acknowledge how power functions between and intra groups. However, all this does not mean we should do away with Galeotti's version of toleration; perhaps what is required is for toleration as recognition to be remodeled in a different theoretical framework.

I argue that care ethics can provide the framework for the conceptual resources that Locke's criteria require, while also being able to reshape and enhance Galeotti's toleration as recognition. The immediate alignment with Galeotti is that care ethics acknowledges the key premise that toleration as recognition rests upon: asymmetrical relations of power exist between individuals and groups. We saw in the previous section that care ethics' fundamental normative claim emerges from the recognition of power asymmetries operating within our relational interdependencies of care. If, as Galeotti says, non-trivial cases of toleration arise through unequal social standing, care ethics will not be difference-blind to this occurrence. Care theorists highlight the need to be attentive to the specific needs of others, which inherently requires the recognition of difference. To do so otherwise undermines the values of attentiveness and responsiveness at base. Moreover, if the task of the care theorist is to promote caring relations to allow civic capabilities and societal cooperation to occur in the political context, Galeotti shares the same goal of mitigating social tyranny to allow minorities full access to their citizenship. Galeotti can thus recommend to the care theorist that toleration as recognition can help fulfil these aims: to promote caring relations, minority groups ought to have their contested practice legitimized through a symbolic state gesture.

However, where care ethics moves to subsume toleration as recognition into its normative framework is its provision of the conceptual resources that justify the attitudinal shifts that wide recognition requires—resources that Galeotti implicitly relies on. Galeotti does not provide a justification that obligates a majority group to shift their mindset away from mistrust, despite a state's symbolic recognition of a contested minority practice. Within the care ethical framework, such obligation is grounded in the need to support and maintain good caring relations. This obligation entails a state and its residents developing an ethos of trustworthiness and civility, which can in part be practically achieved via the social education policy examples outlined in the previous section. Given that we are obligated to exemplify the values of care in the fulfilment of successful caring practices, state recognition of a minority practice should be met not with mistrust but a need for both majority and minority groups to be sincerely attentive and responsive with each other. In turn, this builds trustworthiness and the space for sincere discourse and/or praxis to take place. This move, which remoulds and justifies toleration as recognition within the care ethical framework, grants its remodelling into a theory of toleration as care.

For toleration as care, the purpose of toleration is to build a space intended to support and maintain good caring relations. This entails several important acknowledgements that, while enabling toleration as a recognition, also provide further nuance to Galeotti's argument. First, though moral disagreement may persist within a pluralistic society, all sides are obligated to contribute to building an ethos of trustworthiness and civility. As Locke saw it, it is possible to morally disagree while still behaving civilly toward one another with love and respect. Without this, we fail in our moral obligation to support and maintain good caring relations. Importantly, in the recognition of the power asymmetry between minority and majority groups, the burden falls on majority groups to contribute more to promoting this ethos. Minority groups, whose practices are already under suspicion from majority groups, will typically bear the most weight defending their practices with minimal resources available to them. This creates real harm over time to individuals within those groups—racial trauma or race-based traumatic stress from a cultural majority's intolerance, for instance, erodes one's "sense of self-worth," leading to "anxiety, depression, chronic stress, high blood pressure [and] even symptoms of PTSD" [49].

Second, the space that toleration as care creates must recognize discourse beyond debate, discussion, or other argumentative speech-based forms. Iris Marion Young writes that restricting democratic discussion to critical argument assumes a culturally biased conception of discussion that tends to silence or devalue some people or groups. Parliamentary debates or arguments in court are not simply free and open public forums; "speech that is assertive and confrontational is here more valued than speech that is tentative, exploratory, or conciliatory" [50] (p. 124). Such norms of deliberation create mechanisms of exclusion, reinforcing class divisions of "better-educated" and culturally specific expressions. Young argues that additional modes of communication are required to broaden the space for communicative understanding across difference: greeting ("care-taking, deferential, polite acknowledgement of the Otherness of others"); rhetoric ("constructs speaker, audience, and occasion by invoking or creating specific meanings, connotations, and symbols"); and storytelling ("narrative fosters understanding across such difference without making those who are different symmetrical") [50] (p. 129–132). Speech is not necessitated by these modes of communication, yet they provide important mediums for groups to strengthen the bond of society.

Third, toleration as care, supported by attentiveness toward difference, recognizes that intragroup identities also exist. Essentializing groups and cultural identities creates moral and legal problems for how groups can shift, merge, and change over time. The irony of this problem, Caroline Dick states, is that "identity-driven rights frameworks succeed in suppressing differences within minority cultural groups while masking the differences of the dominant cultural group that order institutional norms and legal rules" [51] (p. 195). To combat this, the focus of analysis should shift away from specifying a group's "authentic" practices to the "control of identity and cultural reproduction" [51] (p. 195). Dick's intragroup difference framework offers a solution to navigate this issue, recognizing that individuals can hold multiple group memberships that may experience oppression in one context but not another. This anti-essentialist stance must frame toleration as care, providing nuance to symbolic public recognition and recognizing who is actually participating in the construction of cultural, political, and legal codes. As Baumeister pointed out, Galeotti does not specify the significance of the politics of intragroup identities; toleration as care, supported by the values of care, does have the conceptual resources to do so.

The above responsibilities that derive from toleration as care may give the impression that care ethics is a demanding moral theory, in which the values of care obligate persons to evaluate and strengthen their personal and political caring relations. But care ethics may only look demanding because of its comparison to a non-interfering liberal state that does not demand these kinds of efforts from its residents. Indeed, this apparent moral demandingness is the main benefit of toleration as care: that the goal of toleration is toward promoting good caring relations that are foundational to the flourishing of a good life and society. For care theorists, the point of toleration is not for setting up a neutral space for (uncomfortable) peaceful coexistence. Instead, the point is to set up a space that enables trust building and good caring relations to develop.

With this argument now in place, we can revisit how toleration as care is preferable to the other approach to toleration within the care ethical framework: Engster's argument from minimal care, discussed at this paper's outset. We saw there that Engster's argument is indifferent to symbolic recognition, with the example given of same-sex marriage. How does toleration as care compare? First, toleration as care has the conceptual resources to not only recognize the cultural power asymmetry between heterosexual and same-sex relationships, but critically evaluate how this power asymmetry emerged in the first place. Attentiveness as a caring value morally obligates us to pay attention to historical and potential current biases that have impacted the way a given society perceives same-sex marriage. In the build up to the Supreme Court's Obergefell v. Hodges ruling, these biases typically included homophobic hostility and arguments from tradition. Toleration as care immediately tells us, then, that the symbolism of recognizing and institutionalizing same-sex marriage not only denounces relations of domination and hostility from historic heterosexual norms (religious-based or otherwise) but also acts as a celebration of enabling same-sex couples to exemplify caring values more strongly in their relation through consensual marriage. In doing so, same-sex marriage is recognized as a viable option open to society and reinforces good caring relations.

The question then concerns shifting attitudinal mindsets to accept this symbolic recognition. In this paper's introduction, I gave the example of certain Legislatures in the United States refusing to acknowledge the Obergefell v. Hodges ruling. Fundamentally, this was a refusal of the religious right to shift their attitudinal mindsets to acknowledge same-sex marriage as a viable option open to society. Given that toleration as care requires a state and its residents to create the space for good caring relations to flourish, we are obligated to explore methods for how to best do so for recognizing same-sex marriage—especially the heterosexual societal majority, which bears the burden of obligation for such recognition. Brian Harrison and Melissa Michelson provide such methods, discovering that the best canvassers on behalf of same-sex marriage are those who share an identity with the intolerant audience—otherwise known as in-group messengers [52]. These canvassers do not need to necessarily be religious leaders; they could be in-group influencers, as was the case with a Catholic elderly couple's viral video that urged the yes vote for same-sex marriage in Ireland. Toleration as care builds the space for sincere discourse and/or praxis, which establishes toleration as the first step toward stronger societal caring relations and overarching acceptance.

Before concluding, a potential criticism of toleration as care should be addressed. This criticism is presented by John Horton, who argues that a *modus vivendi* may be the best option a tolerant society ever has when dealing with hard cases of genuinely irreconcilable differences between groups. Horton writes that, "to ask people to be more inclusive and mandate indifference is to fail to take seriously the less comfortable implications of conflicts between values and ways of life that are mutually antagonistic" [53] (p. 299). A tolerant society "may not be a very comfortable society," but it is one that at least recognizes the authenticity of differing negative viewpoints [53] (p. 303). How can toleration as care respond to this objection?

Care ethics never denies that antagonistic relations exist; the purpose of toleration as care is to foster an environment in which antagonistic relations can function peacefully within society toward the goal of building and sustaining good caring relations over time. It may be that some differences are truly irreconcilable between groups. However, this limbo does not prevent caring relations from developing, at least in part. Relations are not one-dimensional (according to a single moral disagreement) but multifaceted. It is quite possible for toleration as care to allow people with moral disagreements to begin cooperating successfully within various civic associations, despite their differences. The trust building that develops here is a basic one of each side learning that they will not be undermined by the other. But through this basic trust as a foundation, stronger displays of civic attentiveness can begin to be made over time with support from a wider caring society. It is to begin with the minimal basis of a risk-assessment account of trustworthiness, with the view that a will-based account is always possible to strive for. If the choice is between a society that only believes a mere *modus vivendi* is possible (by Horton's own admission, an uncomfortable society) and a society that actively wants to foster caring relations and its associative caring values at the personal and political level, then the latter is preferable. In this way, a theory of toleration can emerge.

5. Conclusions

In this paper, I sought to engage care ethics with Locke's thought on toleration. The purpose for doing so was to build out the foundation for a theory of toleration as care—a theory modeled not only on Locke's criteria for toleration to emerge in society but also Galeotti's toleration as recognition. For toleration as care, the point of toleration is not to set up a neutral space for peaceful coexistence (an albeit uncomfortable *modus vivendi*). Toleration as care goes further, obligating the state and its residents to develop an ethos of trustworthiness and civility, building the space for sincere discourse and/or praxis that supports and maintains good caring relations.

Funding: This research received no external funding.

Data Availability Statement: Not applicable.

Acknowledgments: I am grateful to Elizabeth Brown for her detailed review of an early draft of this paper, which strengthened my arguments in key areas. I also thank Richard Vernon for his graduate course on toleration that deepened my understanding of John Locke's thought, alongside introducing me to contemporary literature on the subject. Last, I am appreciative of the constructive feedback provided by Maggie Fitzgerald and Maurice Hamington, alongside incredibly helpful suggestions provided by two anonymous reviewers.

Conflicts of Interest: The author declares no conflict of interest.

Notes

1.   Robinson's critique is not to downplay the importance of rights-language per se, only that the positive responsibilities of care we have to one another have been often overlooked and discounted by liberal thinkers in favour of negative rights. For further reading, see [8].

2.   The idea of the "relational self" is shared with communitarian political theory. However, there are important differences between communitarianism and feminist philosophy (and care ethics specifically). For a discussion of these differences, see [3] (pp. 10–11).

3.   As seen most explicitly in Locke's *Second Treatise* [19].

4.   Galeotti does not denounce difference-blind toleration as an obsolete approach for resolving individual moral disagreements. As will be shown, all Galeotti means to say is that difference-blind toleration cannot explain what is genuinely at stake between non-trivial cases of toleration arising between groups in pluralistic societies.

References

1. Held, V. *The Ethics of Care: Personal, Political, Global*; Oxford University Press: Oxford, UK, 2006.
2. Galeotti, A.E. *Toleration as Recognition*; Cambridge University Press: Cambridge, UK, 2002.
3. Engster, D. *The Heart of Justice: Care Ethics and Political Theory*; Oxford University Press: Oxford, UK, 2007.
4. Vernon, R. Introduction. In *Locke on Toleration*; Vernon, R., Ed.; Cambridge University Press: Cambridge, UK, 2010; pp. viii–xxxii.
5. Waldron, J. Locke: Toleration and the Rationality of Persecution. In *Justifying Toleration: Conceptual and Historical Approaches*; Mendus, S., Ed.; Cambridge University Press: Cambridge, UK, 1988; pp. 61–86.
6. Rawls, J. *Justice as Fairness: A Restatement*; Kelly, E., Ed.; Belknap Press: Cambridge, MA, USA, 2001.
7. Robinson, F. *The Ethics of Care: A Feminist Approach to Human Security*; Temple University Press: Philadelphia, PA, USA, 2011.
8. Held, V. Care and Human Rights. In *Philosophical Foundations of Human Rights*; Cruft, R., Matthew, L.S., Massim, R., Eds.; Oxford University Press: Oxford, UK, 2015; pp. 624–641.
9. Bhandary, Asha. *Freedom to Care: Liberalism, Dependency Care, and Culture*; Routledge: New York, NY, USA, 2020.
10. Jolley, N. *Toleration and Understanding in Locke*; Oxford University Press: Oxford, UK, 2016.
11. Bejan, T. *Mere Civility: Disagreement and the Limits of Toleration*; Harvard University Press: Cambridge, MA, USA, 2017.
12. Tate, J.W. Locke, Rationality, and Persecution. *Political Stud.* **2010**, *58*, 988–1008. [CrossRef]
13. Locke, J. First Tract on Government (1660). In *Locke: Political Essays*; Goldie, M., Ed.; Cambridge University Press: Cambridge, UK, 1997; pp. 3–53.
14. Locke, J. Second Tract on Government (c. 1662). In *Locke: Political Essays*; Goldie, M., Ed.; Cambridge University Press: Cambridge, UK, 1997; pp. 54–78.
15. Locke, J. An Essay on Toleration (1667). In *Locke: Political Essays*; Goldie, M., Ed.; Cambridge University Press: Cambridge, UK, 1997; pp. 134–159.
16. Locke, J. A Second Letter Concerning Toleration. In *Locke on Toleration*; Vernon, R., Ed.; Cambridge University Press: Cambridge, UK, 2010; pp. 67–107.
17. Locke, J. A Third Letter for Toleration. In *Locke on Toleration*; Vernon, R., Ed.; Cambridge University Press: Cambridge, UK, 2010; pp. 123–163.
18. Locke, J. *Some Thoughts Concerning Education and of the Conduct of the Understanding*; Grant, R., Tarcov, N., Eds.; Hackett: Indianapolis, IN, USA, 1996.
19. Locke, J. Second Treatise. In *Two Treatises of Government. In Locke on Toleration*; Vernon, R., Ed.; Cambridge University Press: Cambridge, UK, 2010; pp. 47–49.
20. Hobbes, T. *Leviathan*; Curley, E., Ed.; Hackett: Indianapolis, IN, USA, 1994.
21. Collins, J. *The Allegiance of Thomas Hobbes*; Oxford University Press: Oxford, UK, 2005.
22. Locke, J. A Letter Concerning Toleration. In *John Locke: A Letter Concerning Toleration and Other Writings*; Goldie, M., Ed.; Liberty Fund: Indianapolis, IN, USA, 2010; pp. 1–68.
23. Casson, D.J. *Liberating Judgment: Fanatics, Skeptics, and John Locke's Politics of Probability*; Princeton University Press: Princeton, NJ, USA, 2011.
24. Locke, J. *The Correspondence of John Locke: Volume 3*; De Beer, E.S., Ed.; Clarendon Press: Oxford, UK, 1978.
25. Locke, J. Essays on the Law of Nature (1663–1634). In *Locke: Political Essays*; Goldie, M., Ed.; Cambridge University Press: Cambridge, UK, 1997; pp. 79–133.
26. Kant, I. *The Metaphysics of Morals*; Gregor, M.J., Translator; Cambridge University Press: Cambridge, UK, 1996.
27. Keller, J. Autonomy, Relationality, and Feminist Ethics. *Hypatia* **1997**, *12*, 152–165. [CrossRef]
28. Mackenzie, C. Personal Identity, Narrative Integration and Embodiment. In *Embodiment and Agency*; Campbell, S., Meynell, L., Sherwin, S., Eds.; The Pennsylvania State University Press: University Park, TX, USA, 2009; pp. 100–125.
29. Noddings, N. *Caring: A Feminine Approach to Ethics and Moral Education*; University of California Press: Berkeley, CA, USA, 1984.
30. Tronto, J.C. *Moral Boundaries: A Political Argument for an Ethic of Care*; Routledge: New York, NY, USA, 1993.
31. Kittay, E.F. Human Dependency and Rawlsian Equality. In *Feminists Rethink the Self*; Meyers, D.T., Ed.; Westview Press: Boulder, CO, USA, 1997; pp. 219–266.
32. Sullivan, L.; Niker, F. Relational Autonomy, Paternalism, and Maternalism. *Ethical Theory Moral Pract.* **2018**, *21*, 649–667. [CrossRef]
33. Slote, M. *An Ethics of Care and Empathy*; Routledge: London, UK, 2007.
34. Randall, T. Justifying Partiality in Care Ethics. *Res Publica* **2020**, *26*, 67–87. [CrossRef]
35. Baier, A. Demoralization, Trust, and the Virtues. In *Setting the Moral Compass*; Calhoun, C., Ed.; Oxford University Press: Oxford, UK, 2004; pp. 176–188.
36. Jones, K. Second-Hand Moral Knowledge. *J. Philos.* **1999**, *96*, 55–78. [CrossRef]
37. Dagger, R. *Civic Virtues*; Oxford University Press: Oxford, UK, 1997.
38. Nussbaum, M. *Upheavals of Thought: The Intelligence of the Emotions*; Cambridge University Press: Cambridge, UK, 2001.
39. Rothstein, B.; Uslaner, E. All for All: Equality, Corruption, and Social Trust. *World Politics* **2005**, *58*, 41–72. [CrossRef]
40. Cohen, J. (Ed.) *Caring Classrooms/Intelligent Schools: The Social Emotional Education of Young Children*; Teachers College Press: New York, NY, USA, 2001.

41. West, R. The Right to Care. In *The Subject of Care: Feminist Perspectives on Dependency*; Kittay, E.F., Ed.; Rowman and Littlefield: Lanham, MD, USA, 2002; pp. 88–114.
42. Sinopoli, R.C. Liberalism and Contested Conceptions of the Good: The Limits of Neutrality. *J. Politics* **1993**, *55*, 644–663. [CrossRef]
43. Galeotti, A.E. The Range of Toleration: From Toleration as Recognition Back to Disrespectful Tolerance. *Philos. Soc. Crit.* **2015**, *41*, 93–110. [CrossRef]
44. Lægaard, S. Galeotti on Recognition as Inclusion. *Crit. Rev. Int. Soc. Political Philos.* **2008**, *11*, 291–314. [CrossRef]
45. Jones, p. Toleration, Recognition, and Identity. *J. Political Philos.* **2006**, *14*, 123–143. [CrossRef]
46. Baumeister, A. Diversity and Equality: 'Toleration as Recognition' Reconsidered. In *Democracy, Religious Pluralism and the Liberal Dilemma of Accommodation*; Mookherjee, M., Ed.; Springer: New York, NY, USA, 2011; pp. 103–117.
47. Seglow, J. Theorizing Recognition. In *Multiculturalism, Identity and Rights*; Haddock, B., Sutch, P., Eds.; Routledge: London, UK, 2003; pp. 78–93.
48. Oliver, K. Witnessing, Recognition, and Response Ethics. *Philos. Rhetor.* **2015**, *48*, 473–493. [CrossRef]
49. Racism and Mental Health. HelpGuide. Available online: https://www.helpguide.org/articles/ptsd-trauma/racism-and-mental-health.htm (accessed on 17 April 2022).
50. Young, I.M. Communication and the Other: Beyond Deliberative Democracy. In *Democracy and Difference: Contesting the Boundaries of the Political*; Benhabib, S., Ed.; Princeton University Press: Princeton, NJ, USA, 1996; pp. 120–135.
51. Dick, C. *The Perils of Identity: Group Rights and the Politics of Intragroup Difference*; UBC Press: Vancouver, BC, Canada, 2011.
52. Harrison, B.; Michelson, M. *Listen, We Need to Talk: How to Change Attitudes about LGBT Rights*; Oxford University Press: Oxford, UK, 2017.
53. Horton, J. Why the Traditional Conception of Toleration Still Matters. *Crit. Rev. Int. Soc. Political Philos.* **2011**, *14*, 289–305. [CrossRef]

Article

Thinking about the Institutionalization of Care with Hannah Arendt: A Nonsense Filiation?

Catherine Chaberty and Christine Noel Lemaitre *

Institut d'Histoire de la Philosophie, UR 3276, Aix Marseille University, 13621 Aix en Provence, France; catherine.chaberty@univ-amu.fr
* Correspondence: christine.lemaitre@univ-amu.fr

Abstract: In recent decades, some feminists have turned to the writings of Hannah Arendt in order to propose a truly emancipatory ethic of care or to find the principles that could lead to the political institutionalization of care. Nevertheless, the feminist interpretations of Hannah Arendt are particularly contrasted. According to Sophie Bourgault, this recourse to Hannah Arendt is deeply problematic, mainly because of her strong distinction between the private and public spheres. This article discusses the relevance of using Arendt's concepts to think about the institutionalization of care by Joan Tronto. Indeed, the most recent analyses developed on the politics of care are shaped by Arendt's concepts such as power, amor mundi or by her conception of politics as a relationship.

Keywords: care; commun world; plurality

Citation: Chaberty, C.; Lemaitre, C.N. Thinking about the Institutionalization of Care with Hannah Arendt: A Nonsense Filiation? *Philosophies* **2022**, *7*, 51. https://doi.org/10.3390/philosophies7030051

Academic Editors: Maurice Hamington and Maggie FitzGerald

Received: 29 March 2022
Accepted: 12 May 2022
Published: 16 May 2022

Publisher's Note: MDPI stays neutral with regard to jurisdictional claims in published maps and institutional affiliations.

1. Introduction

Paradoxically, several recent works highlight the relevance of Hannah Arendt's philosophy [1] for understanding and renewing the theory of care [2]. This is the case with Ella Myers [3], who proposes to draw from Arendt's philosophy the foundations of a democratic ethics of care, and Bonnie Mann [4] or Jess Kyle [5], among many others. It is possible to speak of a paradox insofar as Arendt's thought was very badly received in feminist circles in the 1980s. For Adrienne Rich [6], *The Human Condition* [7] is a "condescending and lame book [which] embodies the tragedy of a woman's spirit steeped in masculine ideology" [6] (pp. 211–212). Mary O'Brien turns Arendt into a female adept of male supremacy [8] (p. 9).

It is true that the philosopher Hannah Arendt never carried the voice of feminism. She abhorred any form of identity-based labeling that tends to obscure and distort political discourse. However, this does not mean that Arendt never raised the question of women's place in society. Diane Lamoureux [9] identifies three texts in the Arendtian corpus where the philosopher raises the question of the feminine condition. First, in a book review written in 1933 and published in *Essays in Understanding* [10], the philosopher underlines the fact that the women's emancipation through work places them in a situation of double constraint on their husbands since they continue to ensure the main domestic and household responsibilities, and in their employment, where they are often poorly remunerated [10] (pp. 70–71). Then, the work she devotes to Rahel Varnhagen emphasizes her dual status as a woman and a Jew [11]. Finally, in *Men in Dark Times* [12], Arendt's interest in Rosa Luxembourg evokes the difference that constitutes her femininity in order to understand her trajectory. However, one would look in vain for a developed reflection on feminism in the writings of Hannah Arendt.

The connection between the concepts forged by Hannah Arendt and care would be eminently problematic for Sophie Bourgault [2]. Indeed, she rightly points out that one of the salient features of the German philosopher's political thought is the strict separation between the public domain, the place of politics, the private domain and the place of satisfaction of needs. This separation leads Arendt to denounce the social rise as likely to deviate political action and would invalidate any rapprochement between feminist theories of care and Arendt's political theory [11] (p. 14). Indeed, for Arendt, the confusion of the politics and the social risks undermining freedom because the prominence of the social might orient politics towards an instrumental logic of satisfying needs. Such logic tends to transform action into doing.

Our contribution aims to propose a reading of the key concepts of Arendt's political philosophy (world, plurality and power) that could shed light on the conciliation between her thought and the theory of care. Several works have indeed outlined bridges between Arendt's concepts and the theory of care. This is the case of those led by Lamoureux [9]. After having restored the difficulties of the dialogue between Arendt and feminism, we will synthesize the main foundations of the theory of care, and then we will see in what way the Arendtian concepts can be particularly relevant for thinking about care.

2. Arendt and Feminism: The Reasons for a Misunderstanding

Hannah Arendt's distinction between the private and the public and her rejection of social interference in political action are at the heart of the criticisms that feminist currents have made of the philosopher's thought.

2.1. The Private and the Public in Their Political Articulation

Arendt [7] is firmly convinced that politics requires a strict separation between the public sphere (*der öffentliche Raum*) and the private sphere of the household (*der Haushalt*) devoted to the preservation of life. If the first is the place of freedom and action, the second is that of necessity and life. Man thus belongs to two distinct orders of existence: the private domain with what is proper to him (*idion*) and the public domain that he shares with other men (*koinon*). Arendt draws from the Greek tradition the opposition between the private as domestic sphere (*oikos*) and the public around the city (*polis*). As for Aristotle, Arendt characterizes the economy by the management of domestic affairs. The domestic sphere is that of the satisfaction of vital needs. Additionally, the economy is a means towards the "good life" within the *oikos*.

The private sphere and the public sphere are opposed on several points. The public sphere implies a free meeting of citizens and an equally free exchange of speech. The private sphere is subject to the sphere of necessity and unequal relationships. The public sphere is the place of light, a necessary medium for dialogue and exchanges. The private domain is a place of darkness and intimacy, necessary for birth and death. It also shelters what is considered shameful but necessary, the bodily needs and their satisfaction. It is finally the home of love. Françoise Collin [13] underlines that Arendtian love is to the private domain, what respect is to the public domain. Love is a kind of respect for the other that reveals the identity of the lover because insofar as only a disinterested impulse can be qualified as love, it leads the lover to reveal himself. Love distances those who experience it from the public sphere, and they are only reintroduced to it by giving life, by the child they introduce into the world.

The public domain is public in a double sense: it is the field of appearance, and it is also what is common to all. Public gardens are places where my actions are exposed to the eyes of all. I appear in all transparency. This quality of the public space to be the place where the men show up confers him an objectivity, which results from the multitude of the perceptions printed by subjectivities also multiple. The public domain provides a space where citizens can appear in a face-to-face manner that forbids anonymity and engages their responsibility. Perceived by all, the public domain constitutes a collective memory, which guarantees the permanence of meaning. Public gardens are also public in the sense that they belong to everyone and are not my property. Additionally, they are common because they proceed from men without ever belonging to any man in particular.

The division operated by Arendt between private space and public space was strongly attacked by the feminists. They denounce, in particular, the inoperative character of this distinction. Jean Bethke Elshtain [14] reproaches Arendt for her reductive interpretation of the private space. Arendt turns a blind eye to the fact that, historically, the private sphere, understood as the domestic sphere, has been attributed to women as a place to ensure the vital necessities and to take care of the family. With access to the public sphere being forbidden to them, they could only exist in the hushed space of what was out of public view. For men, the home was a place of refuge from public life. According to Elshtain, the Arendtian distinction would lead to the endorsement of patriarchal and, more generally, masculine domination. For Couture [15], the example of the struggle of the women in the workers' movements reveals all the fragility of Arendt's thought on this point. Arendt criticizes feminist struggles for not being political struggles. Based on natural attributes (the distinction between man/woman), the feminist claims are directed on social questions—thus not political. Against Arendt, Couture suggests that these struggles proceed precisely from a wrenching from nature. Indeed, if workers fight against the alienation of work, women struggle against their confinement to the role of reproducer and guardian of the home. Feminist demands for emancipation from a so-called "feminine essence" associated with a division of political functions are opposed to an essentialist-inspired alienation. The women thus claim the freedom to belong to the world. Their struggle is, in this sense, fully political, contrary to what Hannah Arendt seems to think. Benhabib [16] comes to wonder if the concept of public space that Arendt borrows from Antiquity is tailored to fit the complexity of the institutions and problems of the modern world. For her part, Hanna Fenichel Pitkin [17] argues that the distinction between public space and private life is totally inoperative in the modern age.

For Clarke [18], it is possible to confront two interpretations of the Arendtian separation of the private and the public. A first register of interpretation would see Arendt as an adversary of modernity associated with the drifts of a productivist and consumerist mass society. A second register of interpretation would underline the influence of Rosa Luxembourg's thought on Arendt. This influence would be revealed in the interest expressed by Arendt for participatory democracy and in particular for workers' councils.

2.2. The Rejecation of the Social

The social is defined by Arendt as a space in which men are linked by needs. It derives from the cycle of the vital process of production–consumption. The modern world, which was built around the sacralization of work, saw the emergence of the social question, which Arendt refers to as "the advent of the social". Arendt paid little attention in her work to issues of social justice. In two of her texts, the essay *On Revolution* [19] and her *Reflections on Little Rock* [20], she separates political action from the temptation to use power for social justice purposes. For Arendt, the French Revolution would have failed precisely because its initial goals of citizen political emancipation quickly faded away in favor of the overriding goal of solving the "social question." Providing a solution to manage the misery of the popular masses became a priority concern that overshadowed the initial political considerations of the French Revolution. The Jacobins very quickly sacrificed the search for stabilizing political emancipation to focus on improving the material conditions of life of the people. This urgency would have justified the terror and the confiscation of the democratic word under the pretext that the social question imposed beforehand a collective submission: "It was necessity, the urgent needs of the people that unleashed the terror and sent the Revolution to its doom" [19] (p. 60). For this reason, Arendt contrasts the failure of the French Revolution with the success of the American Revolution in the essay *On Revolution* [19].

For Arendt, the social is thus the place of a kind of deviation from the political. It is even the opposite of politics. While politics is deliberation, plurality and freedom, the social is only the claim of a starving people, the product of vital requirements that send man back to animality. Already in *The Human Condition* [7], Hannah Arendt deplored that the modern era signs the victory of the social over the political. The modern advent of the social makes activities that were once devolved into the private sphere, such as productive work or professional relations, the center of the collective interest. Arendt does not hesitate to interpret this narrowing of the political field as an erasure of what makes man specific.

With the advent of the social, the public becomes a function of the private, and the latter becomes the one and only common concern.

In her reciprocal delimitation of the social and the politics, Arendt faced many criticisms and reactions. Her lifelong friend, Mary McCarthy, questioned her at a conference Hannah Arendt held in Toronto in 1972: "What are we supposed to do on the public stage, in the public space, when we are not concerned ourselves with the social? What is left...? If all the questions of economics, human welfare, racial diversity, if everything that belongs to the social sphere is excluded from the political scene, then I am fooled. All that is left is war and speeches. But speeches cannot be just speeches. They must have an object purpose". According to Mary McCarthy, Arendt's response to this objection did not satisfy most of the philosopher's opponents. In an attempt to counter her friend's argument, Arendt used the question of the right to adequate housing as an example. In her view, each question has a dual political aspect and a social aspect, which should be separated. In the question of the right to decent housing, the political aspect consists of determining whether the granting of decent housing allows the person to be released in order to participate in political life. The social aspect is related to a question of justice, which implies that all members of a community should have the same right to access decent housing. The right to decent housing, understood in this sense, should not be debated because it is self-evident. Politics can only emerge when man is freed from vital necessities. This may have led Margaret Canovan [21] to assert that Arendt's political theory would be elitist because of its lack of consideration for questions of economic and social equality. Indeed, in Arendt's thought, politics can only concern those who can fully dedicate themselves to action by being freed from material constraints.

Hannah Arendt was convinced of the weakness of the notion of the social, affirming that it is an obstacle to the political apprehension of the human condition. According to her, the notion of social is conceived as a vague encompassing which brings together the discourses which attempt to rationalize a posteriori the advent of modern society. For Arendt, the social is the index of the penetration of the private sphere in the political sphere. The economic production becomes the priority of societies, and this prioritization comes to pervert the public space. To speak about the social is to mix the public and the private, to conceive the public space as the place of a gigantic household of which it is advisable to discuss the mode of management. Linked to work and to its organization, the social gangrenes the politics. This is why it evacuates any question relating to the social question by considering that social problems are "matters of administration to be put into the hands of experts, rather than issues which could be settled by the twofold process of decision and persuasion" [19] (p. 91).

How can we explain such a disinterest of Hannah Arendt for these questions that she evacuates out of politics? A rapprochement with the Kantian perspective can be sketched. For Kant, the State's function is to civilize inter-individual relations, not to ensure the happiness of its members. Happiness is to be understood as psychological and ideological aspiration of individuals. For Kant, happiness is the satisfaction of all our inclinations, both extensive, in terms of their variety, and intensive, in terms of degree, and also in terms of their duration. Such subjective aspirations cannot give rise to valid legal rules because the law is not so much determined by content as by the constraint attached to it. The public will must not be oriented by a eudemonic motive. Indeed, the conception of happiness is so relative and indeterminate that, despite the desire shared by every man to achieve happiness, no one can ever say exactly what would make him happy. Happiness is an ideal of the imagination, not an ideal of reason. Therefore, the legislator who claims to act for the happiness of his people should be omniscient. The aim of the legislator is not thus to achieve sensible well-being but rather rational well-being, which is none other than the preservation of the State through obedience to the laws. The paternal or despotic government considers its subjects as minors incapable of making a decision by themselves. It therefore tries to impose its vision of happiness on them. A government that would be founded on the principle of benevolence towards the people, like a father towards his children, is a paternalistic government, where the subjects are obliged to behave in a simply passive way, like minor children, incapable of distinguishing what is really useful or harmful to them and who must expect simply from the judgment of the head of state the way in which they must be happy, and simply from his goodness that he also wants it, is the greatest despotism that one can conceive [22–24].

The State, as thought by Kant, must not have a social function; it must not intervene outside the preservation of the fundamental political liberties of which it is the guarantor. The interference of the state in private affairs would slow down the vitality of the economic community. Kant is opposed to the institution of permanent state relief for the poorest because it would lead to making "poverty a profession for the lazy". State intervention in the social sphere destabilizes the normal play (game) of the economy and favors the emergence of perverse effects such as a form of encouragement for the lazy. Arendt's repugnance for the social question reflects Kantian distrust and the link he makes between political interest in material equality and despotism.

Arendt is hostile to the violence by which the social question manifests itself when it bursts into the political sphere. "Wherever the breakdown of a traditional authority set the poor of the earth on the march, where they left the obscurity of their misfortune and stream upon the market-place, their furor seemed as irresistible as the motion of the stars, a torrent rushing forward with elemental force and engulfing a whole world" [19] (p. 113). We must not forget that it was hunger and misery that led in part to the Bolshevik Revolution, to the Jacobin terror, as well as to many bloody revolutions. Misery can only lead to explosions of anger that swallow up politics with them.

3. From an Ethic of Care to a Political Theory of Care

Since the 1980s, care has become an ethic in its own right. Whether an ethic of the solicitude—concern for others and attention to the other—or an ethic of the ordinary [25], it attempts to understand the concrete moral norms that characterize a society or the community. For representatives of care ethics, morality cannot be constructed in the abstract from hypothetical moral dilemmas or on the basis of universal moral principles [26]. In contrast to a conception of the selfish and rational, autonomous and independent individual carried by deontological and consequentialist moral theory care emphasizes interdependence and places vulnerability as an "original condition" in response to the Rawlsian "original position" [27]. (Rawls proposes to imagine that people determine together, behind a veil of ignorance, the principles of society organization. The veil of ignorance prevents them from favoring their own interests because they know nothing about the position they will occupy in society). "Throughout our lives, all of us go through varying degrees of dependence and independence, of autonomy and vulnerability" [28] (p. 135). Vulnerability is then constitutive of the person, and the ideal of autonomy becomes fictional.

The moral foundations of concern for others, nurturance, compassion, attentiveness and responsibility constitute the values of caring for Tronto [28]. They are opposed to rational ethics and justice. Rather than relying on principles and rules of a theory of justice's law that apply to all in an indeterminate way, care focuses on the responsibilities and relationships understanding of particular beings in their singularity.

Long focused on gender, care or work issues, care is introduced in the field of politics, especially with Joan Tronto in the 1990s. For the author, feminist theories have not led to political change. Arguments based on the "women's morality" have proved ineffective, and care has remained confined to the private sphere, a domain attributed to women, and relegated to the rank of secondary activities. For Tronto, moral arguments are to be understood in a political context, in the sense that this latter may affect their acceptability: "Widely accepted moral values constitute the context within which we interpret all "moral arguments" [28] (p. 6) and certain interpretations may limit the consideration of other ideas of morality. This awareness helps us understand how these boundaries shape morality. In *Moral Boundaries, A Political Argument for an Ethic of Care*, Tronto [28] analyzes these "moral boundaries" that have kept care values peripheral. The theory of care she proposes invites not only an ethical practice but a moral reflection by redefining the existing moral boundaries that she has identified between morality and politics—the "moral point of view"—and between public and private life.

3.1. Moral Boundaries, Excluding Care

In recent Western and liberal thought, politics is situated outside the domain of morality. Moral values are introduced into politics only insofar as they correspond to its requirements. The history of the relationship between morality and politics is indeed characterized by two opposite tendencies. For a first tradition, eminently illustrated by Platonic philosophy, politics must be conceived as a kind of enlarged morality. The aim of the politician must be to guarantee an organization that favors fair behavior. For a second tradition initiated by Machiavelli, politics and morality are distinguished both by their ends and by their means.

Fitting into the Aristotelian framework where morality and politics are closely intertwined, Tronto proposes a concept of care having the capacity to describe a moral and political version of the good life, thus escaping the dilemmas of seeing these as separate spheres. "When the world is rigidly divided between the realms of power and of virtue, we lose sight of the facts that power requires a moral base, and more importantly for our present boundary between politics and morality prevents us from seeing that moral theory conveys power and privilege" [28] (p. 93). Tronto then considers that care can play both the role of a moral value and politically found a good society. Politics and morality are no longer thought of in an instrumental relationship or considered as areas of life that should be kept separate but become closely linked. "The practice of care that I have developed and described … , can itself be understood not only as a moral concept, but as a political concept as well. Because the practice of care is also a political idea, I do not face the problem of trying to import a moral concept into a political order" [28] (p. 161). Tronto is going to make the crossing between the moral concept and political concept the starting point of his reflection, which is all the more interesting if we note the conception of morality understood as referring to the conduct of our relations with others [29], a domain that Arendt attributes to politics, morality, for the latter, concerning the individual in his singularity. [29] (p. 195).

By describing care as an attitude or disposition, it is easier to assign it to the private sphere. Thus considered by the patriarchal model as a female prerogative, care suffers from traditional gender roles. The idea of a gender-differentiated morality reinforces some of the existing moral boundaries and makes it difficult to transform our conceptions of politics, morality and gender roles.

This association of care/women or care/providers of care, reflecting gender, class and racial inequalities, obscures its complexity and the fact that it is inextricably linked to all aspects of life in general. We can also analyze the exclusion of care from the public sphere in terms of power [30]. The public sphere is of far greater importance than the private sphere, and since political life is identified with public life, care's reference to a private life means that it is beyond or below political concerns [28] (p. 96).

Considering it also as a practice, Tronto [28] allows care to escape this confinement. By embodying a moral ideal, while already being a contextualized practice, care is no longer excluded from the public sphere and can therefore require a political theory: without a policy of care, the ethics of care is not only deprived of its means, as if the relationship between morality and politics were external, but also of its conditions of existence [30].

Without going back entirely to the moral sentiments of the Scottish Enlightenment, Tronto notes that the care ethic presents certain similarities with this thought because this philosophy gives a different idea of the relationship between moral life and political life than the Kantian approach. Since the end of the 18th century, the Kantian model has been almost completely unchallenged: the Kantian moral theory excludes any analysis of morality that appeals to the emotions, to the circumstances of everyday life, as Kant considers them irrational. Morality, then, becomes a domain located beyond the world of emotions and feelings and belongs only to reason. It appears universal, where the actor is presented as objective, autonomous, distant and guided by abstract rationality [28] (p. 37). These assumptions divert our attention from the value of care in our lives. However, by being part of a "movement to rehabilitate emotions and feelings in moral theory, the ethics of care legitimizes what is in the private domain as political" [31] (p. 321). Arendt and Tronto agree on the "Kantian error". For Tronto [28], Kant's conception of ethical life sets the boundaries around morality understood as an autonomous sphere of human life. There is a requirement that morality be derived from human reason in the form of universal, abstract and formal principles. These boundaries request that the relation of the social and the political to morality is not central to morality itself and suppose that morality is rigorously separated from self-interest, thus reflecting "the moral point of view", i.e., a set of universalizable, objective principles describing what is right. Kant's fundamental error, in the Arendtian thought, is that his moral philosophy misses the moral judgment in its confrontation with a singular situation [29] (p. 197). Kantian morality is then a morality of impotence in the sense that man remains a prisoner of himself and is unable to take the risk of commitment with others because he is stuck to an ethic of conviction. However, action implies a constant engagement with others; it is "relationship with". Kantian morality would therefore be unsuitable for thinking about human action because it forbids any relationship with the plurality of individuals (although Kant is not hostile to plurality) [29] (p. 196).

3.2. Conditions including Care

For Tronto, taking account of otherness but also the fact of thinking care in relation to the world and considering it as power are necessary conditions for a political conception of care.

3.2.1. From Otherness to Plurality

Tronto argues that unless we abandon our current way of thinking about the boundary between morality and politics, it is not possible to honestly approach the problem of otherness. To think about otherness, then, is to rethink the boundary between morality and politics [28] (p. 63), in particular, by recognizing the partiality (of gender, of class) at work in universalist morality. It is crucial, in any feminist theory, to take into account questions of distance and otherness, to consider what our relations with others, close and distant, should be, to pay attention to situating in a general context the conditions in which others are found, in contrast to "treat morally distant others who we think are similar to ourselves" [28] (p. 13). For Tronto [28], otherness is an inherent element of the human condition in our Western societies, citing Simone de Beauvoir: "Otherness is a fundamental category of human thought. Thus is it that no group ever sets itself up as the One without at once setting up the Other over against itself..." [28] (p. 70). Kohlberg's theory "masks" this otherness and "makes it impossible for moral reasoners to deal with the 'others' they have created" [29] (p. 72), a process that Tronto calls assimilation, for it assumes that people are interchangeable. Tronto suggests considering the position of the other as he himself expresses it: thus, one is engaged from the standpoint of the other, but not simply by presuming that the other is exactly like the self. From such a perspective, "we may well imagine that questions of otherness would be more adequately addressed than they are in current moral frameworks that presume that people are interchangeable" [28] (p. 136). For the author, only the otherness thus conceived makes it possible to think about plurality.

Plurality in Arendt's philosophy allows confrontation with otherness: a plurality of equal beings and yet radically different from each other [32]. Otherness is an important aspect of plurality; it is because of it that we are unable to say what one thing is without distinguishing it from something else. Plurality is properly the condition of the human existence, and it introduces the otherness by which we can go out of ourself: a human being exists only from the other human beings. World's plurality is elaborated in this acceptance of the uniqueness of each one. "Plurality is the condition of human action because we are all the same, that is, human in such a way that nobody is ever the same as anyone else who ever lived, lives or will live" [7] (p. 7).

For Arendt, politics, the place of action, is based on a fact: human plurality. Plurality is therefore a factual condition of action. "Action, the only activity that goes on directly between men without the intermediary of things or matter, corresponds to the human condition of plurality" [7] (p. 176). In *The Life of the Mind*, she writes, "plurality is the law of the Earth" [33] (p. 19). Politics, for Arendt, originates in the space-that-is-between-men and is constituted as a relation; it is what happens between men in the plural [34]. The institution of the common world is realized through action, thought of as activity and as an entire dimension of life—what man does—as a pragmatic requirement of action. The constitution of this intermediate space, this public space, is the precondition for the emergence of politics, so action takes place in the public space.

For Tronto, including care in the public space is a necessary condition for the emergence of a politics of care. Care, as a central aspect of human life, requires a universalist principle: "we must care for those around us and those with whom we form society, this principle requiring a political commitment that contributes to valuing care and reshaping institutions" [28] (p. 178).

Could care, through its practice of otherness and its desire for pluralism, allow the emergence of political action in the Arendtian sense of the term as a co-action in a common world? A world no longer described as the juxtaposition of autonomous and independent individuals, distant and similar, but described in the form of the relationship to the other.

3.2.2. Care as a Relationship to the World

However, Tronto does not limit the care to the interactions that humans have with others, she will include the possibility that it can also apply to objects and especially to the environment. It is a holistic vision that Berenice Fisher and Joan Tronto will propose by defining care "as a species activity that includes everything that we do to maintain, continue, and repair our 'world' so that we can live in it as well as possible. That world includes not only our identities but also our environment, all of which we seek to interweave in a complex, life-sustaining web" [28] (p. 103). Involving our "environment", Tronto and Fisher [35] go further than Gilligan, who limited care to the individual needs. Care is here more than a relationship to oneself and to another; care defines the relationship with the world. It thus extends beyond the private sphere and becomes part of the public space; care as a political concept in its own right will allow a parallel with the Arendtian's care of the world. The latter is a central concept in Arendt, everything brings her back to the "care of the world", to the "love of the world": the *Amor Mundi*, defined by taking up and transforming Saint Augustin's concept of love of neighbor. Equals in front of their destiny, men are also interdependent since they cannot live outside a chain of reciprocal exchanges. *Amor Mundi* determines the being-together of men, that is to say, their worldly and political consciousness. Through love of the world, society becomes the vector of self-conscious solidarity and is no longer an organization based on reasoned exchange. Only by belonging to the world can we feel at home on this earth; it is the moral responsibility of every human being to care for the world in which he or she lives. As such, could the ethics of care allow action and be reconciled with politics?

3.2.3. Care as "The Power of the Weak"

As opposed to life, the world is a stable environment. This stability is essential for Arendt. However, does not the vulnerability of man refer to the vulnerability of the world [36]. If the concern for life is, in Jonas, immediately a concern for the world, the concern for the world in Arendt is a distancing from the question of life, in the sense that the process of maintaining life can be carried out without the assistance of a human world.

For Arendt, there is a rupture between human existence and the biological activity of the living, and politics must remain free of the tasks of biological life [7]. The submission of the society to the condition of the life is done to the detriment of belonging to the world and of plurality.

Care can appear as the privileged site of a government of life, especially if we consider it as a power. Because it is one of the promises of care for Tronto [28] (p. 122). She defines it as one of the "powers of the weak" [28] (p. 122) because it supports life by responding to the satisfaction of needs. In this sense, one might think of care as fundamentally anti-Arendtian politically: the satisfaction of needs, but it is, above all, profoundly anti-capitalist: the satisfaction of needs rather than the pursuit of profit [28] (p. 175). In a world where politics is described as limited to the protection of interests, there is a redefinition of power here, as well as a different conception of the notion of needs. The error of universalism, for Tronto, is to consider needs as commodities [28] (p. 138). For Nussbaum, the concept of need is very useful: By introducing the specific vocabulary of "capabilities" introduced by Sen, she opens a framework in which it is possible to form more objective judgments about needs [28] (p. 140).

Arendt's idea of power, especially in its distinction between power and domination, can be relevant to the political conception of power and its attributes: power, force, authority and violence. Political domination, which denies any individuality, goes against a power in common, as a capacity to live and act together, the only power able to build a common world. A power that derives from the characteristics of the action: plurality and concertation.

Indeed, Tronto insists on the conditions of a democratic and pluralist care. She affirms that "care, is only viable as a political ideal in the context of liberal, pluralistic, democratic institutions" [28] (p. 158). The socio-economic relations constitute, from now on, an irreducible dimension of the social with an extension of the merchandising to all the fields of life. Power as a capacity to be and to act fades away in front of the preponderance of the market and in favor of a politics reduced to the preservation of wealth, contributing to maintaining vast relations of domination, gender, class and race, inherent to the care and current conditions of power. Ultimately limiting the human being or even the environment to its economic component.

To replace needs in a political context, as Tronto proposes, is not only to extract them from the economic sphere, but also to operate a shift in our conceptions of human nature: connecting "our notion of 'interests' with the broader cultural concern with 'needs'". A shift from production utility to the care: care of others, care of the planet, care of the world.

4. Thinking about Care with Arendt?

It is important to differentiate care from life, to relate it to the individual, to the other and to the world; to consider it politically, by going beyond the moral boundaries highlighted by feminist studies, by questioning interdependence, plurality and caring, in a theory that could question the Arendtian social, notably in the definition of the "needs" of our current world; and finally to propose a rereading of power, thought in terms of relationship.

4.1. World

The notion of world is central to understanding the specific originality of Tronto's conception of care. Care, which consists of being concerned about (caring about), must lead us to want to "repair our world". Let us recall the definition given by Tronto and Fischer: "On the most general level, we suggest that caring be viewed as a species activity that includes everything that we do to maintain, continue and repair our 'world', so that we can live in it as well as possible. This world includes our bodies, ourselves, and our environment, all of which we seek to interweave in a complex, life-sustaining web" [28] (p. 103).

Several elements of this definition are worth emphasizing. First, care is not limited to the interactions of humans with each other. It applies not only to others but also to inert objects and the environment [28] (p. 103). Furthermore, care is not a dual or inter-individual relationship. Care is too often described as a dyadic relationship, modeled on the relationship that unites a mother to her child; this representation presupposes that care is an individual dimension, even though few societies in the world have been able to conceive of the education of children as the sole responsibility of the mother. This dyadic reading of care could only lead to the rejection of the different ways in which care can intervene on the political level. Care, far from being a purely individual disposition, would characterize any practice whose aim is the maintenance, perpetuation or repair of our world.

The notion of the world is also central to Arendt's philosophy through the notion of "world building", "care of the world" or Amor Mundi. The world is what allows us to be human. Indeed, Arendt's analysis of the totalitarian phenomenon led her to be convinced of the absence of human nature [37]. Man is characterized by his condition, i.e., his inscription in a state of the world, which is constituted by the whole of the durable works that he makes. The world is thus populated by artificial objects that allow men to go beyond the strict determinations of their biological inscription and to be part of sustainability. The excess of the human claims offered by the technique induces a radical transformation of the world of that we forget to take care of. The world extends between men, and this "between"—much more than (as it is often thought) men or man—is today the object of the greatest concern". The world is not reduced in Arendt to a simple set of functional objects. It becomes a world by offering man a durable living place, in which he can find reference points for his own existence and in which we can find an echo of our own voice. The world is not human because it was made by men, and it does not become human because the voice of man resounds in it but only when it has become an object of dialogue. This notion of world has received recent attention from researchers in philosophy in relation to the Anthropocene. Following the analyses of Arendt, Hyvönen [38] elaborates a notion of "material culture of care" as a modality of mediating human interactions with non-human nature. If the notion of care, as it is used by Arendt, is "usually associated with the narrowly political activities of democratic participation" [38] (p. 98), according to Hyvönen, care should also be applied to human's relationship with nature.

4.2. Plurality

A second point of intersection between Tronto and Arendt is found in the concept of otherness. For Tronto, care defined as a political ideal can only be viable in the context of free, pluralistic and democratic institutions [28] (p. 158). Tronto is more likely to refer to the concept of otherness than to that of plurality, but the two are intimately linked. For Arendt, it is the recognition of human plurality that conditions the very possibility of an authentic political project. In a letter to Jaspers on 4 March 1951, Arendt states: "Western philosophy has never had and has not been able to have a precise idea of politics because it was necessarily addressed to man—that is, to abstract man, in his general idea, not in his particular conditions—and held the reality of plurality to be incidental" [33]. To take into account human plurality, to pose it as a condition of politics, consists not only in recognizing that men are united by their diversity. The world we inhabit is not populated by clones whose essence would be fixed once and for all by their nature. It is populated by unique and irreplaceable identities. According to Arendt, by acting and speaking, men make it clear who they are, actively reveal their unique personal identities, and thus make their appearance in the human world [7]. Politics must be defined as that which allows each man to reveal what makes him irreplaceable. In *The eggs speak up*, Arendt takes up the popular formula according to which "you can't make an omelette without breaking eggs" and proposes analyzing the foundations that it reflects. The image of the omelet refers to the transformation of society into a shapeless mass of substitutable individuals, impossible to distinguish in their singular identity. In the omelet, the eggs are always present, but they are no longer identifiable as such [10]. The society thought from the metaphor of the omelet designates the society directed by the tyrant and made homogeneous by terror, as the democratic "mass society", governed in appearance by the principle of freedom, but which does not escape an effort of standardization in which the individuals deny their specificities and are reduced to masses of conforming and interchangeable individuals. This is the case in societies where politics has given way to the economy and its laws of operation, which demand standardization of behavior. Indeed, the individual reduced to a simple *homo economicus* is thought of as a rational being, totally predictable and anticipable from the understanding of his short-, medium- or long-term interests.

However, politics, according to Arendt, should always be primarily concerned with eggs, with taking care of them, rather than having as a priority objective the success of omelets. We find here a convergence with the principle of care and the very possibility of its extension from the moral to the political field. Instead of aligning individual singularities with a sanitized and homogenized mass, politics must strive to preserve singularity because only it can allow for plurality, the condition of a human world.

4.3. Power

For Sintomer [39] (p. 117), the notion of power constitutes the keystone of Arendtian political philosophy. Power is conceived by Arendt as the result of the interaction between men; it is an intersubjective notion. Power springs up among men when they act together and falls back as soon as they disperse. Power does not translate into the "I want" of an isolated individual but into the "I can", which results from an action carried out in common. It is thus always uncertain and dependent on a common potential that may disperse. This allows us to understand why power is not owned since it results from common actions. It can either decrease or increase again and again.

Tronto's conception of power emphasizes the idea of strength and power. The powerful accumulate power, which is distributed in a very unequal way in contemporary societies. Thus, women are virtually excluded from power in political, economic and cultural institutions in the United States.

Yet, moral theories are not designed to take account of these inequalities of power. The "every man for himself" doctrines of universal moral judgment mask the inequalities of resources, power, and privilege that have made it possible for some to succeed and not others [28] (p. 147). These resource inequalities, by preventing citizens from accessing equal power, are important political issues, not mere theoretical dilemmas.

For Tronto, feminist theory stems from the attempt to end the marginal status of women in society, which leads to them remaining on the periphery of power circles. To share the power of those who are central, there are few options, for if it is difficult to be admitted among the powerful, it is even more difficult to cause their fall. To try to obtain power from the margins of the system, to persuade those who have power to share it, the feminist theory puts forward only two options: the recourse to the logic of identity (to become similar to those who are already in place) or to the logic of difference (to have something interesting to offer them) [28] (p. 20). On the other hand, Tronto will propose to see the world differently: to ensure that the activities that legitimize the accumulation of power among the powerful are less valued and that those that could legitimize a sharing of power with the "outsiders" are more appreciated. This process will consist, for Tronto, "in recognizing that the current boundaries of moral and political life are drawn such that the concerns and activities of the relatively powerless are omitted from the central concerns of society" [28] (p. 20).

Referring to the theories of moral sentiments, she shows that care, extended to concern for oneself and others and no longer reduced to the sole experience of women, could be morally central until the 19th century. With the development of trade, the place of care in societies will be increasingly reduced. "Moral sense" proposed by Hutcheson or Smithian "sympathy" will give way to rationality and moral universalism, where only independence and autonomy constitute the essence of life. (Hutcheson, following Shaftesbury, thinks that we can account for our moral evaluations by postulating the existence of a "moral sense"; they conceived the moral sense as a natural organ of moral perception, Hume and Smith argue that this concept has no explanatory value and that our moral judgments are best explained from our propensity to "sympathize" with the feelings of others. They argue that our moral evaluations are generated primarily by our social and affective interactions.)

Tronto wishes to reconnect with this Scottish Enlightenment centrality of care and to present it as the "organization of our world". Considered as "support for life" (without care, children would not become adults, men would not have children to inherit their fortune, etc.), it constitutes a power, and it is one of the promises of care as the "power of the weak".

To define power, Tronto will refer to the analyses of C. Wright Mills: "my notion is partly Marxian, that the powerful are members of the class that can command resources, and party Weberian, that the powerful are those who occupy status positions and can command resources in society" [40] (p. 185). In other words, the decisions that will most affect people's lives are made by those, and only those, who are in the "command positions" of society. From the perspective of care, Tronto will point out that the institutions that are thus placed at the center of power (government, armies and firms) will shape those that are pushed to the periphery (family, schools, churches, hospitals). Moreover, the conquest and the satisfaction of needs thanks to the work of the subjugated classes will come to serve the political and economic interests of the ruling classes [28] (p. 227). There will be a publicization of the questions of care, notably during the 20th century under the action of the government and the market [28] (p. 173): certain private, family-related care will be assumed by the State (social assistance, etc.), the latter taking on the role of "head of the household" [28] (p. 173), and non-family-related care will be provided by the market. From now on, only those who can afford it receive care independently of their needs, thus creating an unequal distribution of resources. Care thus reveals the ways in which the powerful monopolize resources. If care is used by the powerful to manifest their power and to preserve it, it also makes it possible to reveal these relations of power and to grasp how it is distributed or not in the society, from where the necessity of its confinement: "It is the enormous real power of care that makes its containment necessary" [28] (p. 122).

For Tronto, the ultimate strength of care as a political concept lies in its ability to serve as a basis for political change. Consisting of "a way to collect the "powers of the weak "" [28] (p. 177), care could go beyond a simple renunciation by the powerful of some of their power. Tronto's differentiation between the weak and the powerful can be understood as that between the dominated and the dominants. According to Arendt, power is not an individual property; one does not possess power. Moreover, domination is individual, and power is common. Arendt's distinction between power and domination could shed new light on the integration of women as political actors and, more generally, of people who are generally excluded—one of the most important issues to be resolved for care. Power derives from the characteristics of action: plurality and concertation. In this Arendtian conception, the power of care could be born from the concerted action of the "weak" and of the "dominated", whose aim would be to found a new political body that would allow them to maintain their freedom and power through collective action. It is because the source and the objective of power are identical, namely action in its plurality, that power appears to be its own principle. This conception of power anchored in plurality implies the fact that it cannot be only identified with the powerful. It is an opposition to the values of individualism in the very conception of power. In this sense, the notions of weak and powerful would become obsolete in a politics of care, considered as an alternative to violent domination, a politics at the service of the good of the world.

5. Conclusions

The reconstitution of the reasons that led some feminist authors to reject Hannah Arendt's thought makes it possible to understand the reasons for this misunderstanding. The question of gender is not considered a relevant question by the philosopher, and its separation of the private and public spheres has been considered an element invalidating any claim to reconcile the fundamental principles of care with Arendt's conception of the politics. Nevertheless, a careful reading of Arendt's corpus allows us to assert that a rapprochement between her own philosophy and the theories of care, in particular the thought of Tronto on which we have focused in this study, is possible. If we have underlined the extent to which the notions of world, power or plurality, as conceived by

Arendt, resonated in the perspective of care as opened by Tronto, another rapprochement would deserve to be made in particular on the notion of interest.

By making care exist in the public space, Tronto allows it to become a power, no longer thought of in terms of domination but defined as the capacity of humans to act together. Arendt then allows us to conceive a politics of care linked to action, putting people in relation with a view to a common world, sharing the conditions of plurality in a world populated by interdependent beings. Arendt and Tronto make it possible to pose the concern for the other and the concern for the world at a political level, in a society vector of solidarity and no longer based on the defense of individual interests, thus shifting the notion of interest toward care: the care of the world in the Arendtian sense of Amor Mundi.

Author Contributions: Conceptualization, C.N.L. and C.C.; writing—original draft preparation, C.N.L. and C.C.; writing—review and editing: C.N.L. and C.C. All authors have read and agreed to the published version of the manuscript.

Funding: This research received no external funding.

Institutional Review Board Statement: Not applicable.

Informed Consent Statement: Not applicable.

Conflicts of Interest: The authors declare no conflict of interest.

References

1. Honig, B. *Feminist Interpretation of Hannah Arendt*; Penn State Press: University Park, PA, USA, 1995.
2. Bourgault, S. Le féminisme du care et la pensée politique d'Hannah Arendt. *Rech. Féministes* **2015**, *28*, 11–27. [CrossRef]
3. Myers, E. *Worthly Ethics: Democratic Politics and Care for the World, Durham*; Duke University Press: Durham, NC, USA, 2003.
4. Mann, B. Dependence on Place, Dependence in Place. In *The Subject of Care. Feminist Perspectives on Dependency*; Kittay, E.F., Feder, E.K., Eds.; Rowman & Littlefield: Oxford, UK, 2002; pp. 348–368.
5. Kyle, J. Protecting the World: Military Humanitarian Intervention and the Ethics of Care. *Hypatia* **2013**, *28*, 257–273. [CrossRef]
6. Rich, A. *La Contrainte à l'Hétérosexualité et Autres Essais*; Mamamélis—Nouvelles Questions Féministes: Genève-Lausanne, Switzerland, 2010.
7. Arendt, H. *The Human Condition*; University of Chicago Press: Chicago, IL, USA, 1958.
8. O'Brien, M. *The Politics of Reproduction*; Routledge and Kegan Paul: Boston, MA, USA, 1981.
9. Lamoureux, D. Hannah Arendt: Agir le Donné. In *Sous les Sciences Sociales, le Genre. Relectures Critiques, de Max Weber à Bruno Latour*; Chabaud-Rychter, D., Ed.; La Découverte: Paris, France, 2010; pp. 471–484. [CrossRef]
10. Arendt, H. *Essays in Understanding: 1930–1954*; Kohn, J., Ed.; Harcourt Brace & Company: New York, NY, USA, 1994.
11. Arendt, H. *Rahel Varnhagen: The Life of a Jewish Woman*; Winston, R., Winston, C., Eds.; Harcourt Brace Jovanovich: New York, NY, USA, 1974; Critical edition Edited by Liliane Weissberg; Johns Hopkins University Press: Baltimore, MA, USA, 1997.
12. Arendt, H. *Men in Dark Times*; Harcourt Brace Jovanovich: New York, NY, USA, 1968.
13. Collin, F. Du privé au public. *Les Cah. Du CRIF* **1986**, *33*, 37–48. [CrossRef]
14. Bethke Elsthain, J. *Public Man. Private Woman*; Princeton University Press: Princeton, NJ, USA, 1981.
15. Couture, J.P. La Politique Comme Arrachement à la Nature. Essai Sur le Concept de Social. Montréal. Master Thesis, University of Quebec, Montréal, QC, Canada, 2000.
16. Benhabib, S. Feminist theory and Hannah Arendt's concept of public space. *Hist. Hum. Sci.* **1993**, *6*, 97. [CrossRef]
17. Pitkin Fenichel, H. *The Attack of the Blob. Hannah Arendt's Concept of the Social*; Chicago Press: Chicago, IL, USA, 1994.
18. Clarke, J.P. Social justice and political freedom: Revisiting Hannah Arendt's conception of need. *Philos. Soc. Crit.* **1993**, *19*, 333–347. [CrossRef]
19. Arendt, H. *On Revolution*; Penguin Books: New York, NY, USA, 1990.
20. Arendt, H. *Responsibility and Judgment*; Kohn, J., Ed.; Schocken Books: New York, NY, USA, 2003.
21. Canovan, M. *The Political Thought of Hannah Arendt*; Harcourt Brace Jovanovich: London, UK, 1974.
22. Kant, E. *Critique of Practical Reason*; Cambridge University Press: Cambridge, UK, 1997.
23. Kant, E. *Theory and Practice*; translated by Humphrey, T.; Springer: Berlin/Heidelberg, Germany, 1988.
24. Kant, E. *The Metaphysics of Moral*; Cambridge University Press: Cambridge, UK, 2006.
25. Laugier, S. L'éthique comme politique de l'ordinaire. *Multitudes* **2009**, *37–38*, 80–88. [CrossRef]
26. Gilligan, C. *In a Different Voice*; Harvard University Press: Cambridge, MA, USA, 1983.
27. Noddings, N. *A Feminine Approach to Ethics and Moral Education*; University of California Press: Berkeley, CA, USA, 1984.
28. Tronto, J. Moral Boundaries. In *A Political Argument for an Ethic of Care*; Routledge: New York, NY, USA, 1993.
29. Noel Lemaitre, C. *Arendt, Pas à Pas*; Ellipses: Paris, France, 2019.
30. Molinier, P.; Laugier, S.; Paperman, P. *Qu'Est ce Que le Care? Souci des Autres, Sensibilité, Responsabilité*; Payot: Paris, France, 2009.

31. Paperman, P.; Laugier, S. *Ethique et Politique du Care. Le Souci des Autres, (nouvelle édition augmentée)*; EHESS, coll. Raisons Pratiques: Paris, France, 2011.
32. Arendt, H. *The Promise of Politics*; Kohn, J., Ed.; Schocken Books: New York, NY, USA, 2005.
33. Arendt, H. *The Life of the Mind, One, Thinking*; Houghton Mifflin Hercourt: New York, NY, USA, 1978.
34. Bragantini, A. Pluralité et être en commun. Arendt Confrontée à Heidegger. *AUC Interpret.* **2013**, *3*, 63–83.
35. Fischer, B.; Tronto, J. Towards a Feminist Theory of Care. In *Circles of Care: Work and Identity in Women's Lives*; Abel, E., Nelson, M., Eds.; State University of New York Press: New York, NY, USA, 1991.
36. Jonas, H. *The Imperative of Responsibility: In Search of an Ethics for the Technological Age*; University of Chicago Press: Chicago, IL, USA, 1984.
37. Arendt, H. *The Origins of Totalitarianism*, Revised ed.; First published 1951. (Includes All the Prefaces and Additions from the 1958, 1968 and 1972 Editions); Schocken Books: New York, NY, USA, 2004.
38. Hyvönen, A.-E. Amor Tellus? For a Material Culture of Care. *HannahArendt.net* **2022**, *11*. Available online: https://www.hannaharendt.net/index.php/han/article/view/460 (accessed on 10 May 2022). [CrossRef]
39. Sintomer, Y. Pouvoir et autorité chez Hannah Arendt. *L'Homme Société* **1994**, *113*, 117–131. [CrossRef]
40. Mills, C.W. *The New Men of Power*; Kelley: New York, NY, USA, 1971.

 *philosophies*

Article

Care Ethics and Paternalism: A Beauvoirian Approach

Deniz Durmuş

Department of Philosophy, John Carroll University, University Heights, OH 44118, USA; ddurmus@jcu.edu

Abstract: Feminist care ethics has become a prominent ethical theory that influenced theoretical and practical discussions in a variety of disciplines and institutions on a global scale. However, it has been criticized by transnational feminist scholars for operating with Western-centric assumptions and registers, especially by universalizing care as it is practiced in the Global North. It has also been criticized for prioritizing gender over other categories of intersectionality and hence for not being truly intersectional. Given the imperialist and colonial legacies embedded into the unequal distribution of care work across the globe, a Western-centric approach may also carry the danger of paternalism. Hence, a critical approach to care ethics would require reckoning with these challenges. The aim of this article is first to unfold these discussions and the responses to them from care ethics scholars and then to present resources in Beauvoir's existentialist ethics, specifically the tenet of treating the other as freedom, as productive tools for countering the Western-centric and paternalistic aspects of care practices.

Keywords: critical care ethics; existentialism; Simone de Beauvoir; feminism; paternalism; existentialist ethics; western-centric approaches in care; transnational feminism

Citation: Durmuş, D. Care Ethics and Paternalism: A Beauvoirian Approach. *Philosophies* **2022**, *7*, 53. https://doi.org/10.3390/philosophies7030053

Academic Editors: Maurice Hamington and Maggie FitzGerald

Received: 28 February 2022
Accepted: 28 April 2022
Published: 18 May 2022

Publisher's Note: MDPI stays neutral with regard to jurisdictional claims in published maps and institutional affiliations.

1. Introduction

Care ethics has succeeded in establishing itself as a well-developed and well-recognized ethical discipline in its own right. In response to criticisms from feminist scholars charging it with essentialism and focusing only on the personal, the field has reworked itself to incorporate social and political dimensions and applications of caring relationships. Transformation of the early formulations of care ethics into a more political account of care ethics provided fruitful discussions addressing the hard-pressing questions related to the imperialist and colonial legacies embedded into the unequal distribution of care work across the globe. However, despite its critique of Western-centric ethical theories, care ethics has received criticisms for operating—mostly unintentionally—with Western-centric assumptions and registers by way of universalizing care as practiced in the Global North. This tendency can be understood as an indirect manifestation of the dominant Eurocentric perspective that undergirds most of the Western thinking and theorizing. Ethnocentrism is still present as a significant challenge in feminist theories developed in the Global North and care ethics is not immune to that challenge. In addition to the lack of engagement, the Eurocentric approach may also present itself in the form of paternalism when there is some level of engagement with different localities.

Transnational feminist ethics scholarship presents rich discussions showing why care questions have to be dislocated from the Global North and relocated in different geographies by taking intersectional oppressions into account. While transnational feminist ethics provides us with these well-justified concerns regarding care ethics, the discussion needs to be complemented with accounts of how to resolve such concerns and challenges with respect to care ethics, especially given the historical reality of paternalism in Western encounters with non-Western geographies. For this reason, I introduce Beauvoir's existentialist ethics and argue that it has immense resources to offer in helping us map out how to decenter care ethics from the Global North to more diverse locations without falling into the trap of paternalism. I argue that Beauvoir's notion of treating the other as freedom enables us to better perceive the incongruity between the ideas of the one caring and that of the one cared; for regarding the best way of care as the Beauvoirian notion of treating the other as freedom helps us acknowledge the one cared for as a subject who has their unique set of needs and projects distinct and sometimes even in conflict with the one caring.

Using Beauvoir's existentialist ethics in combination with transnational feminist critiques of care ethics to argue against the paternalism that is assumed by Eurocentric conceptions of care ethics and to formulate a more viable form of care ethics may sound counterproductive given that Beauvoir herself is situated in the Western philosophical tradition. However, Beauvoir's writings and activism responds to the problem of coloniality in unique and productive ways. Her interventions in the Algerian decolonization war are exemplary of how one can be mindful of their privileged and colonialist position in interacting with the colonized.

In her interventions, Beauvoir was particularly conscious of her background as a middle-class French intellectual who was benefitting from the Algerian colonization. She treaded lightly the line between indifference and paternalism, only showing care for the oppressed who was a victim of the French imperialist policies. As she explains in *Force of Circumstance*, indifference to the Algerian War could no longer be an option for her, given her complicity in the oppression Algerian people have been enduring [1] (pp. 369, 371, 384, 652). Intervening in a paternalistic manner, on the other hand, would go against the principles of existentialist ethics she laid out in her philosophical works [2,3]. For her, engagement with the Algerian girl, Djamila Boupacha, who was raped by the French military, was like walking a razor blade, not only because she could taint the care for her with her own subject-position's values, but also because she had to avoid the pitfall of philosophical imperialism herself, painting her act as a paternalistic charity of a hypocrite [4,5].

That is why, while she deployed every tool at her disposal to help free Djamila Boupacha, she abstained from intervening when Boupacha asked for help to resist the FLN's (*Front de Libération Nationale*) request that she goes back to Algeria [6] (p. 529). Such nuances in Beauvoir's interactions will be central to my claim that Beauvoirian existentialist ethics may provide fruitful resources to counteract transnational feminist criticisms leveled against care ethics [7–10], despite the seeming fact that Beauvoir, too, is a Western, privileged, middle-class feminist. However, it is important to note that care ethics has been one of the few ethical theories with a Western origin that both welcomed and encouraged criticisms especially to open up space for inclusion and contributions from all over the world. For example, Cree scholar Katherine Walker's work [11] and Tula Brannelly and Amohia Boulton's scholarship [12] on the interlacement of care ethics and Māori thought attests to meaningful engagement with indigenous thought in care ethics. In addition, Vrinda Dalmiya's comparative reading of care ethics and the Mahābhārata [13], Chenyang Li's comparative study of Confucianism and care ethics [14], Thaddeus Metz's suggestions for care ethics based on her articulation of sub-Saharan communitarian morality [15], Sarah Munawar's work that develops an Islamic ethics of care [16], and Hil Malatino's theorization of trans care work [17] are only a few more examples from a rapidly growing literature that engages care ethics not only with different localities, but also with marginalized groups in the Western geographies.

The affinity existentialist ethics shares with care ethics regarding their fundamental principles and their critique of traditional ethical theories also constitutes another reason to turn to Beauvoir in discussing the questions above. Unfortunately, the compatibility between existentialism and care ethics has not been fully explored by feminists yet. Kristana Arp points out some of these commonalities, yet she limits her analysis to Beauvoir's *The Ethics of Ambiguity* and only focuses on the phenomenological aspects of Beauvoir's work [18]. Maurice Hamington and Anya Daly also join the efforts to find a grounding for feminist ethics in the phenomenological tradition. Hamington explores Maurice Merleau-Ponty's theory of embodiment as a grounding for care ethics [19] (pp. 5, 39). Daly, on the other hand, appeals to Merleau-Ponty's non-dualist ontology in order to answer the questions of why one should care [20] (pp. 12–14). Based on Merleau-Ponty's thesis of interdependence and the reciprocity and empathy that ensues from this interdependence, Daly argues that Merleau-Ponty's phenomenology explains the potential of and tendency to care [21] (p. 290). Tove Pettersen is another care ethics scholar who focuses on the relational ontology of care ethics as its most significant feature [22] (p. 52). While these phenomenological aspects provide significant grounding for care, existential and postcolonial aspects of Beauvoir's work compound this process of grounding in meaningful ways.

In my analysis, I delineate that the notion of interconnectedness of people and the accompanying notion of the relational self are central to both care ethics and Beauvoir's thinking. My goal in bringing in Beauvoir's notion of existentialist freedom is to offer a model to counter the detrimental effects of ethnocentric thinking in care ethics and help relate to different localities across the globe. Ethnocentric thinking has also historically produced paternalistic interventions directed at the women of the Global South by Western women. As I have shown elsewhere, Beauvoir skillfully practiced existentialist tenets in her colonial and postcolonial engagements, and she avoided the dangers of paternalism in her interventions in the Algerian decolonization movement [23]. Here, I analyze her writings and activism in the hopes that it will shed some light on the discussions of paternalism in care ethics.

In the first section, I explain Beauvoir's existential ethics with an emphasis on her notion of freedom. In the second section, I trace the development of care ethics in relation to its critique of traditional Western ethical theories. In the third section, I locate discussions of care in postcolonial literature. The work in this section aims to address the social inequalities surrounding caring practices between the Global North and the Global South. By exploring the local contingencies in defining care practices with an emphasis on their relation to the global political structures, I show the colonial and paternalistic aspects of care practices and call for a more exhaustive account of these aspects in discussions of care ethics. In the last section, I discuss how Beauvoir's existentialist notion of freedom may prove to be useful for addressing the colonial and paternalistic aspects of care practices.

2. Beauvoir's Existentialist Ethics

A close examination of Beauvoir's life and writings reveals that her existentialist ethics was developed within the context of her political activism against the oppressive structures of her time [23]. While her existentialist ethics is centered around the notion of freedom, the notion of freedom she develops in her ethics is closely informed by an account of oppression. We first experience ourselves as dominated by our situation in our childhood where we are confined with meanings created by others. Only after childhood we can materially challenge those meanings and endorse ours. However, since we are born into those meanings and they constitute our facticity, our freedom is shaped in relation to those meanings [3] (pp. 38–39). Hence, our projects and choices are shaped by others from the very beginning. In *The Prime of Life*, Beauvoir writes,

> An individual, I thought, only receives a human dimension by recognizing the existence of others. Yet, in my essay, coexistence appears as a sort of accident that each individual should somehow surmount; he would begin by creating his project in isolation, and only then ask the community to endorse its validity. In truth, society shapes me from the day of my birth and it is within that society, and through my close relationship with it, that I decide who I am to be [24] (p. 456).

For Beauvoir, then, the self is constructed through the projects one undertakes based on those meanings. Moreover, these projects can neither be realized nor take on any meaning without others. Hence, instead of considering the need for others as an egoistic move, we should understand it as a contingent fact of our human condition. She writes, " … the individual is defined only by his relationship to the world and to other individuals; he exists only by transcending himself, and his freedom can be achieved only through the freedom of others" [3] (p. 156).

According to Beauvoir, we are free, yet we also choose to will ourselves free or not. In *The Ethics of Ambiguity*, she appeals to a distinction between natural freedom and ethical freedom to resolve the seeming contradiction on being free and choosing that freedom at the same time [3] (p. 24). Instead of postulating these freedoms as two separate types of freedoms, drawing from Edward Fullbrook and Kate Fullbrook, I argue that they are two dimensions of our freedom [25] (p. 106). In other words, they are two different experiences of our freedom. According to Beauvoir, one's experience of herself as a consciousness is the same as one's experience of herself as freedom. In that sense, being free is the essential mode of all human beings, which she describes as natural freedom. However, in our interactions with others we are also objects and we need recognition of others to create and pursue our projects. In creating our genuine projects and demanding attention of others to our projects we are exercising ethical freedom. Beauvoir states that "To will oneself free is to effect the transition from nature to morality by establishing a genuine freedom on the original upsurge of our existence" [3] (p. 25).

One may abstain from exercising her ethical freedom because of different reasons. Two main reasons Beauvoir considers for failure in exercising ethical freedom are being in bad faith and being under oppressive conditions. In " . . . laziness, heedlessness, capriciousness, cowardice, impatience . . . " one is escaping from the anguish her freedom brings about [3] (p. 25). In situations of oppression, on the other hand, the oppressed is not given the chance to assume her ethical freedom and act upon it. If the oppressive condition ceases to exist and the oppressed still does not claim her freedom, then she would be in bad faith.

While these two modes of failure in exercising ethical freedom, i.e., being in bad faith and being under oppressive conditions, are separate, they can overlap. An example of an overlap would be acting in bad faith and choosing to remain in immanence in response to an oppressive situation. Beauvoir calls this being complicit with one's oppression in *The Second Sex*. In a patriarchal society, Beauvoir argues, many women admit or even welcome their object-like status defined by men for the purposes of rewards and protection. Nevertheless, acting in bad faith in a situation like this should be differentiated from acting in bad faith in the absence of oppressive conditions. As Beauvoir states " . . . she [*women*] discovers and choses herself in a world where men force her to assume herself as Other: an attempt is made to freeze her as an object and doom her to immanence, since her transcendence will be forever transcended by another essential and sovereign consciousness." [26] (p. 17).

Furthermore, one can exercise their ethical freedom in an oppressive way and hence act in bad faith. For example, oppressors act in bad faith by declaring themselves as sovereign and independent subjects, because they refuse to acknowledge the interconnected nature of human freedom and existence. While they may be acting towards transcendence, their actions undermine the actualization of freedom of others. Beauvoir's analysis of patriarchal oppression as practiced by men provides a good example of being bad faith in that regard.

The notions of natural freedom and ethical freedom are as well referred to as ontological (metaphysical) freedom and moral freedom, respectively. For Beauvoir, ontological freedom is the very condition of moral freedom, yet moral freedom can be realized only through the "conscious affirmation of one's ontological freedom" [27] (p. 2). We can choose among ethical action and unethical action only because we potentially have ontological freedom. Beauvoir uses the notion of natural freedom to refer to the conditions of the possibility of willing one's self free. Based on this possibility and depending on the conditions the individual finds herself in, she may or may not choose to will herself free. For Beauvoir "There is ethics only if ethical action is not present" [3] (p. 24). In some situations, on the other hand, the subject may not even have the option to will herself free, although theoretically she has this potential. Beauvoir analyzes those options under the notion of ethical freedom. Natural freedom refers to our ability to act in the world within the limits of our facticity and ethical freedom is overcoming the limits of natural freedom and being involved in the actions which transcends our facticity; thus, natural freedom is the condition of ethical freedom. Here, Beauvoir defines ethics as "the triumph of freedom over facticity" [3] (p. 44).

Notions of transcendence and immanence play a pivotal role in Beauvoir's existential ethics as well. Transcendence and immanence can refer to states of being or adjectives for actions. In other words, an action can be considered as transcendent or immanent or a human being can be in a state that is closer to transcendence or immanence. If a person is only interested in activities that reproduce daily life (activities that are described as immanent) and does not engage in any further activities that create meaning in the world (activities that are described as transcendent) then this person would be closer to a state of immanence rather than transcendence. According to Beauvoir, we are always somewhere between transcendence and immanence and most of our activities involve elements of both [3] (p. 82).

Beauvoir exalts activities of transcendence such as inventions, industries, and books as she thinks they " ... people the world concretely and open concrete possibilities to men" [3] (pp. 80–81). Her treatment of activities of immanence on the other hand takes a negative tone. In *The Second Sex*, for example, she talks about care-related activities as drudgery. She writes,

> Few tasks are more like the torture of Sisyphus than housework, with its endless repetition. The clean becomes soiled, the soiled is made clean, over and over, day after day. The housewife wears herself out marking time: she makes nothing, simply perpetuates the present [26] (p. 451).

Some may read Beauvoir's treatment of care-related activities as immanence as an act of undervaluing such activities. And, understandably, they might question the turn to a Beauvoirian notion of freedom for a grounding of care ethics. To analyze this possible criticism let us turn to Iris Marion Young, who criticized Beauvoir's treatment of domestic work in a negative light. Young reads Beauvoir as devaluing housework by associating it with immanence. In doing so, she argues Beauvoir " ... misses the creatively human aspects of women's traditional household work, in activities I call preservation" [28] (p. 124). Since for Beauvoir activities that shape human history are marked as transcendence and activities that only serve perpetuation of life are marked as immanence, and domestic work is associated with immanence, Young argues that Beauvoir fails to see the creative potential in activities that maintain daily life [28] (p. 138).

While I do not agree with Young that Beauvoir thinks it is impossible for domestic work to be carried out in creative and transcendent ways, this discussion is beyond the scope of this paper. However, based on what I laid out about Beauvoir's existentialist ethics so far, I argue that Beauvoir accords utmost value to activities of immanence because they create the environment in which activities of transcendence can be practiced. In other words, they are prerequisite for any transcendent activity and hence for any type of expression of freedom. However, their relegation to a certain group creates oppressive conditions. Since Beauvoir aimed to change this oppressive structure or at least disturb it, her insistence on talking about the negative aspects of that type of work for women can be understood as a strategy to underscore the impediments created for women by the systemic patriarchal oppression. Yet, that does not necessarily translate into devaluing of such activities per se.

Young also seems to read Beauvoir's position as one of endorsing a dichotomy between the notions of transcendence and immanence and favoring transcendence over immanence. I do not agree with Young on two grounds. First, some of our activities incorporate elements of both immanence and transcendence and therefore do not allow us to maintain this dichotomy. Second, given Beauvoir's emphasis on providing people with the resources to be able to choose and pursue their own projects, the relationship between transcendence and immanence is best described not as a dichotomous one, but rather, an interdependent one, as activities of immanence create the conditions for the possibility of transcendent activities. Many activities that fall under the category of care work, for example, can be creative and transcendent as well as well as mundane and immanent. Andrea Veltman identifies four characteristics of transcendent activity based on Beauvoir's description. They are as follows: producing something durable, enabling individual self-expression, transforming or annexing the world, and contributing to the constructive endeavors of the human race [29] (p. 123). Based on these characteristics, cooking by trying different recipes, experimenting with them, and inventing new ones can become a creative and non-repetitive caring activity. Nevertheless, no matter how creative and self-expressive I am in my cooking, I can only produce a recipe that is durable but not a meal. Hence, cooking falls in between transcendence and immanence. On the other hand, most of the work the factory workers perform produces a great deal of durable objects, while that type of work lacks both self-expression and creativity. There are many activities which fall in the gray zone between transcendence and immanence.

Moreover, in a 1971 interview, Beauvoir acknowledged the socially constructed gender association of care while also recognizing the value of care: In an interview she maintains that "There is often, in women, a kind of caring for others that is inculcated in them by education, and which should be eliminated when it takes the form of slavery. But caring about others, the ability to give to others, to give of your time, your intelligence—this is something women should keep, and something that men should learn to acquire." [30] (p. 191).

Existence in the world necessitates action and interaction with other freedoms. One may well escape assuming others' freedom, treat others as immanence, and avoid seeing how their projects need others to have a meaning, yet they cannot escape their natural freedom in doing so; they would only escape their ethical freedom. Because for Beauvoir "to will oneself moral and to will oneself free are one and the same decision" [3] (p. 24). This is how the distinction Beauvoir makes between natural freedom and ethical freedom in connection with the notions of immanence and transcendence establishes the bond between freedom and ethics and makes an account of oppression possible in the framework of existentialist ethics [3] (p. 24). To be ethical is the same as acting on one's facticity by exercising one's freedom. Nevertheless, ethics is not limited to the subject's freedom. Since our freedoms are interconnected in such a way that we cannot conceive of a subject being free without at the same time conceiving her fellows as free, to be ethical means also to be concerned with Other's freedom. This inherent interconnectedness between individuals then necessitates a continuous interaction between us to help determine the best ways to be involved in promoting each other's freedoms.

Existentialism has been widely criticized for not providing any guidelines for action. While the claim is true, Beauvoir presents this aspect of existentialism not as a weakness but as a strength of the theory. In *The Ethics of Ambiguity*, she emphasizes that the ethics she is proposing does not provide any recipes for action [3] (p. 134). Her justification is that each experience is unique, and it is impossible to determine the best way to respond to it beforehand. Beauvoir writes, "the good of an individual or a group of individuals requires that it be taken as an absolute end of our action; but we are not authorized to decide upon this end a priori" [3] (p. 142). Hence, for Beauvoir, establishing a priori rules of moral action is a useless attempt as no ethical dilemma replicates itself in the exact same manner. Therefore, the individual is called upon to reinvent the rules for action every single time. However, despite nullifying every justification that can be drawn from society, history, or culture, Beauvoir leaves us with one single precept: "to treat the other […] as a freedom," [3] (p. 142). Treating the other as a freedom means treating them as a subject who has their unique projects to pursue. The most significant aspect of this approach is not to conflate one's own ideals, aspirations, and desires with the other person being engaged with. While we cannot know a priori how one should care for the other, as it will depend on the particularities of the case, and treating someone as a freedom is not the only type of care, surely it is one necessary form of care in care's fullest expression. Thus, caring for another has to include treating them as a freedom. If I care for another person but do so in a way that does not treat her/him as a freedom, I am only caring for her in a diminished, and perhaps even harmful way.

For Beauvoir, an action that enhances one's (and others') freedom would be an authentic action, whereas an action that diminishes or undermines one's and/or others' freedom would be deemed inauthentic. Although Beauvoir did not use the terms "care" or "caring" in relation to her notion of freedom, the idea of interconnectedness of freedoms presents caring for the other as an authentic mode of being in existentialism. We can argue that for Beauvoir, one of the most important ways that we should care for others is to care for and help enable them as a freedom. In other words, one cannot treat a person as a freedom without caring for him/her in some respect, and vice versa. Thus, caring for another must include treating them as a freedom. If this aspect is missing, the type of care provided would be a diminished way of caring. The type of care men provide women with in patriarchal and paternalizing white middle class heterosexual relationships presents a good example of a diminished way of caring. By treating women as fragile, and unable to protect themselves, men put women on a pedestal as dependent on them, which eventually diminishes women's freedom. Since freedom is the basic value for existentialism, any demand or act that proves to be undermining the other's freedom or that which oppresses should be denied.

Some Beauvoir scholars have highlighted Beauvoir's attention to caring for others as well. For example, Karen Vintges writes that "For Beauvoir . . . the whole point of ethics is our choosing to become connected and emotionally involved with other people, and of course care for others is much more prominent in this view" [31] (p. 176). Moreover, every performance of care navigates the antimony between social controls and individual autonomy but is grounded in the freedom and disclosure of the other. As Beauvoir describes, "One can reveal the world only on a basis revealed by other men" [3] (p. 71). A project becomes possible and meaningful only on the background of former projects and through its engagement with other past and current projects. Accordingly, ethics cannot evade moral ambiguity by appealing to transcendent justification but must do the hard work of finding common projects with other subjects that foster moral freedom. One's freedom is possible only through others' freedoms.

We can observe that Beauvoir utilized this tenet in her political activism specifically in her involvement with Djamila Boupacha's case. Her discomfort with the colonial violence exercised in Algeria and her reckoning with her indirect implicatedness in this violence as a French citizen who benefits from the colonial regime show her awareness of the interconnectedness between her freedom and the freedoms of Algerian people [1] (pp. 369, 652). Her meticulous attention to recount Boupacha's story as truthfully as possible and her careful abstention from imposing her own values on the case attest to her acknowledgment of the dangers of paternalistic care [23]. Yet, such dangers and the possibility of doing harm while the intention is enhancing others' freedoms do not deter Beauvoir from taking action. Her activism around the Boupacha case exemplifies the need to act when another's moral freedom is threatened despite the ambiguities involved.

I will explain in more detail in the last section of this paper how Beauvoir's existentialist ethics may help counter the Western-centric and paternalistic tendencies in the attempts to universalize care. However, in the following sections, I first provide an account of the origin and development of care ethics and then discuss these Western-centric and paternalistic elements prevalent in the current care practices around the globe.

3. Origins and Evolution of Care Ethics

The term "care ethics" encompasses a broad category of literature that includes substantial differences, especially when we consider early formulation of the theory by Carol Gilligan [32] and Nel Noddings [33], and the more contemporary scholarship on care ethics by Fiona Robinson, Virginia Held, and Joan Tronto, to name a few [32–39]. Moreover, care ethics has also been expanding in connection with a variety of theories and disciplines ranging from geography to political science, sociology, economics, and science. In very broad terms, care ethics can be construed as a feminist moral theory that takes human beings' interdependence by way of meeting each other's needs as the theoretical basis for considering moral questions.

Feminist care ethics has criticized traditional ethical theories for two main reasons: (i) under-emphasizing or ignoring caring needs or relegating them to the private sphere and focusing on the public sphere, which historically has undermined women's experiences

Ferrarese's reminder that every life is susceptible to being vulnerable and dependent at any moment in life underscores the fact that "vulnerability exists only in situations" [54] (p. 154). Hence, particular manifestations of care would have to stem from the specific circumstances at hand.

An important step towards countering the challenges stemming from such positionality could be to dislocate care ethics discussions from the Global North, which is what I explore in the next section.

4. (Dis)Locating Care

Care ethics rejects a priori universal ethical principles that guide action and focuses on the particularity of actual situations to determine the course of ethical action. The basis for rejecting such principles is ontological. According to care ethics, we exist in the world as caregivers and caretakers, and are responsible for goodness of care-related actions. It is difficult to define care and show how one is supposed to care for others. It is the subject's responsibility to do this work and no guidance or assurance is available. In practice then, care seems to be open-ended, which makes care practices susceptible to various risks such as malpractice, exploitation, and paternalism. While we cannot change the fact that care is open-ended as this is a human condition, we can find ways to address and minimize these risks.

Since it is impossible to determine the best form of care without knowing the local and particular conditions of the cared for, predetermined universal ethical laws seem inadequate and, at times, even harmful to the care ethicist. The transnational feminist ethicist is also wary of universal ethical laws due to the historical association between such universal ideals and Western imperialism and colonialism. However, feminists (both care ethicists and transnational feminist ethicists) would agree that care is a universal value and practice. As such, one needs to move cautiously while trying to understand what that universality means. While theoretical articulations of care ethics underline the particularity of ethical dilemmas and invite us to closely examine the concrete conditions of the ethical question to determine the course of action, discussions of care ethics in philosophical scholarship still mostly revolve around the Global North. Hence, in its efforts to provide a comprehensive ethical theory that would address care practices across the globe, care ethics by omission carries the danger of being ethnocentric as it is centered in the West. However, as Khader argues, universal normativity does not have to mean Western ethnocentrism [55] (p. 3). Historically speaking, it might be unimaginable to think of universalism independent of imperialism, yet in theory universalism and anti-imperialism are not mutually exclusive [56]. In the same vein, we are able to imagine a universalist yet anti-imperialist care ethics theory, the tenets of which I articulate in this paper by appealing to Beauvoir.

Moreover, the context of care ethics has undergone drastic changes recently. We are experiencing an increasing circulation of care needs, caregivers, and caretakers globally. Such global circulations are highly marked and organized by colonial and postcolonial engagements between different geographies. Due to migration of caregivers globally, the caregivers are increasingly leading "transnational lives" [7] (p. 156). Deployment of care theories in analyzing these transnational care practices has the implication that "the concept [care] has been transferred relatively unreflectively to different parts of the Global South without recognizing that in doing so one inherits the very different histories of development policy, care arrangements and gender regimes that influence the notion of care" [7] (p. 161). Given the colonial and postcolonial entanglements in globalization of care practices, the need for locating care thinking in transnational feminist discussions becomes inevitable. In Raghuram's words "localizing care ethics implies dislocating it from its unspoken but often implicit locatedness in very particular locations and practices of care" [8] (p. 513).

The universalizing and imperialist hegemony of Western culture and history made many care ethicists wary of the potential of care ethics, as a theory that originated in the Western geography, to be implicitly operating under universalizing claims based on Western practices of care. Raghuram mainly criticizes care ethics scholarship for postulating caring as a universal ethical ideal without addressing the differences between regions regarding the multiplicity of care practices when she writes, "by being locationally ambiguous, the Global North became implicit" [8] (p. 521). The erasure of multiplicity of care practices leads to the tacit imposition of care as practiced in the Global North as the norm. Moreover, the ways in which the caring needs of the world are relegated to the different localities determined by colonial and neocolonial relationalities remain underexplored.

Another scholar, Olena Hankivsky, criticizes care ethicists for prioritizing gender and gendered power relations over other identity categories such as race, class, and disability. Since intersectionality vehemently opposes spotlighting one category of difference over others, Hankivsky concludes, these approaches cannot be considered truly intersectional [10] (p. 256).

While Raghuram's and Hankivsky's claims merit significant attention, there are active endeavors among care scholars to address the threat of Western-centric thinking and to foster deeper intersectional analysis in care ethics. Fiona Robinson is one care ethicist who takes both criticisms seriously. While acknowledging the need to be hyper-vigilant in care ethics scholarship to make sure the analysis is truly intersectional and it does not lend itself to Western-centric thinking, she also highlights the amenability of care ethics to address and confront these two threats. She writes, "Far from universalizing care or silencing alternative understandings of care, it could be argued that care ethics provides a basis for contesting racial and neocolonial hierarchies" [35] (p. 16). In another article, Mahon and Robinson emphasize the need for a critical care ethics to be embedded in "the concrete activities of real people in the context of webs of social relations" [57] (p. 2). Robinson also agrees with Raghuram on the need to dislocate care from the "normative white body through which much care is theorized" [35] (p. 21).

Regarding the criticism about intersectionality, the discussions around shifting the register in which we analyze care as a theory and practice away from gender should not be read as claiming gender irrelevant or unimportant. As Nancy Fraser explains, the burden of social reproduction has been mostly carried by women. Capitalist structure does not fully recognize the material and affective labor that goes into maintaining social reproductive practices. Fraser maintains that this type of labor is taken for granted and treated as an infinitely available "gift" [58] (p. 31). The depletion of affective and material capacities of caregivers in our contemporary capitalist society led to what Fraser calls "a crisis of care" [58] (p. 31). She contends: "When a society simultaneously withdraws public support for social reproduction and conscripts the chief providers of it into long and grueling hours of paid work, it depletes the very social capacities on which it depends" [58] (p. 31).

Fraser's emphasis on the inequalities centered on care practices provides us with tools to challenge the dominant structures of gender inequalities, especially apparent in care work. The neocolonial configuration of the global care chain forces us to look into the racialized and classed dynamics in addition to gendered dynamics. The category of gender is as equally important as other categories like class, race, disability, sexual orientation, ethnicity, etc. Yet, I argue that the focus on gender should not come at the expense of these categories. For example, Western feminisms have a long history of advocating for and protecting the rights of middle- and upper-class white women by sacrificing black women's rights, as was the case during the suffragette movement. Focusing on gender has historically meant prioritizing middle- and upper-class white women's needs and rights. Using an intersectional approach by incorporating categories such as class, race, sexual orientation, disability, ethnicity, etc. into the discussion ensures inclusion of the demands of marginalized women in the debate.

Hence, the entanglements of race, class and gender dynamics among women in global care practices remain as one of the most interesting aspects of the feminization of care work on a global scale that is in need of further theorization. Care work is still mainly considered to be women's duty despite feminists' efforts to gender-neutralize it. Since the majority of the states in the Global North do not provide public support for childcare or other forms of care, women's participation in the labor force in the Global North largely depends on their chances of outsourcing such services. While some women prefer and are able to employ full-time care workers who are mostly migrants from the Global South, some others choose or are able to afford partial care services which are, again, provided mostly by women of color. Upper-class women of the Global North, for example, are able to completely delegate their care needs to lower class women who mostly migrate from the Global South [59] (pp. 25–26). While the emotional and intellectual labor of finding, organizing, and supervising paid care remains women's work, being able to afford such paid care work makes holding a demanding full-time job possible for them. The care deficit created by the two-wage families in the Global North has been absorbed by the caregivers migrating from the Global South [60–62]. While this flow of care services might seem like a rational economic transaction for some, as Robinson articulates, it disguises the responsibility of two main parties to the question: men and states [63] (p. 74).

Despite the relentless efforts of feminists to include men in practices of caregiving, research shows that men are still falling considerably behind in participating in meeting care needs in the family [62] (p. 9). In the same vein, by not providing support for childcare, states relegate this work to families, where in turn this work gets relegated to women in the family. Hence, we can say that the upper-class women of the Global North partly owe their career development to the women migrating from the Global South who handle their care needs at low wages. Fraser makes this point succinctly by referencing Sheryl Sandberg's call to women to "lean in": " … it is only possible for her [Sandberg's] readership to envision leaning in at the corporate boardroom in so far as they can lean on the low-paid care workers who clean their toilets and their homes, diaper their children, care for their aging parents, and so on" [58] (p. 34).

While middle- and upper-class women of the Global North are able to choose between the options of hiring care workers or taking a break from work to provide care, the migrating care workers mostly have no place to turn to in order to meet their own caregiving needs in the family. As a result, women in those conditions mostly have to leave their children or elderly unattended, which in turn increases the risk of accidents and emergencies [64] (p. 88). Postcolonial scholars have documented how a middle class well-intentioned Western individual's daily practices are always already implicated in a broader web of exploitative relationships [65] (p. 71). Therefore, as Noxolo et al. maintain, postcolonial politics has to acknowledge the insurmountable gap created between the North and the South, despite " … its anti-colonising impulse in the face of continued inequality and exploitation … " [66] (p. 423).

Khader also underlines this point when she writes, "The facts of militarism, cultural domination, and transnational economic exploitation mean that Western women are complicit in 'other' women's oppression" [55] (p. 2). In spite of, or perhaps because of this complicity, Western feminism is in dire need of transnational coalitions that are both responsible and caring. However, Western feminists should be wary of the fact that these coalitions themselves carry the danger of replicating practices of cultural domination and economic exploitation. Most importantly, as Noxolo et al. state, responsible caring action "involves an openness and vulnerability to that which most resists European thought: those aspects of the 'other' that are not shared and are not comfortable" [66] (p. 423).

Because of these important aspects of actual care practices (it being racialized, feminized, and relegated to the lower classes) that may get lost in normative theorizations of care ethics, I mostly focus on care as practice rather than as a theory. Caring is always embodied, and it is a combination of both physical and affective labor. As Sander-Staut states, care as ethics and care as practice can never be thought of as distinct from each other [67] (p. 22). My choice to do so is also informed by the long-standing historical struggle of feminists to claim the personal as political. The distribution of household care work is a political and economic process. Hence, even when we are talking about the most intimate types of caring practices, we are not talking about a personal phenomenon, but rather a representation of a historical construction of complex social, political, and cultural norms, values, and practices. Feminist theory, for example, provides invaluable scholarship on the dominant influence of institutions, such as patriarchy and capitalism, in the production of norms, values, and practices that define caring relationships, especially in the household. As Eva Kittay argues and skillfully demonstrates in *Love's Labor* by providing theoretical accounts and examples of lived experiences, care is never personal but rather it is a public matter [68].

Just as care is a public issue, it is also varied in cultural matrices. Scholars of care ethics are not oblivious to the differences in cultures, values, and norms. Robinson, for example, calls for "a critical account of care ethics" that can address "the question of relations among moral agents on a global scale" [34] (p. 114). She also argues that care ethics is a "phenomenology of moral life that recognizes addressing moral problems involves first, an understanding of identities, relationships and contexts" [34] (p. 131). In the same vein, Held points out the necessity to be attuned to the particularity and specificity of the needs of the cared-for instead of universally generalizing and assuming what the needs of the other would be [36] (p. 39). However, caring for the distant others in a global world brings forth new questions, such as what it means to understand these identities, relationships, and contexts.

Although feminist theory has made some progress in terms of being open to and understanding non-Western feminist practices, there are still significant challenges on the way to a pluralistic ideal of feminism. María Lugones' and Elizabeth Spelman's critique of Western ethnocentrism in "Have We Got a Theory for You!" still proves to be highly relevant in that context [69]. Engagement of Western feminists with non-Western feminists continues to be incomplete and devoid of a genuine understanding of the non-Western feminist theories and practices. What is more, the increasing global hegemony of Western cultures obscures the necessity of such engagement and understanding. Raghuram makes a crucial point when she states that being attuned to differences in the practices of care across localities around the globe is an invaluable exercise for feminist thinkers in itself [8] (p. 523). As Lugones and Spelman write, theories can be "disrespectful, ignorant, ethnocentric, imperialistic" [69] (p. 578). The implicit locatedness of care ethics in the Global North means that any conversation on care ethics is actually happening in the language of the Global North. Given the power and privilege the Global North enjoys in comparison to the Global South, this asymmetry left unaddressed creates the bedrock for "disrespectful, ignorant, ethnocentric, imperialistic" theory.

For these reasons, the challenges to feminist coalition-building that have been pointed out by transnational feminist ethicists need to be addressed in discussions of care ethics. One of the main challenges in seeking coalitions around transnational feminist ethics is the erasure of the racial and class dynamics between the Global North and Global South. Many scholars have discussed the harmful consequences of postulating universal sisterhood without accounting for the inequalities created and perpetuated by colonial and neocolonial political engagements. Hence, these attempts have been highly problematic in their selectivity and paternalism. Failing to account for the contributions of the Global North in the creation of these inequalities in the Global South creates an inauthentic and imbalanced relationship between the one-caring and the one-cared for. Such an imbalance may and does lend itself to a paternalistic form of caring. Re-establishing an authentic caring relationship between the Global North and the Global South requires that each party accepts each other as equals. For that reason, in the last section of my paper I will discuss Beauvoir's existentialist ethics as a fruitful source to address the question of paternalism in engagements of care ethics and transnational feminist ethics.

5. Paternalistic Care and Beauvoir

In the previous section, I established that care ethics should be theorized within the global order due to the increasingly global character of caring practices. Considering care ethics within the current global context also requires reckoning with the paternalistic care discourses used in justifying colonialism, as the current global order is the product of past colonial and current neo-colonial practices. Therefore, care ethics has to address and resolve the potential danger of paternalism in applications of care ethics in the colonial and postcolonial configurations of the contemporary world. In this section, I use the theoretical tools existentialism offers to provide possible solutions to the danger of paternalistic care.

As explained in the previous section, care is ontologically an open-ended practice. By rejecting universal, decontextualized, and atemporal ethical laws, care ethicists acknowledge the ambiguity involved in ethical action, including caring practices. Similar to care ethics, existentialism is one of the theories that articulates best the ambiguity and risks involved in ethical action. Daryl Koehn, who is one of the few feminist thinkers who delineates the existential elements in care ethics, puts this succinctly:

> In the care ethic, the moral world is not already "there," fully formed in its rationality. If the world is to be good, the caregiver must make it so through her acts in accordance with her personal ideal of herself as a caring person. Since no one can specify necessary and sufficient conditions for an act to be caring, the caregiver is finally thrown back upon herself to assess the goodness of her acts [70] (pp. 22–23).

However, this ambiguity could easily lend itself to paternalistic and dominating practices. While care ethicists should always be on the lookout to make sure that they are not explicitly or implicitly endorsing tenets that may lead to paternalistic or diminishing care practices, such practices do and probably will continue to take place. If the caregiver, for example, is to determine the best form of care without any conversation with the one cared-for, they may intentionally or unintentionally choose to care in ways that diminish the agency and freedom of the one cared-for. They may impose what they think would be best for the one cared-for or they may care for the other in ways that will promote their own agenda. Due to these risks, the caring action should always be decided in an ongoing conversation with the one cared-for. The focus in care ethics on the extraordinary challenge of decentering the self so as to try and center the other and moving from a care-giver/cared-for framing to a relational one shows that care ethicists are aware of these problems arising in care practices [35,71–73].

The way Beauvoir articulates the relational and interconnected self includes a receptive mode of the ethical subject, an emphasis on action, and a considerable responsibility for one's actions. Receptivity is to be open to see, hear, and feel what the other has to convey to me. Receptivity involves both activity and passivity in that, while I am opening myself up to the other, I am also letting go of myself and my attempts to control the other. The opening up is active, yet neither manipulative nor assimilative [33] (p. 146). Explicitly drawing from existentialist philosophy, Noddings places the receptive mode at the heart of human existence. She concurs to the existential way of existence which requires a constant awareness and questioning of one's values and actions within the context of interactions with others. In this process of questioning, Noddings explains, the caregiver sees the demand of the other and remains with two options; proceeding "in a state of truth" which refers to acknowledging the call of the other or denying "what I have received and talk myself into feeling comfortable with the denial" [33] (p. 35). Existentialist ethics also requires the ethical subject to be attentive to the other's call and genuinely respond to it. For Beauvoir the good of others should be "taken as an absolute end of our action" [3] (p. 142). Taking the good of others as the main goal of our action implies the ethical necessity to be receptive to the others' call. Refusing to do so would mean being in bad faith, which is an unethical stance to take.

Responding to this call inevitably requires some action. Care ethics also entails a moral obligation to act in a way that addresses the demands for care. Caring for someone without being moved to action would not be considered as genuine caring. Noddings maintains that authentic care signifies the existence of a genuine concern for the other's wellbeing. In addition, we expect this concern to translate into actions that would endorse their wellbeing. This relationship to the other solidified in practical action again reminds us of Beauvoir's notion of interconnectedness of human freedoms and the ethical obligation to endorse others' freedoms. Noddings reminds us of "our fundamental relatedness, of our dependence upon each other. We are both free—that which I do, I do—and bound—I might do far better if you reach out to help me and far, far worse if you abuse, taunt, or ignore me" [33] (p. 49). For Beauvoir, we need the approval and support of others for the projects we take up. Hence, we demand their support and, depending on their response, we may do "far better" or "far worse".

Relationships play a fundamental part in the constitution of the self in care ethics. We experience ourselves as independent as well, yet a self that totally perceives herself as separated from others—as in the case of justice-oriented ethical approaches—does not constitute an ideal ethical starting point for care ethicists. Recognition of one's connection to other people also constitutes the source of moral obligation to others. As Gilligan states, in a care-based approach to morality "an awareness of the connection between people gives rise to a recognition of responsibility for one another, a perception of the need for response" [32] (p. 30).

As such, endorsement of a conception of the self as relational constitutes one of the main principles of care ethics. Care ethicists converge on the claim that human beings are ontologically related to each other and posit this aspect of our ontology as the basis of an ethics. However, in postulating this relational ontology, one has to be attentive to unjust and oppressive forms of relationality as well. To that end, both care ethics and existentialist ethics emphasize the necessity of a meticulous evaluation of the case in question before the caring action can take place. They both highlight the fact that some cases may require the subject to look for or even create alternative solutions that may not be immediately available at the beginning of the inquiry.

The notion of the relational self gains even more significance in discussions of ethics given the increasing circulation of labor and the world's resources, usually in the form of a flow from less affluent nations to more affluent ones. The same contextual sensitivity is at the core of existentialist ethics as well. Existentialist ethics establishes the subject as fundamentally interconnected and interrelated through each one's freedoms and projects. The subject has an ethical responsibility to analyze her actions within the context of a web of relations. Awareness of the need for others in order to realize one's projects forms the basis for the ethical responsibility to respond to others' needs to be successful in their projects.

Others' infinite demands on us constitute one of the main sources of anguish in our lives. On the one hand, we are fundamentally connected to others and we cannot simply ignore their call; on the other hand, we are tempted to and free to ignore their call. When we do the latter, we immediately turn to reestablish our relatedness. There is an ever-present potentiality for both reciprocity and conflict in human relations. We have the potential to treat others as subjects or objects and we are susceptible to be treated as such. Therefore, we always work through these questions and constantly decide on how to treat others and assess how others treat us. Hence, the anguish arising out of this deliberation is bound to be a part of our lives. Since our freedoms are fundamentally dependent on others' freedoms, one cannot simply treat the other as an object and avoid feelings of anguish. One justifies her existence in promoting others' freedoms; "I concern others and they concern me" [3] (p. 72).

Caring for the other as a freedom then entails a variety of considerations that may guide the caring person in their deliberations on how to care. First, the caring person should be attentive to the context by analyzing the background of the individual or group being cared for. Being attentive to the context should also involve an honest evaluation of one's own subject-position in the power matrix. Second, the caring person should aim for creating the conditions for the possibility of free action and transcendence for the person or group being cared for. By focusing on creating such conditions, as opposed to imposing their own vision of what freedom and free action would entail, the caring person would also be able to have a clear picture of the genuine needs, goals, and desires of the person or group being cared for. These considerations, if embraced and practiced conscientiously, would significantly reduce the possibility of engaging in paternalistic care unwittingly.

In this article, I engaged with critiques of care ethics regarding its Western ethnocentrism and paternalism mostly raised by transnational feminist scholarship. While the critiques are viable and constitute serious challenges to the care ethics theory, they are not insurmountable, as there are many resources available both in care ethics and existentialist scholarship to address these challenges. Nonetheless, drawing on Beauvoir's political engagements during the Algerian War, I showed that even scholars situated in a Western geography could eschew ethnocentric and paternalistic engagements by deploying a self-critical approach in their caring practices. Moreover, I sought to advance a critical care ethic by using Beauvoirian existentialism and her tenet of treating the other as freedom as fruitful resources against latent paternalism and Western-centric tendencies.

To conclude, care ethics has numerous strengths in its ability to address questions of relationality and responsibility to care in an increasingly unjust global world. However, in doing so it has to be rigorously observant of the historical and geographical power dynamics and the inegalitarian terrain such dynamics have created over time. Western feminists can no longer afford overlooking the Western-centered and paternalistic caring practices presented under the disguise of universalism and responsible caring. Continuing to do so would create crucial obstacles in the way of building transnational feminist coalitions across the globe. Turning to Beauvoir's existentialism helps us generate a robust account of feminist and transnational care ethics without jettisoning the responsibility to care. The possibility of falling into paternalism in caring is not a reason to resign to indifference. On the contrary, as both care ethics and existentialist ethics articulate, any ethical action including the caring act in a messy and ambiguous world is bound to be imperfect. Hence the goal should be to minimize such imperfections as opposed to not to act at all. The existentialist tenet of treating the other as freedom may help us avoid many dangers associated with Western ethnocentrism, including being paternalistic in the process of caring for the other.

Funding: This research received no external funding.

Conflicts of Interest: The author declares no conflict of interest.

References

1. de Beauvoir, S. *Force of Circumstance*, 1965 ed.; Howard, R., Translator; G.P. Putnam's Sons: New York, NY, USA, 1963.
2. de Beauvoir, S. Pyrrhus and Cinéas. In *Philosophical Writings*, 2004 ed.; Timmerman, M., Translator; Simons, M.A., Timmermann, A., Mader, M.B., Eds.; University of Illinois Press: Urbana, IL, USA, 1944; pp. 89–149.
3. de Beauvoir, S. *The Ethics of Ambiguity*, 1976 ed.; Frechtman, S.B., Translator; Citadel Press: New York, NY, USA, 1947.
4. de Beauvoir, S. Pour Djamila Boupacha. *Le Monde*, 2 June 1960.
5. de Beauvoir, S.; Halimi, G. *Djamila Boupacha*; Green, P., Translator; Macmillan Company: New York, NY, USA, 1962.
6. Ptacek, M. Simone de Beauvoir's Algerian war: Torture and the rejection of ethics. *Theory Soc.* **2015**, *44*, 499–535. [CrossRef]
7. Raghuram, P. Global care, local configurations: Challenges to conceptualizations of care. *Glob. Netw.* **2012**, *12*, 155–174. [CrossRef]
8. Raghuram, P. Locating care ethics beyond the global North. *ACME Int. J. Crit. Geogr.* **2016**, *15*, 511–533.
9. Raghuram, P. Race and feminist care ethics: Intersectionality as method. *Gend. Place Cult.* **2019**, *26*, 613–663. [CrossRef]
10. Hankivsky, O. Rethinking care ethics: On the promise and potential of intersectional analysis. *Am. Political Sci. Rev.* **2014**, *108*, 252–264. [CrossRef]
11. Walker, K. Okâwîmâwaskiy: Regenerating a Wholistic Ethics. Ph.D. Dissertation, University of British Columbia, Vancouver, BC, Canada, 2021. Available online: https://open.library.ubc.ca/collections/ubctheses/24/items/1.0398723 (accessed on 12 January 2022).
12. Brannelly, T.; Boulton, A. The ethics of care and transformational research practices in Aotearoa New Zealand. *Qual. Res.* **2017**, *17*, 340–350. [CrossRef]
13. Dalmiya, V. *Caring to Know: Comparative Care Ethics, Feminist Epistemology, and the Mahābhārata*; Oxford University Press: India, 2016.
14. Li, C. The Confucian concept of jen and the feminist ethics of care: A comparative study. *Hypatia* **1994**, *9*, 70–89. [CrossRef]
15. Metz, T. The western ethic of care or an afro-communitarian ethic?: Finding the right relational morality. *J. Glob. Ethics* **2013**, *9*, 77–92. [CrossRef]
16. Munawar, S. In Hajar's Footsteps: A de-Colonial and Islamic Ethic of Care. Ph.D. Dissertation, University of British Columbia, Vancouver, BC, Canada, 2019. Available online: https://open.library.ubc.ca/collections/ubctheses/24/items/1.0387144 (accessed on 14 January 2022).

17. Malatino, H. *Trans Care*; University of Minnesota Press: Minneapolis, MN, USA, 2020.
18. Arp, K. A different voice in the phenomenological tradition: Simone de Beauvoir and the ethic of care. In *Feminist Phenomenology*; Fisher, L., Embree, L., Eds.; Springer: New York, NY, USA, 2010; pp. 71–81.
19. Hamington, M. *Embodied Care: Jane Addams, Maurice Merleau-Ponty, and Feminist Ethics*; University of Illinois Press: Urbana, IL, USA, 2004.
20. Daly, A. A phenomenological grounding of feminist ethics. *J. Br. Soc. Phenomenol.* **2018**, *50*, 1–18. [CrossRef]
21. Daly, A. *Merleau-Ponty and the Ethics of Intersubjectivity*; Palgrave-Macmillan: London, UK, 2016.
22. Pettersen, T. The ethics of care normative structures and ethical implications. *Health Care Anal.* **2011**, *19*, 51–64. [CrossRef]
23. Durmuş, D. Lessons from Beauvoir for a transnational feminist ethics. *Simone Beauvoir Stud.* **2020**, *31*, 47–67. [CrossRef]
24. de Beauvoir, S. *The Prime of Life*, 1994 ed.; Green, P., Translator; Lancer Books: New York, NY, USA, 1961.
25. Fullbrook, E.; Fullbrook, K. *Simone de Beauvoir: A Critical Introduction*; Polity Press: Cambridge, UK; Blackwell: Hoboken, NJ, USA, 1998.
26. de Beauvoir, S. *The Second Sex*, 2012 ed.; Borde, C.; Malovany-Chevallier, S., Translators; Random House: New York, NY, USA, 1949.
27. Arp, K. *The Bonds of Freedom: Simone de Beauvoir's Existential Ethics*; Open Court Publishing: Chicago, IL, USA, 2001.
28. Young, I.M. *On Female Body Experience: "Throwing Like a Girl" and Other Essays*; Oxford University Press: New York, NY, USA, 2005.
29. Veltman, A. The Sisyphean torture of housework: Simone de Beauvoir and inequitable divisions of domestic work in marriage. *Hypatia* **2004**, *19*, 121–143. [CrossRef]
30. Brison, S.J. Beauvoir and feminism: Interview and reflections. In *The Cambridge Companion to Simone de Beauvoir*; Card, C., Ed.; Cambridge University Press: New York, NY, USA, 2003; pp. 189–207.
31. Vintges, K. 'Must we burn Foucault?' ethics as art of living: Simone de Beauvoir and Michel Foucault. *Cont. Philos. Rev.* **2001**, *34*, 165–181. [CrossRef]
32. Gilligan, C. *In a Different Voice: Psychological Theory and Women's Development*; Harvard University Press: Cambridge, MA, USA, 1982.
33. Noddings, N. *Caring: A Feminine Approach to Ethics and Moral Education*, 2003 ed.; University of California Press: Berkeley, CA, USA, 1984.
34. Robinson, F. *Globalizing Care: Ethics, Feminist Theory, and International Relations*; Westview Press: Boulder, CO, USA, 1999.
35. Robinson, F. Resisting hierarchies through relationality in the ethics of care. *Int. J. Care Caring* **2020**, *4*, 11–23. [CrossRef]
36. Held, V. *The Ethics of Care: Personal, Political, and Global*; Oxford University Press: New York, NY, USA, 2006.
37. Tronto, J.C. *Moral Boundaries: A Political Argument for an Ethic of Care*; Psychology Press: London, UK, 1993.
38. Tronto, J.C. Creating caring institutions: Politics, plurality, and purpose. *Ethics Soc. Welf.* **2010**, *4*, 158–171. [CrossRef]
39. Tronto, J.C. *Caring Democracy: Markets, Equality, and Justice*; New York University Press: New York, NY, USA, 2013.
40. Hoagland, S.L. Some thoughts about 'caring'. In *Feminist Ethics*; Card, C., Kan, L., Eds.; University Press of Kansas: Lawrence, KS, USA, 1991; pp. 246–263.
41. Hoagland, S.L. *Lesbian Ethics: Toward New Value*; Institute of Lesbian Studies: Palo Alto, CA, USA, 1988.
42. MacKinnon, C.A. *Toward a Feminist Theory of the State*; Harvard University Press: Cambridge, MA, USA, 1989.
43. Jaggar, A. Feminist ethics: Projects, problems, prospects. In *Feminist Ethics*; Card, C., Ed.; University Press of Kansas: Lawrence, KS, USA, 1991; pp. 78–104.
44. Walker, M.U. Feminist ethics and human conditions. *Tijdschr. Voor Filos.* **2002**, *64*, 433–450.
45. Kittay, E. The moral harm of migrant care work: Realizing a global right to care. *Philos. Top.* **2009**, *37*, 53–75. [CrossRef]
46. Ahmed, L. Western ethnocentrism and perceptions of the Harem. *Fem. Stud.* **1982**, *8*, 521–534. [CrossRef]
47. Narayan, U. Colonialism and its others: Considerations on rights and care discourses. *Hypatia* **1995**, *10*, 133–140. [CrossRef]
48. Al-Saji, A. The racialization of Muslim veils: A philosophical analysis. *J. Philos. Soc. Crit.* **2010**, *36*, 875–902. [CrossRef]
49. Khader, S. The feminist case against relational autonomy. *J. Moral Philos.* **2020**, *17*, 499–526. [CrossRef]
50. Murphy, M. Unsettling care: Troubling transnational itineraries of care in feminist health practices. *Soc. Stud. Sci.* **2015**, *45*, 717–737. [CrossRef] [PubMed]
51. Kittichaisaree, K. *International Human Rights Law and Diplomacy*; Edward Elgar Publishing: Cheltenham, UK, 2020.
52. Durmuş, D. Middle Eastern feminisms: A phenomenological analysis of the Turkish and the Iranian experience. *Comp. Cont. Philos.* **2018**, *10*, 221–237. [CrossRef]
53. Tronto, J.C. Caring democracy: How should concepts travel? In *Care Ethics, Democratic Citizenship and the State*; Urban, P., Ward, L., Eds.; Palgrave Macmillan: London, UK, 2020.
54. Ferrarese, E. Vulnerability: A concept with which to undo the world as it is? *Crit. Horiz.* **2016**, *17*, 149–159. [CrossRef]
55. Khader, S. *Decolonizing Universalism: A Transnational Feminist Ethic*; Oxford University Press: New York, NY, USA, 2019.
56. Chakrabarty, D. *Provincializing Europe: Postcolonial Thought and Historical Difference*; Princeton University Press: New Jersey, NJ, USA, 2000.
57. Mahon, R.; Robinson, F. *Feminist Ethics and Social Policy: Towards a New Global Political Economy of Care*; University of British Columbia Press: Vancouver, BC, Canada, 2011.
58. Fraser, N. Capitalism's crisis of care: A conversation with Nancy Fraser. *Dissent* **2016**, *63*, 30–37. [CrossRef]
59. The Care Collective. *The Care Manifesto: The Politics of Interdependence*; Verso: London, UK, 2020.

60. Hochschild, A.R. The nanny chain: Mothers minding other mothers' children. *Am. Prospect.* **2000**, *11*, 32–36.
61. Parreñas, R.S. *Servants of Globalization: Women, Migration and Domestic Work*; Stanford University Press: Redwood City, CA, USA, 2001.
62. Ehrenreich, B.; Hochschild, A.R. *Global Woman: Nannies, Maids, and Sex Workers in the New Economy*; Henry Holt: New York, NY, USA, 2004.
63. Robinson, F. *The Ethics of Care: A Feminist Approach to Human Security*; Temple University Press: Philadelphia, PA, USA, 2011.
64. Heymann, J.; Fischer, A.; Engelmann, M. *Labor Conditions and the Health of Children, Elderly and Disabled Family Members*; Oxford University Press: New York, NY, USA, 2003.
65. Spivak, G. *Other Asias*; Blackwell Publishing: Oxford, UK, 2008.
66. Noxolo, P.; Raghuram, P.; Madge, C. Unsettling responsibility: Postcolonial interventions. *Trans. Inst. Br. Geogr.* **2012**, *37*, 418–429. [CrossRef]
67. Sander-Staudt, M. The unhappy marriage of care ethics and virtue ethics. *Hypatia* **2006**, *21*, 21–39. [CrossRef]
68. Kittay, E. *Love's Labor: Essays on Women, Equality, and Dependency*; Routledge: New York, NY, USA, 1999.
69. Lugones, M.; Spelman, E. Have we got a theory for you! Feminist theory, cultural imperialism and the demand for 'the woman's voice'. *Women's Stud. Int. Forum* **1983**, *6*, 573–581. [CrossRef]
70. Koehn, D. *Rethinking Feminist Ethics: Care, Trust and Empathy*, 2012 ed.; Routledge: London, UK, 1998.
71. Gilligan, C. *Joining the Resistance*; Polity Press: Cambridge, UK, 2011.
72. Robinson, F. Methods of feminist normative theory: A political ethic of care for international relations. In *Feminist Methodologies for International Relations*; Ackerly, B.A., Stern, M., True, J., Eds.; Cambridge University Press: New York, NY, USA, 2006; pp. 221–240.
73. Confortini, C.C.; Ruane, A.E. Sara Ruddick's *Maternal Thinking* as weaving epistemology for *justpeace*. *J. Int. Political Theory* **2013**, *10*, 70–93. [CrossRef]

philosophies

MDPI

Article

Žižek's Hegel, Feminist Theory, and Care Ethics

Sacha Ghandeharian

Department of Political Science, Carleton University, Ottawa, ON K1S 5B6, Canada;
sacha.ghandeharian@carleton.ca

Abstract: This article presents conceptual bridges that exist between the philosophy of G.W.F Hegel and a feminist ethics of care. To do so, it engages with Slavoj Žižek's contemporary reading of Hegel in concert with existing feminist interpretations of Hegel's thought. The goal of doing so is to demonstrate how both Žižek and a selection of critical feminist thinkers interpret Hegel's perspective on the nature of subjectivity, intersubjective relations and the relationship between the subject and the world it inhabits, in a way that can further our thinking on the feminist ethics of care as a relational and contextualist ethics that foregrounds vulnerability as a condition of existence. These readings of Hegel highlight the radical contingency of human subjectivity, as well as the relationship between human subjectivity and the external world, in a way that is compatible with the feminist ethics of care's emphasis on the particularity, fluidity, and interdependency of human relationships. I argue that this confrontation between care ethics and mainstream philosophy is valuable because it offers mutual contributions to both care ethics as a moral and political theory and the philosophy of Hegel and Žižek.

Keywords: Slavoj Žižek; feminist theory; care ethics; G.W.F. Hegel; subjectivity; vulnerability; relationality

Citation: Ghandeharian, S. Žižek's Hegel, Feminist Theory, and Care Ethics. *Philosophies* **2022**, *7*, 59. https://doi.org/10.3390/philosophies7030059

Academic Editors: Maurice Hamington and Maggie FitzGerald

Received: 21 March 2022
Accepted: 18 May 2022
Published: 31 May 2022

Publisher's Note: MDPI stays neutral with regard to jurisdictional claims in published maps and institutional affiliations.

1. Introduction

This article presents conceptual bridges that exist between the philosophy of G.W.F Hegel and a feminist ethics of care. To do so, it engages with Slavoj Žižek's contemporary reading of Hegel in concert with existing feminist interpretations of Hegel's thought. The goal of doing so is to demonstrate how both Žižek and a selection of critical feminist thinkers interpret Hegel's perspective on the nature of subjectivity, intersubjective relations and the relationship between the subject and the world it inhabits, in a way that can further our thinking on the feminist ethics of care as a "relational" and contextualist ethics that foregrounds vulnerability as a condition of existence [1] (p. 13). These readings of Hegel highlight the radical contingency of human subjectivity, as well as the relationship between human subjectivity and the external world, in a way that is arguably compatible with the feminist ethics of care's emphasis on the particularity, fluidity, and interdependency that is characteristic of human relationships, a reality which care ethicists argue should be better reflected in our dominant moral and political theories (see Gilligan [2]; Held [3]; Tronto [4]). To put it simply, the picture of Hegel's thought that emerges out of the specific interpretations presented in this article is one that acknowledges that human beings, the relationships that they engage in, and their relationship with the natural world, are never static, always changing, and thus require an ethics that recognizes and responds to the fluid contexts of mutually vulnerable individuals and their always evolving relationships with others and the external world.

Hegel was a 19th century German philosopher known for his immense philosophical contributions related to the tradition now known as German Idealism. It is important to note that the reading of Hegel presented in this article is a very particular one and is primarily focused on his theory of the self as discussed in the preface to the *Phenomenology of Spirit* (1807) [5]. Furthermore, this reading of Hegel draws heavily upon the prominent Slovenian political philosopher and cultural theorist, Slavoj Žižek, who provides a contemporary interpretation of Hegel's theory of the self. As such, the goal is not to provide an exhaustive assessment of Hegel's thought, and many scholars may disagree with the picture of Hegel which is foregrounded in this article. Despite this, I argue that Hegel is a thinker who lends himself to a variety of legitimate interpretations, and the main purpose of this article is to demonstrate how certain readings of Hegel are amenable to the kind of critical ethics that is evident in the ethics of care, an approach to feminist moral philosophy which focuses on the ethical and political implications of our relational and vulnerable being (see Robinson [1]). Žižek's reading of Hegel presents the subject (i.e., the human self) as a source of radical contingency and, I argue, this picture of the subject can be mobilized to counter the dominant vision of the independent and rationalistic self from the Enlightenment tradition of liberalism. The mainstream liberal conception of the human self is equally a target for the feminist ethics of care which juxtaposes said vision of the subject with a picture of the relational and vulnerable self. The objective is thus not to provide a definitive interpretation of Hegel's thought, it is to demonstrate how Hegel, Žižek, and critical feminist theory can intersect in ways that can further our thinking on the ethics of care and the theory of the self—as relational—that is a key piece of care ethics.

The article proceeds with three main sections. The first presents some of the basic elements of Žižek's reading of Hegel that are relevant to this article; the second foregrounds a selection of feminist interpretations of Hegel that are compatible with Žižek's reading and which will in turn be compatible with the relational ontology of the ethics of care; the final main section identifies some of the points of intersection between these readings of Hegel and recent literature related to the ethics of care, with particular emphasis on the conceptual themes of relationality and vulnerability. Specifically, I will argue that Žižek's emphasis on the instability (or contingency) that characterizes Hegel's model of the human self can be usefully expanded when it is reimagined with the conceptual themes of relationality and vulnerability. At the same time, Žižek's reading of Hegel—in addition to the other feminist thinkers discussed—can offer care ethics additional philosophical resources in further developing its own theory of subjectivity and intersubjective relations which is consistent with its commitment to foregrounding the centrality of relationships at the ontological, epistemological, and ethical levels. The conversation between these approaches contributes to a wider philosophical effort to understand the nature of the human self and intersubjective relations in a way that is rooted in concrete relationships and contexts, and yet, does not presuppose that either individual subjects, nor the relationships that shape them, have, or will take on, any essential or fixed characteristics.

2. Žižek's Reading of Hegel's Theory of the Self

Before turning to Žižek and his reading of Hegel, I would like to offer some brief words on two related approaches in contemporary ethics which seek to foreground the importance of intersubjectivity (i.e., of relationships), and which can also be traced back to aspects of Hegel's thought; these two approaches are the communitarian tradition and the politics of recognition. A brief mention of these approaches, as well as why some feminists see them as inadequate attempts to develop a relational moral and/or political theory, not only provides some additional philosophical context, but also helps to identify how Žižek's approach to Hegel (developed in the rest of the section) differs in a way that can potentially avoid the feminist criticisms levelled at communitarian and recognition-based theories and, based on these differences, can then be potentially compatible with care ethics, especially when reimagined with the themes of relationality and vulnerability.

The communitarian tradition holds a prominent place in the fields of contemporary ethics and political theory because of its thoroughgoing critique of liberalism and the theory of the self as a rational autonomous individual that is a core piece of the liberal tradition. The emphasis on communal life and its role in the development of the human self can be traced back to Hegel's theory of the state and its role in the development of self-consciousness. One of the most well-known examples of the communitarian critique of liberalism comes from Charles Taylor [6], who argues that liberal conceptions of political society, premised on a belief in the "primacy of rights," reveal a misguided conception of human society, the human self, and the relationship between the two. The liberal tradition, for Taylor [6], can be labelled as a form of "atomism," which neglects the ways in which the human self depends on their community in order to sustain itself and to then develop the "capacities" that are traditionally seen as defining what it is to be human and key to living a fulfilling human life in a particular political community.

One can observe a clear affinity with care ethics here, in the sense that care ethics is also premised on a relational understanding of the self and society, where the self cannot be separated from its relations with others in either theory or practice; furthermore, this relational understanding of the self and social relations also acts as a key premise upon which care ethics makes its own critique of liberalism. However, feminist theorists like Susan Hekman [7] demonstrate why a focus on community, on its own, may not be the best avenue to critiquing liberalism if we wish to take seriously the specific concerns of feminists (including those in the care-tradition) as it relates to the various hierarchical relations of power which can exist between different groups within the same community (pp. 1115–1117). As Hekman helpfully writes:

> One of the central claims of the communitarian critique of liberalism is the assertion that the disembodied "I" of liberalism must be replaced by the embodied "we" of community. A feminist appraisal of this critique, then, must question the constitution of this "we" that the communitarians espouse. [7] (p. 1108)

A related and influential attempt to develop a moral and political theory suitable to a relational understanding of persons and the reality that modern communities, especially in the West, will have within them a range of group differences is that which is commonly known as the "politics of recognition" [8]. The scope of this article does not allow for an extended discussion of a moral or political theory of recognition; however, it is important to note that this approach is influenced by the well-known dialectic of lordship and bondage found in Hegel's *Phenomenology* which is used to convey Hegel's insight that the development of human subjectivity is, at its core, an intersubjective process which requires mutual recognition between human selves. By extension, for theorists of recognition such as Taylor [8], we should extend this ideal of mutual recognition to relations between different social groups within a given society to foster greater levels of social and political inclusivity. This is again an effort to develop an ethics and/or politics which is consistent with a relational understanding of persons, which is attentive to the contexts of individuals within social relations, and where there is a normative concern that these relationships (broadly conceived) allow individuals to flourish within their societies. However, and like Hekman's [7] critique of communitarianism, there are concerns that can be raised about this related recognition-based approach to subjectivity and social relations. For instance, Kelly Oliver [9] raises the concern that a politics of recognition stops short of challenging the unequal relations of power that make the act of recognition necessary; furthermore, Oliver argues that the possible unintended consequence of making recognition the core of a politics of inclusion is that it reaffirms unequal relations of power in which certain social groups are in need of recognition and the dominant group(s) have the power to grant it [9] (p. 474, 476–477).

Both above approaches can be seen as efforts to think of ethics and politics according to a vision of human beings as relational, and, where the development of subjectivity is an inter-subjective process. An emphasis on the importance of community and mutual recognition are ideas that could also be compatible with an ethics of care. However, as demonstrated by the corresponding feminist critique of each of these approaches, there is a risk that certain problematic relations of power can go unchallenged. In other words, the response from communitarians, or recognition theorists, to the realities of human interdependency and difference do not go far enough in some sense. This is where a more Žižekian reading of Hegel's subject can be helpful. As will be demonstrated below, Žižek's reading of Hegel on the subject has a more anti-foundational tone, where the subject is not defined by any thick or positive characteristics. In fact, for Žižek, subjectivity itself can be thought of as a source of contingency or instability, and this means that intersubjective relations, all the way up to political communities, could also be characterized as contingent. This is important for two reasons. First, Žižek's reading of Hegel is one that is less likely to fall into the idealization of communities, and, in fact, suggests that any community is always vulnerable to change because of the nature of human subjectivity; second, it also allows us to think of human subjectivity in a way that is non-essentializing. I argue that placing these insights into conversation with the themes of relationality and vulnerability, which are important to care ethics, provides mutual benefits. It allows the care ethics perspectives to further reflect on the implications of taking a wholly relational approach to ontology, epistemology, and ethics, and to do so in a way that can further avoid the charges of essentialism (gendered and otherwise) from its critics; on the other hand, reading Žižek alongside feminist theorists (including care theorists) can make this particular approach to Hegelian subjectivity more concrete and relational in ways that are philosophically and politically interesting. Overall, the conversation I stage in this article between Žižek's Hegel, feminist theory, and care ethics, seeks to demonstrate that the contingency, or what Žižek will below call the "negativity," of the subject, is perhaps more a property of relationships as such, rather than of an individual/abstract subject, and that relationships (in all their forms) are always vulnerable to change because to a certain extent they embody the possibility of change. In this sense, the *subject* as the source of *negativity* becomes reimagined as *relationships* as the embodiment of *vulnerability*. I will unpack this line of thought in what follows, beginning with a discussion of Žižek and his reading of Hegel on the nature of subjectivity.

2.1. Intellectual Context of Žižek's Thought

Žižek has engaged extensively in efforts to synthesize the work of G.W.F. Hegel, Karl Marx and Jacques Lacan in order to investigate the co-constitutive relationship between subjectivity and ideology. This section will deal exclusively with Žižek's reading of Hegel, which is arguably more important to Žižek's broader oeuvre than the influence of Marx. However, Žižek's own reading of Hegel very much takes place through a Lacanian lens, and, therefore, it is more so the influence of Lacan that needs to be briefly elaborated on here. Lacanian psychoanalysis is a vast field of study, and it is outside of the scope of this article to engage in a thorough analysis of said field; therefore, I concern myself mostly with the aspects of Lacan's thought which orient Žižek's reading of Hegel.

The most significant aspect of Lacanian thought is his tripartite schematic of human experience [10]. Human experience, for Lacan, is constituted by three zones—"Imaginary," "Symbolic," and "Real." The "Imaginary" is the zone we inhabit in our everyday life and characterizes how what we commonly call "reality" is experienced at the level of appearance. The "Symbolic" zone is constituted by "the 'big Other', the invisible order that structures our experience of reality, the complex network of rules and meanings which makes us see what we see the way we see it (and what we don't see the way we don't see it)" [11] (p. 119); see also [12] (p. 336). This brings us to the last zone of human experience in the Lacanian structure—the "Real." We can read "the Real" as representing that ultimate remainder which is "'impossible': something which can neither be directly experienced nor symbolized – like a traumatic encounter of extreme violence which destabilizes our entire universe of meaning. As such, the Real can only be discerned in its traces, effects or aftershocks" [11] (p. 120). Of course, the claim here is not that those who experience violence lack an awareness that such violence took place, or that they were implicated in the violent act. It is rather that there are some aspects of human existence which can escape linguistic signification, and because signification structures experience, these events escape total incorporation into a meaningful symbolic whole that can be perceived in conscious thought. In other words, there can be acts of violence that are so extreme, and/or, traumatic, that they cannot be reconciled within our everyday frameworks of meaning. But this does not only go for violence; it also applies to other aspects of reality that escape signification into everyday frameworks of meaning, some of which are rendered inarticulable because of socio-political practices of exclusion.

This notion of the "real" will come into play as I now turn to an exegetical reading of the preface of Hegel's *Phenomenology of Spirit* in tandem with one of Žižek's key texts on Hegel, *Less Than Nothing: Hegel and the Shadow of Dialectical Materialism* (2012). The key aspect of said reading will be the way that human subjectivity is framed as a source of "negativity" and "spontaneity."

2.2. Subjectivity, Negativity, and Hegel's Phenomenology

The thematic core of how we will understand subjectivity—i.e., the nature of the human self—in this section is helpfully encapsulated by comparing the two passages below, one from Žižek on some of the defining features of German Idealism as a philosophical tradition, and the other from Hegel who is the central figure of that tradition. It is also helpful to remember that this reading of Hegel ultimately seeks to foreground the perspective that there is a certain contingency (or lack of stability) at the root of all human existence that, when read in line with critical feminist theory and the ethics of care, bolsters an understanding of the self (and the world) as relational, fluid, and vulnerable (understood as both a physical and social vulnerability):

> Two features which cannot but appear opposed characterize the modern subject as it was conceptualized by German Idealism: (1) the subject is the power of "spontaneous" (i.e., autonomous, stating-in-itself, irreducible to a prior cause) synthetic activity. . . . (2) the subject is the power of negativity, of introducing a gap/cut into the given-immediate substantial unity. [12] (p. 106)

> In essence, Spirit is the result of its own activity: its activity is the transcending of what is immediately there, by negating it and returning into itself. We can compare it with the seed of a plant: the plant begins with the seed, but the seed is also the result of the plant's entire life. [13] (p. 82)

The distinct feature of Žižek's reading of Hegel is the emphasis that is placed on the "negativity" of the subject. When I refer to contingency above, it is considering this "negativity," which, as Žižek reads it, implies a kind of existential inability, or vulnerability, the result of which is that individual persons can never achieve a completely fixed or stable identity (or sense of their own subjectivity), and, by extension, a society at large can never achieve a fixed or stable ideological framework to bind subjects to a social framework. By the end of this article, I propose that this idea of "negativity" can be applied to the context of human relationships more broadly, and moreover, I suggest that this points to the simple fact that relationships as such—including relations of care—are always contingent, open, and in fact, vulnerable, to change.

If we are to accept this picture of the subject and its relations with the world (including other subjects) developed through Žižek's reading of Hegel, ahistorical ethical theories premised on the universalizability of moral rules and principles, and the impartial perspective, become theoretically and practically inadequate because they cannot attend to the particularities and fluidity of concrete human experience. On the other hand, a contextualist ethics such as the ethics of care, founded upon an explicitly relational ontology, is well-suited to deal with the simple fact that to be a human subject is to be in a constant state of flux which extends to various networks of relations that range from the personal to the political dimensions. If we are to stay true to the relational ontology of the ethics of care, we would be better served to understand the "negativity" at the root of Hegel's vision of the subject (as Žižek reads him) as a form of ontological and subjective vulnerability. Furthermore, this "negativity" should be understood as not merely a characteristic of an individual subject, but rather as the result of the dynamics of intersubjectivity itself—the fact that any human relationship is a constantly evolving entity resulting from the "negativity," or vulnerability, of the self-in-relation. With this broad perspective in mind, it is helpful to now look more closely at both Hegel's and Žižek's texts which develop these ideas.

2.3. Hegel's Subject

Here, I present a Žižek-inspired reading of Hegel's subject as radically contingent, i.e., as having no determined foundation, and thus as characterized by what Žižek above termed as a kind of spontaneity. This picture of Hegel's subject will be drawn out of a key passage from the preface to his *Phenomenology of Spirit* [5] and will anchor itself in Hegel's metaphorical illustration of human subjectivity as a kind of self-moving-void—i.e., as "the power of negativity" (see Žižek [12] and Kojève [14]).

From the beginning of Hegel's *Phenomenology*, the stated aim is to demonstrate how the history of philosophy has been "the progressive unfolding of truth … moments of an organic unity" [5] (p. 2) [1]. At risk of simplification—but perhaps a useful one for the purposes of this article—one of the main goals of Hegel's *Phenomenology* is to reflect upon, and spell out, the journey that human consciousness, the subject, or the "I," took to gain knowledge of itself, to become self-aware or self-conscious. This path that Hegel describes involves a set of progressive stages which include both an engagement with the natural world and with other human beings, or what will come to be recognized as other subjects. In this sense, it demonstrates how subjectivity is always relational, or, intersubjective; subjectivity is importantly constituted in its relationship with otherness. The influence of this aspect of Hegel's thought, especially as encapsulated in the dialectic of "lordship" and "bondage", is deep and pervasive and is perhaps most reflected in contemporary theories of recognition (e.g., Charles Taylor and Axel Honneth); however, it has also been mobilized by feminist theorists (e.g., Oliver [15]) as is explored further below. In either case, the interesting question from which varied readings of Hegel can emerge is whether the specific developmental journey that consciousness goes on was a pre-determined, necessary, or inevitable one? Or was it, to use Žižek's terms, "spontaneous"?

Žižek has elsewhere characterized Hegel's well-known *Philosophy of History* (1840) [13], in which he traces the emergence of self-consciousness throughout world history, as a "retroactive necessity," meaning that it only becomes 'necessary' as we reflect back on the journey, rather than being a necessary journey from the start; this idea of "retroactive necessity" is therefore also applicable to the description of consciousness in the *Phenomenology* [16] (pp. 129–131). Hegel himself emphasizes that it is only in 'remembering' the different stages that consciousness becomes self-consciousness and this characterization is at play in his initial discussion of "subject" (i.e., the human self) and "substance" (e.g., nature) – specifically regarding their eventual unity:

> [T]he living Substance is being which is in truth *Subject*, or, what is the same, is in truth actual only in so far as it is the movement of positing itself, or is the mediation of its self-othering with itself. This Substance is, as Subject, pure, *simple negativity* . . . . Only this self-*restoring* sameness, or this reflection in otherness within itself—not an *original* or *immediate* unity as such—is the True. It is the process of its own becoming, the circle that presupposes its end as its goal, having its end also as its beginning; and only by being worked out to its end, is it actual. [5] (p. 18)

The Žižekian interpretation can be distinguished from readings of Hegel that interpret the development of self-conscious reason through history as a kind of pre-determined and/or dominating force. For instance, such a "panlogical" reading of Hegel, which as Todd McGowan [17] notes is the defining common feature of his critics, would qualify as an example of what Leo Strauss saw as a characteristic of modern (political) philosophy more broadly and which is representative of what Strauss calls the "conquest of nature" [18] (p. 23, 27); what Strauss means here is that human rationality—as shaped by modern science—takes on an inflated and domineering role in both human and non-human existence. As opposed to this reading, thinkers like Žižek, and sympathetic readers of Žižek such as McGowan, emphasize the side of Hegel that describes the development of human consciousness and/or reason in history as "the spontaneous becoming of itself" [5] (p. 20), where "[i]t is reflection that makes the True a result" [5] (p. 21). As Žižek succinctly explains, "the process of becoming is not in itself necessary, but is the *becoming* (the gradual contingent emergence) *of necessity itself*" [12] (p. 231).

The following passage from the preface of the *Phenomenology* is central to the picture of human subjectivity that is the focus of this article and, I would argue, to the Žižekian reading of Hegel's subject, and takes place within the broader conversation regarding the dialectical process whereby human subjectivity emerges in the world by distinguishing itself from what is other than it:

> The disparity which exists in consciousness between the 'I' and the substance which is its object is the distinction between them, the *negative* in general. This can be regarded as the *defect* of both, though it is their soul, or that which moves them. That is why some of the ancients conceived the *void* as the principle of motion, for they rightly saw the moving principle as the *negative*, though they did not as yet grasp that the negative is the self. [5] (p. 37)

Hegel's evocation of the image of "the *void* as the principle of motion" is significant here since it suggests that it is the radical contingency of the "I" and its ability to take itself as other, as object, which moves consciousness forward along its path. It is the tension ("disparity") within consciousness itself—between the formal/abstract "I" as the void, or the negative, and its existence in/encounter with the world as positive content that leads to a series of progressive moments in the ongoing process of attempting to unify what Hegel labels "subject" and "substance" within/by the human self. In this sense, the self becomes both the "principle of motion" as well as that which is moved: it is self-positing, self-negating, self-moving.

The above passage is, in a sense, the key to answering the question as to whether the path taken by consciousness was a spontaneous one. By characterizing the self as, at bottom, constituted by the negative, or the void, and attributing to this void the property of motion, the result becomes a productive tension (between the "I" and the world) that plays out over historical time, but one that is not inevitable since the moving force remains characterized by contingency—i.e., having no positive determination as its foundation. To reiterate, the void becomes the "principle of motion" when it is faced with a positive externality which must then be reconciled with it; however, one can argue that the void or the negative (for Hegel) remains throughout such acts of reconciliation, and it is this dynamic that makes it possible to develop an interpretation of Hegel's subject as the seat of radical contingency in the world. It is here where the influence of Lacan on Žižek is relevant to Žižek's reading of Hegel, as the void, the "negativity" of subjectivity, becomes associated with the Lacanian notion of "the real" mentioned above [12] (p. 646, 841, 959–960, 963, 967).

The connection between Hegel's subject as a void, or the negative, and Lacan's notion of the "real" revolves around the idea that there is a "gap" in human subjectivity that can never be permanently filled by a particular social identity, system of meaning, or political ideology connecting the subject to the external human and non-human world [12] (p. 963). Žižek reads Hegel's notion of "Absolute Knowing" not as a kind of positive "total knowledge," but as a naming of our inherent "limitation" (we might even say vulnerability), the reality "that there is no external point of reference from which we could perceive the relativity of our own "merely subjective" standpoint" [12] (p. 393). McGowan [17] effectively argues that Žižek's reading of Hegel goes beyond merely positing that there is an "epistemological" gap existing in human subjectivity, and that this gap, or "self-division," is also an "ontological" condition of the human and non-human world.

This theme of the contingency, "negativity," and/or limitations of Hegel's subject—as interpreted by thinkers like Žižek—can also be seen as the focus of the critical feminist readings of Hegel that are discussed in the following section. This discussion thus continues us down a road that involves rethinking the *negativity* of the subject as the *vulnerability* of the subject in such a way where, I argue, a philosophically interesting connection can be made between Žižek's Hegel and a feminist ethics of care. This connection also involves conceiving of Hegel's subject in more explicitly embodied and relational terms, a task which is effectively accomplished by feminist ethicists.

3. Critical Feminist Interpretations of Hegel

In this second main section, I present some important examples of feminist engagements with Hegel's thought; in particular, I am interested in demonstrating how such readings bear a resemblance to more open readings of Hegel's ontological and epistemological claims, akin to Žižek's reading. The discussion in this section is relevant to an understanding of how Hegel has been interpreted by critical feminist thinkers and also serves as a bridge to the third and final main section in which I suggest that these readings of Hegel can contribute to the ethics of care literature and its focus on the philosophical and practical importance of human vulnerability.

First, in Kimberly Hutchings [19,20] and Tuija Pulkkinen [21] we find an explicitly feminist reading of Hegel's formulation of "substance as subject." Second, Judith Butler [22–25] and Luce Irigaray [26,27] are both influenced by a reading of Hegel which foregrounds "the power of negativity," in relation to subjectivity—in the way described by Hutchings and Pulkinnen, in addition to Žižek [12]—while rethinking this "negativity" in a way which foregrounds its concrete embodiment and constitution within particularistic social relations. It is this shift to theorizing subject, substance, and the negative in more embodied and relational terms that make Hegel's notions more amenable to care ethics as a moral and political theory which foregrounds vulnerability as a key aspect of human and social life.

The feminist reading of Hegel I mobilize here is guided by two key aspects related to the *Phenomenology's* [5] equation of "substance as subject." These two aspects are: (a) the possible feminist ontology and epistemology which can be drawn out of Hegel's philosophy as representative of what Hutchings [19] calls "the simultaneous identity and non-identity of being and truth"; and (b) an emphasis on what Pulkkinen [21] calls the "substance-subject," which refers to an open-ended and inherently self-reflexive ontology that can aid in the effort to overcome traditional conceptual oppositions. Hutchings and Pulkkinen indicate that readings of Hegel usually fall within one of three approaches—"closed," "open," and/or "deconstructive." The distinction between "open" and "closed" has to do with whether Hegel's philosophical system implies a fixed and definitive endpoint vis-à-vis the development of human consciousness in relation to the world, or an open epistemology and ontology which is non-totalizing and always open to a revision of its claims. The "deconstructive" approach seeks to destabilize Hegel's philosophy from within the text itself. Feminist readings of Hegel are typically "open" and "deconstructive" [28] (pp. 4–5). This article primarily follows the "open" interpretation as it is the one that aligns with Žižek's Hegel and, as I argue, is compatible with care ethics. That said, it is worth noting that Butler and Irigaray both incorporate elements of a "deconstructive" reading.

For Hutchings, and calling to mind the point by Žižek [12] above:

> The paradox of Hegel's 'absolute knowledge' is that, if we take it seriously, it is historically contingent and necessarily provisional. A feminist ontology and epistemology which took its cue from Hegel would be one which was premised on the idea of the simultaneous identity and non-identity of being and truth. The identity between being and truth refers to the dependence of thought on experienced being. The non-identity of being and truth refers to the limitlessness of being and the necessary partiality of any particular claim to truth. [19] (pp. 105–106)

Hutchings reads Hegel as foregrounding contingency in his epistemology, ontology, and ethics in the sense that the "limitlessness of being" refers to the hypothesis that human subjects, external nature, the relationship between human subjects and external nature, and the relationships between human subjects themselves, are "self-changing" according to an open-ended and dialectical process situated in historical time. Therefore, we can read a certain degree of spontaneity vis-à-vis "subject-knowers," "objects of knowledge," and relations among "subject-knowers" at the root of Hutchings's reading of Hegel and the "simultaneous identity and non-identity of being and truth" [19] (p. 109). For Hutchings, via Hegel, we can never know reality because reality, and us as part of reality, are in continuous motion: "Hegel's notion of grasping the truth both as substance and subject expresses this idea of the solid but self-moving medium within which claims are both made and judged" [19] (p. 110).

This understanding of substance and subject, according to Hutchings, is a step towards moving past a number of conceptual antinomies as human subjects are constantly in a co-constitutive relationship with nature, and one another. As Nadine Changfoot succinctly puts it:

> He [Hegel] does not see the subject as an instrument or a medium for ascertaining knowledge and its truth. Instead, he attempts to overcome the dualism of subject and object by viewing knowledge as a phenomenon where subject and object are inseparable with their own history and development. [29] (p. 490)

Furthermore, for Hutchings's Hegel, like Žižek's, this is not a teleological process determined by a transcendent order, or preestablished structures in human cognition. The lesson of Hegel's *Phenomenology* is "to think differently" and beyond "mutually exclusive oppositions" [20] (p. 87). It is worth noting that along with the above distinction between "open," "closed," and "deconstructive" interpretations of Hegel, there is also a division between "realist" and "non-realist" accounts of his ontology. As Allison Stone identifies, Hutchings falls within the "non-realist" camp in that she does not presume that "Hegel believes that reality has a determinate character independent of human practices, and that we can gain knowledge about this reality because it is, in itself, organized conceptually" [30] (p. 304).

One of the key components of Hutchings's reading of Hegel which makes it potentially compatible with care ethics is the inclusion of the theme of human vulnerability. On Hutchings's [20] reading, it is mortality and the presence of death which inscribes a kind of ultimate limitation within the "simultaneous identity and non-identity of being and truth." As Hutchings explains:

> When Hegel distinguishes the realm of animal nature from that of spirit he does so in terms of a distinction between determinate and self-determining being. Self-determination involves a self-conscious recognition of limitation, fundamentally an awareness of death that is not available to animal species. Without this recognition of the mutual dependence of nature and spirit, spirit paradoxically becomes reduced to nature by treating itself as a natural kind that transcends the death of its members. As such, Hegel is cutting the ground from under the feet of any moral authority that … takes itself to be grounded in a unique, self-legislating, self-legitimating force that somehow transcends either nature or ethical life. [20] (pp. 93–94)

Thus, Hutchings links our corporeal vulnerability and dependence to the vulnerability and dependence of judgment (see [31]) in a way that calls to mind the broader tradition of standpoint theory in feminist thought (e.g., Haraway [32]; Hartsock [33]; Hekman [34]). The endpoint for Hutchings—who explicitly frames her discussion in reference to Butler and Irigaray—is a form of "Hegelian, heteronomous ethics" [20] (p. 88) wherein

> ethics becomes a painful and incremental process in which reckoning with our own incapacity to know ourselves is intrinsically linked to our potential capacity to communicate with others in ways that are non-assimilative and may make it possible to forge inter-subjective links in the construction of new forms of ethical life. [20] (p. 103) (see also [35])

This is especially relevant for feminism in the sense that, contra Nancy Hartsock [33], the inability to finally "know thyself" can equally be seen as a resistance to "gender codification" [29] (pp. 479, 485–486).

Pulkkinen claims that much of the hesitance in feminist theory when it comes to Hegel – apart from instances of sexism in his work—is with the humanistic portrayal which neglects that side of Hegel concerned with embodied and non-human being. The Hegel exemplified by Hutchings and Pulkkinen, on the other hand, maintains a fuller understanding of the dialectical nature of subjectivity where "[t]he substance-subject is sheer (self-reflecting) activity and a perpetual motion in concepts: it is thought" [21] (p. 25). Yet, thought is inherent to the material world, and the material world to thought, and this is why Pulkkinen uses the notion of a "substance-subject," rather than replicating readings of Hegel which privilege "subject" over and against "substance" in order to uncover "the essential characteristics of a universalized human being" [21] (p. 24). This also further supports Žižek's reading of Hegel and why sympathetic commentators like McGowan [17] emphasize that the importance of Žižek—in juxtaposition to other radical or emancipatory readings of Hegel—is in the way that he not only focuses on the epistemological dimension of Hegel's reflections on the nature of subjectivity, but also on the implications that this has for a Hegelian perspective on the nature of being as such. The suggestion is that the nature of being is, in some sense, incomplete, and this sense of "incompleteness" [17] (p. 7), I argue, is posited for similar reasons as to why Hutchings [19] (p. 106) uses the language of "limitlessness" in their reading of Hegel. Both of these readings of Hegel suggest that the human subject, the natural world, and/or the relationship between subject and world, is always potentially unstable and open to change; furthermore, this is why Hutchings is most interested in the ethical implications of Hegel's thought, since such an understanding of both self and being seems to undermine dominant approaches to ethics premised upon the notion of the independent, rational, and autonomous self whose humanity is defined in opposition to material being—both human and non-human (e.g., Kantianism).

The above is important in the ways in which these feminist readings of Hegel foreground a relational ontology and epistemology, and, how this relational ontology and epistemology implies vulnerability and interdependency. Ontologically, human subjects are engaged in an ongoing relational process with other human subjects, as well as material being—both human and non-human. This is then seen as inseparable from the dimension of consciousness, meaning ongoing relations between subjects, as well as between the subject and material being—both human and non-human. Rather than reading Hegel's ontology and epistemology as a dialectical process moving towards a fixed endpoint, Hutchings and Pulkkinen instead focus on the ways in which the relational character of this ontology and epistemology suggests an inherent vulnerability and interdependency both at the level of embodied being, but also at the level of thought, which, in fact, are seen as inseparable. This reading of Hegel is more compatible with the reading of Hegel offered by Žižek, which, as McGowan pointed out, is in juxtaposition to the "panlogical" readings which are often "closed" and/or "realist" to use some of the categories cited above. A brief examination of how this reading of Hegel is also discernable in the work of two important contemporary feminist theorists, who have a significant presence in the above feminist interpretations of Hegel—Judith Butler and Luce Irigaray—further demonstrates the above claims and serve as a point of transition to the final main section of this article where I suggest that this particular reading of Hegel can be put into a philosophically interesting conversation with care ethics because it aligns with the latter's emphasis on relationality, embodiment and vulnerability.

Judith Butler's sustained interrogation of the politics of identity can be traced back to their engagement with the Hegelian themes of "desire" and "recognition" as found in *Subjects of Desire* [23]. Butler's presentation and mobilization of Hegel's *Phenomenology of Spirit* in their earliest work is characterized by an "open" and "non-realist" reading (see above), which sees the development of consciousness as "neither static nor teleological" and one "that knows no closure" [23] (p. x) (see also [25] (pp. 172–174) [36,37]). Furthermore, the Hegelian subject is described as exemplifying "a critical mobility" [23] (p. xiv) and "is an ek-static one, a subject who constantly finds itself outside itself" [23] (p. xv). The theme of movement is crucial here, and the notion of "ek-stasis" suggests that the "substance-subject" is characterized by a perpetual moving from inside to "outside," or, between "identity" and "non-identity" [19]. For Butler, it is the interplay between desire and recognition that constitutes the source of this movement, and this is at the root of their theory of "subjection" which is subsequently developed in *The Psychic Life of Power* [22].

The foregrounding of desire in Butler's reading of Hegel is both theoretically and practically significant in the sense that it participates in an investigation of subjectivity which sees human subjects as a product of our embodiment and as being implicated in social relations of power. As such, it contributes to the larger goals of critical feminist theory by offering tools for social and political analysis and critique. If desire can be shown to be a crucial element in the development of a self-conscious substance-subject, then this means that we could be said to "reason in our most spontaneous of yearnings" [23] (p. 2) such that desire and rationality are co-constitutive and co-implicated in a way which undermines traditional humanist conceptions of the subject as "unified" and/or "coherent" [23] (pp. 2–6). Desire in this sense is not something that the subject is said to possess but is rather that movement which "perpetually displaces the subject" [23] (p. xv). We can see an affinity with the interpretation of Hegel offered thus far in the sense that the defining feature of Hegel's subject, for these thinkers, is that it is a subject that is constantly vulnerable to a "self-loss," and, in fact, moves itself towards this loss [23] (p. xv).

Desire, in Butler's reading of Hegel, is the source of motion which characterizes the relationship between subject and substance and constitutes said relation as the basis for the ontology of "substance-subject" elaborated by Pulkkinen [21]. Epistemologically, desire becomes the desire for self-knowledge in relation to a world that at first appears external:

> Insofar as desire is this principle of consciousness' reflexivity, desire can be said to be satisfied when a relation to something external to consciousness is discovered to be constitutive of the subject itself. On the other hand, desire's dissatisfaction always signifies ontological rupture, the insurpassability of external difference. [23] (p. 8)

In this sense, "desire is an interrogative mode of being, a corporeal questioning of identity and place" [23] (p. 9). "Negativity," as discussed above, is further rendered "corporeal" in the sense that it becomes an embodied approach towards difference, which participates in a "cyclical process" in an unending effort to gain a better "sense of self" [23] (pp. 18, 13). Butler's reading of Hegel therefore puts desire front and center in the constitution of the substance-subject. Subject and substance are enmeshed because of the movement caused by desire, "the subject's desire to discover itself as substance" [23] (p. 21). This point is crucial in the sense that it serves as the grounds for the "heteronomous ethics" which Hutchings [20] seeks to mobilize for critical feminist theory. "Desire" is that which moves the substance-subject towards, and in relation to, "otherness." The limitations of the human body, which house a physically and socially situated consciousness, simultaneously provides the grounds for an "infinite" engagement with the other propelled by a desire to recognize the self as a broader relation of "interdependence" with said other, as another self-conscious substance-subject [23] (pp. 35–41). In other words, for Butler, negativity becomes a founding element within the "embodied identity" [23] (p. 43) of a self-conscious substance-subject which recognizes its limitation, and thus "dependence" on, others as the means through which to fulfill its "desire-for-another-desire," and thus, a more "self-sustaining picture of himself." But, at the same time, the desire for recognition means that one's sense of identity is constantly open to being destabilized and shaped by the otherness it confronts [23] (p. 42).

This exposition of Butler's early reading of Hegel is significant for two reasons: first, is the idea that "desire" compels "an act of willful self-estrangement" in its effort to satisfy its desire as mediated by alterity [23] (pp. 49–50), which is helpful in terms of thinking about the notion of the "void" in Hegel's theory of the self as presented in the previous section; and second, this means that "desire" is connected to "recognition," as the desire to be "recognized" within an eventual "community" of "desiring" others [23] (pp. 57–58). This interplay, or tension, between desire and recognition as embodying negativity is key to understanding Butler's notion of "subjection" [22] as that which simultaneously constrains and enables human agency and again foregrounds the themes of vulnerability and interdependency in terms of how subjects come to be self-conscious of themselves in relation to a world of other subjects and external nature; however, Butler's work following

Subjects of Desire demonstrates how such relations, including relations of "recognition," are always implicated in multifaceted relations of power and, importantly, as situated in the realm of discourse. Butler also alludes to the relationship between the ontological and epistemological dimensions of Hegel's thought in a way reminiscent of both Žižek's and Hutchings's reading of the Hegelian notion of "universality," and in which Butler foregrounds the role of difference and opposition:

> And just as Hegel insists on revising several times his very definition of 'universality', so he makes plain that the categories by which the world becomes available to us are continually remade by the encounter with the world that they facilitate. We do not remain the same, and neither do our cognitive categories, as we enter into a knowing encounter with the world. Both the knowing subject and the world are undone and redone by the act of knowledge. [24] (p. 20)

Again, the focus on these approaches to Hegel seek to foreground the relational nature of both his epistemology and ontology in a way that can intersect with a feminist ethics of care. Indeed, this focus is the reason why this article presents commentaries on Hegel's work which suggest a picture of the Hegelian subject as being engaged in an open and fluid relationship with the external world at the level of both knowledge (subject) and being (substance), as well as in the interconnection between these two levels.

Irigaray, the final thinker I discuss in this section, is somewhat distinct in that they have been more openly hostile to Hegel's philosophy in their substantive body of work; however, the notion of "sexual difference" in Irigaray's work, at the very least, can be seen as incorporating the notion of "negativity" discussed throughout this article in a way which continues the effort to think about an ontological and epistemological perspective—coming out of Hegel's philosophy—which foregrounds an image of the self and world as always unfinished due to the fact that difference and opposition are an inherent feature of both social and natural reality. Therefore, they could be classified as ultimately offering more of a "deconstructive" reading of Hegel according to Hutchings and Pulkkinen [28], but perhaps this is only because Irigaray, unlike Žižek, Hutchings, and/or Butler, begins with a more "panlogical" [17], "closed" [28], and/or "realist" [30] interpretation of Hegel.

The moments in Irigaray's texts where Hegel is explicitly rejected all refer back to a particular reading of Hegel's "absolute spirit" [27] (p. 13) which does not easily align with the "open" and "non-realist" [30] readings offered by Hutchings [19,20], Pulkkinen [21], or Butler [23–25]. This leads to a reading which, somewhat similar to Simone De Beauvoir [38] and Hartsock [33], sees *spirit-conscious-of-itself* ("absolute spirit") as the paradigmatic form of abstract masculinity which has underpinned the traditional humanist subject and relegated women as its constitutive outside in both theory (particular vs. universal) and practice (family vs. politics). As Tina Chanter helpfully puts it:

> For Hegel the relation between family and state is both necessary and hierarchically arranged in terms of the relation between parts and whole. Similarly, the relation between the sexes is ordered according to the same schema, the female representing the family, and the male representing the state . . . . In a negative movement, the individual departs from his familial grounding in the organic life, supported by the mother, and moves on out into the public sphere, the realm in which his individuality acquires a higher, universal, communal significance. [39] (p. 108)

Therefore, even while Hegel acknowledges the integral role of the family to the broader political community in a way that is relatively progressive for his time, this apparent valorization of mothers, to some of his feminist critics, is nevertheless left open to the charge of maintaining traces of "phallocentric" logic [40] and traditional (gendered) hierarchies between nature-culture, subjective-objective, particular-universal, and family-state (among others) [19] (p. 53), [27] (pp. 22-25), [39] (pp. 88–95), [41] (p. 29), [42].

That being said, in both the *Ethics of Sexual Difference* (1993) and *I Love to You* (1996) we see Irigaray explicitly, and simultaneously, mobilize "Hegelian logic" in which "the relation between the universal and the particular becomes a central theme" [43] (pp. 457–458) in thinking through the ontological, epistemological and ethical consequences of taking seriously the "real" of "sexual difference." In this aspect of their work, Irigaray does not actually seem as distant from the interpreters of Hegel explored in this article. While the historical account of Hegel's social and political thought offered by Irigaray is one that (rightly) identifies the appearance of gendered hierarchies and logics in both the *Phenomenology of Spirit* [5] and *Elements of the Philosophy of Right* (1821) [44], they also mobilize aspects of Hegel's ontology and epistemology, as encapsulated in the notion of a self-moving substance-subject with negativity as its principle of motion. As we see below, there are substantive amendments to these aspects of Hegel's thought (as interpreted by feminist re-readings above); however, it nevertheless ultimately engages in an "open" and "deconstructive" reading of Hegel to aid in the task of thinking the "real" as "sexual difference" as both subject and substance. Taking this into account is helpful not only for clarifying the ontological status of sexed difference in Irigaray, but simultaneously opens the door to the possibility of reading Irigaray and Butler alongside the ethics of care, despite their differing theoretical commitments.

Given that Irigaray's positing of an ontology of "sexual difference" is often controversial [45] (pp. 26–27), it is perhaps helpful to first return to Žižek's notion of "retroactive necessity"—discussed above in the context of his reading of Hegel—and approach Irigaray's notion of sexual difference in a similar fashion. The notion of "retroactive necessity" may be a useful framing device in the sense that it allows us to see "sexual difference" as simultaneously socially constructed, on the one hand, but nevertheless a now stable feature of our (social) ontology with real and lasting effects, on the other. These two aspects of sexual difference are primary features of Irigaray's ethics. A focus on sexual difference takes seriously an analysis of historical relations of (gendered) power, but, at the same time, also points to a future which is more open to difference. It is in this reading of Irigaray's notion of sexual difference—as a retroactive necessity—which also provides the possibility of overlap with Butler's thought. As Hutchings remarks, Irigaray and Butler, despite their differences, both exemplify a form of "Hegelian, heteronomous ethics" [20] (p. 88). These ideas can be demonstrated by a brief look at Irigaray's *Ethics of Sexual Difference* [26] and *I Love to You* [27] in light of the analysis provided up to this point.

"Sexual difference" constitutes the "real" for Irigaray in the sense that "subjectivity" is co-constitutive with the embodied reality of being "sexed" [27] (pp. 13–14), [26] (p. 6). As Irigaray puts it: "It is evident that female and male corporeal morphology are not the same and it therefore follows that their way of experiencing the sensible and of constructing the spiritual is not the same" [27] (p. 38). In other words, everyone is born first as a natural being, a body, and this basic and simple reality is posited as an inescapable and ontological fact for Irigaray. Consequently, there is never just one "nature," "culture," and/or, "subjectivity," but rather, always a "sexed" nature, culture, and/or, subjectivity, and this means each realm is always "at least two" [27] (pp. 13–14, 35–37), [26] (p. 68). I argue that this fundamental aspect of Irigaray's thought aligns them with the above readings of Hegel which emphasize that human consciousness (subject) and human and non-human matter (substance) are always partial, incomplete, and fluid, especially in the sense that these two realms are engaged in an interconnected and ongoing relationship as described most vividly by Hutchings's idea of the "limitlessness of being" and the "identity and non-identity of being and truth" [19] (pp. 105–106).

The further ontological significance of this fact is the way it allows Irigaray to read Hegel's notion of "negativity" as rooted in nature and our status as embodied and "sexed" beings. Whether this claim is open to a charge of biological essentialism is up for debate, but I argue that we can read this ontological claim as one of "retroactive necessity" rather than biological essence; in other words, the claim here is that socio-historical conditions have made it so that the initial source of social identification vis-à-vis a subject is determined along the relational axis of sex—one is born a male or female. Depending on how one's embodied being is designated according to this relational axis significantly impacts the trajectory of their further development as a subject in relation to other subjects, and leads to the two "sexes," each taken as a group, having their own "history" [27] (p. 47). These relations are always socially constructed, and one of Irigaray's main goals is to demonstrate how the history of patriarchy has predominately suppressed the development of a specifically "female identity" and culture, thus simultaneously limiting the relation between the sexes themselves in a way which is detrimental to society as a whole, and any truly emancipatory or egalitarian goals it may have [27] (pp. 5, 15, 27, 38), [26] (pp. 67–68).

Irigaray even frames this notion of sexual difference in the language of "negativity" when they write:

> The negative in sexual difference means an acceptance of the limits of my gender and recognition of the irreducibility of the other. It cannot be overcome, but it gives a positive access—neither instinctual nor drive-related—to the other. [27] (p. 13)

Irigaray distinguishes this from the "negative" presented in Hegel, which Irigaray sees as representative of a (masculinist) "mastery of consciousness" [27] (pp. 13, 35). There is sometimes an apparent disjuncture between Irigaray's assessment of Hegel's philosophy, and then their mobilization of particular aspects of Hegel's ontology. This seems to be primarily a result of Irigaray's reading of Hegel's absolute spirit as teleological, oriented towards a fixed "universal" end which subsumes all particularity, and thus excludes women from shaping human culture (e.g., [27] (p. 21), [26] (p. 108)). I argue that Irigaray is actually more aligned with Hutchings [19,20], Pulkkinen [21] and Butler [23–25] than it seems on first glance, and this is primarily based on the way that Irigaray incorporates a Hegelian understanding of "negativity" within their ontology and ethics of sexual difference; this involves the epistemological claim that human subjectivity is an always unfinished project, which is related to the ontological claim that concrete reality, or nature, is always self-divided, and we can see this self-division most clearly in the fact of sexual difference.

Therefore, the recognition of sexual difference is equally the recognition of difference as a fundamental attribute of nature, and thus presents consciousness with inescapable limitations vis-à-vis any effort to totalize or establish closure in either nature or culture. In other words, male and female—as rooted in relational and differential embodiment—are "irreducible" one to the other and represent within nature, as in human consciousness, its own *limit-point* [27] (pp. 39–41). Somewhat ironically, given that on the surface Butler and Irigaray begin with opposing orientations towards the whole of Hegel's thought, Irigaray mobilizes a re-conceptualization of the category of universality that is reminiscent of Butler's in their account of how particular instantiations of gender relate to its universal form:

> The universal thus results from a retroactive and non-projective constitution. From a return to reality and not from an artificial construct. I belong to the universal in recognizing that I am a woman. This woman's singularity is in having a particular genealogy and history. But belonging to a gender represents a universal that exists prior to me. I have to accomplish it in relation to my particular destiny. [27] (p. 39) (see also (p. 27))

At first glance, a reader may presume a divergence with Butler's approach given the assertion of "a return to reality" in the above passage. However, if we understand sexual difference as a retroactive necessity, and in relation to substance-subject, then the passage becomes surprisingly compatible with a reading of gender as constituted through "subjection" [22] and a relation to universality as theorized by Butler [24,25], at least at the level of formal logic. For both Butler and Irigaray, there is a co-constitutive relationship between thought and matter, which then entails that any notion of universality is one that has to be reproduced in concrete social practices. Furthermore, universality and particularity stand in a relation such that universality can be a space for change [24,25].

Therefore, Irigaray ultimately roots the idea of universality in embodied being by equating it with sexual difference:

> Without doubt, the most appropriate content for the universal is sexual difference. Indeed, this content is both real and universal. Sexual difference is an immediate natural given and it is a real and irreducible component of the universal. [27] (p. 47)

In the above passage, Irigaray identifies universality with negativity; sexual difference, as universal, is the name for the "irreducible" relation of difference that exists between the sexes as such. For Irigaray, sexual difference is that which concretely organizes subjectivity and the societies in which those subjectivities are located across socio-political time-space. It further reconciles universality and particularity within a non-totalizing "dialectic" of "sexual difference" in the sense that universality stands for the "irreducible" relation that exists between concrete beings who themselves each embody both universality-particularity as singular beings who nonetheless relate to one of the two universals of "gender" [27] (p. 51). Furthermore, even though Žižek's engagement with "French Feminism" has not been extensive or particularly collaborative [46], we could read Irigaray as positing that the "void," "gap" and/or "negativity" at the heart of Žižek's Hegel-inspired conceptualization of the subject becomes the more relational and productive gap that exists between the "competing universalities" [25] of masculine and feminine, where there is no possibility of reconciliation into a singular, unified notion of universal humanity. In other words, for Irigaray, it becomes the event of reconciliation without a "unisex[ual]" "absolute spirit" as its result [27] (p. 105), where each "I belong[s] to a gender, which means to a sexed universal and to a relation between two universals" [27] (p. 106). Lastly, if we read these claims as retroactively necessary—i.e., that they could have been otherwise–and according to the insight that external reality (nature as substance) and the way that nature is conceptually understood and posited by consciousness (as subject) is a co-constitutive relation, then we can interpret sexual difference as both really-existing but not essentially determined in any way.

So, why is Irigaray important to a discussion of how Hegel—especially as interpreted through Žižek—can be shown as having points of intersection with feminist philosophy, and ultimately with the ethics of care? The first reason is the simple fact that, despite being critical of Hegel's philosophy, Irigaray's notion of sexual difference can be seen as at least partly constituted by a "deconstructive" reading of some of Hegel's ideas; furthermore, and as demonstrated above, there are some interesting connections that can be made between Irigaray's philosophy and Žižek's reading of Hegel. The second reason is that Irigaray's re-mobilization of some of these Hegelian themes further explores an understanding of Hegel's epistemology and ontology as a relational epistemology and ontology that is a potentially interesting addition to thinking about the ethics of care as a contextualist ethics which contains important assumptions about the nature of the human self and its relation to being (including other embodied selves) and which foregrounds the dimensions of vulnerability and interdependency. Irigaray focuses on the idea of "sexual difference" to present these themes, and demonstrates the implications of sexual difference for thinking about both nature and culture; however, we can extrapolate from this fundamental difference (as Irigaray sees it), and apply it to thinking about relationships and differences more broadly (i.e., beyond differences between the "sexes") and, of particular import for this argument, to the relational ontology of the ethics of care which highlights the particularity of contexts when it comes to individual human selves—understood as embodied and vulnerable—and their relationships with others and the socio-political realm at large. In the final main section, I situate some of these ideas and lines of thought in the context of some contemporary literature on the ethics of care as a moral and political theory. I suggest that we continue to make Hegel's philosophy of the subject increasingly relational and embodied, and therefore more compatible with a care ethics' perspective, by rethinking the theme of *negativity* in the language of *vulnerability*.

4. Points of Intersection between Žižek, Feminist Readings of Hegel, and Care Ethics

In this final main section, I discuss some possible points of intersection between the readings of Hegel offered above and recent literature related to the ethics of care. As noted previously, the ethics of care is concerned with vulnerability, relationality, and interdependency as core features of the human experience which must be accounted for in our moral and political theories (e.g., Robinson [1]); these themes are central to thinking about the fluid nature of human relationships, and, by extension, points to the strengths of a contextualist ethics of care which explicitly seeks to bring to light and further develop forms of moral reasoning which attend to concrete others and the particulars of their various and multifaceted relationships in both the so-called "private" *and* "public" realms (see Gilligan [2]; Held [3]; Tronto [4]). The argument is that Žižek's focus on Hegel's picture of the human subject as a continuously unfinished project, and the relational ontology and epistemology that is connected to this claim and which is developed by feminist readers of Hegel, is even more compelling when we reflect on the nature of human vulnerability and relationship more broadly. For instance, the descriptions so far often call to mind the image of an abstract individual encountering the world, or a singular individual encountering another. In Žižek's discussion, as well as feminist interpretations of Hegel, it always came back to the notion of the human self as an incomplete entity or actively destabilizing force. However, what if we scale this up, and, consistent with the ethics of care's emphasis on concrete relationships, what if it is relationship—as such—rather than the abstract self, which is the source of the *negative*? Furthermore, what if we rethink the notion of *negativity* as reflective of our inherent *vulnerability*—both material and immaterial? Care ethics is key to this rethinking and this philosophical conversation–between Žižek's Hegel, feminist theory, and care ethics—can subsequently expand both the understanding and possible application of the concept of vulnerability within the ethics of care, as well as other related traditions which foreground a relational understanding of subjectivity.

The above also encourages us not to privilege the philosophical model of relationship as an encounter between only two abstract selves, and instead, to think about *relationship* as its own philosophical category which foregrounds the concrete vulnerability of *subjects-in-relation* in a way that can be scaled all the way up to political communities. Doing so would demonstrate the relational nature of all the different aspects of our lives, but also demonstrate their instability and thus fluidity. It is also in this sense that the ethics of care is not just an ethics for personal intimate relationships but is a broader moral and political theory capable of wide-ranging social analysis and critique all the way up to the sphere of international and/or global relations (e.g., Held [3], Robinson [47]). I unpack some of these lines of thought below by engaging with feminist reflections on vulnerability and Carol Gilligan and Naomi Snider's [48] recent reflections on the status of "relationship" in contemporary North American society.

4.1. Vulnerability

As Danielle Petherbridge notes, the notion of vulnerability has often appeared in the history of philosophy, and its remobilization is a key component of contemporary feminist ethics:

> Images of vulnerability have populated the philosophical landscape from Plato to Hobbes, Fichte to Hegel, Levinas to Foucault, often designating a sense of corporeal susceptibility to injury, or of being threatened or wounded and therefore have been predominantly associated with violence, finitude, or mortality . . . . However, more recently, feminist theorists have begun to rethink corporeal vulnerability as a critical or ethical category, one based on our primary interdependence and intercorporeality as human beings. [49] (p. 589)

Importantly, the foregrounding of vulnerability on the part of feminist ethicists, including care ethicists, takes the form of an "existential" premise. Hanna-Kaisa Hoppania and Tiina Vaittinen's understanding of the different meanings of "care" is helpful here:

> We understand care as a "corporeal relation". This is a form of political relatedness
> that does not begin with the practices of caring, as care ethicists would argue, but
> from the living organism of the vulnerable body that makes the labour of care
> an absolute necessity for human beings as individuals and as a species. Since all
> bodies are vulnerable to decay and disease and *no* body can exist without the aid
> of other bodies, the vulnerable body belongs to each and every one of us. This is,
> indeed, an existential fact of human life. [50] (p. 74)

For Sandra Laugier, our everyday practices of care come to reflect this sense of vulnerability
as a key facet of our "ordinary reality," and requires a contextualist ethics that responds to
the unique manifestations of this generalized vulnerability [51] (p. 219).

Vulnerability as explicated within a relational ontology of care also breaks down the
binary between nature (substance) and society (subject), which, as explored in the previous
analysis of the substance-subject, is an important component to feminist theory. Estelle
Ferrarese [52] explains this point in a way that is reminiscent of Hutchings [19,20] and
Pulkkinen [21] on Hegel when they write:

> Endowing oneself with a new ontology—that is by positing the idea that we are
> all vulnerable—might only confirm that there exist two levels of reality, each of
> them hermetic, or *pure*. In these conditions, the political, as representing the social
> world, would be distinguished—possibly to combine with it—from a natural
> world, which would be that of vulnerability. Instead, our view is that the task
> of theories of vulnerability is to recall the inseparability of the two levels. It is
> worth dwelling on the fact that it is the nature-social circle, and the impossibility
> of opening it, which makes it possible to grasp the stakes of vulnerability . . . .
> Social world and ontological vulnerability must be thought of conjointly in order
> to show how they engender and co-produce each other. [52] (p. 154)

Furthermore, in making this "ontological" claim there is an equally related epistemological claim in the sense that when the nature of (social) being is posited as vulnerable
and relational, it follows that knowledge about said world must be equally vulnerable and
relational. Knowledge claims are necessarily made within the fluid contexts of relations,
relations which involve differing perspectives and positions of power, and this applies to
both intimate relations of care, but also broader social, political and/or economic relations.
As such, ethics should involve always being mindful of the vulnerability of human and
non-human materiality, but also of the vulnerability of conscious thought, and by extension,
of judgment [31].

4.2. Relationality

In *Why Does Patriarchy Persist?*, Gilligan and Snider [48] provide a relevant and contemporary examination of how vulnerability and interdependency, as philosophical concepts,
take a very particular shape in contemporary Western society. The situatedness of this
analysis is important, as we can think of the idea of a relational ontology as a formal principle communicating that life is relational, but then see that this relationality takes on many
different forms in concrete social life, and some of these forms can even reflect a denial of
our vulnerability and relationality as ontological features of our existence. Furthermore,
Gilligan and Snider [48] demonstrate how a communal state of denial of our existential
vulnerability and relationality can reflect gendered structures of power, and are especially
constitutive of patriarchal norms. This is a helpful example of how we might think of
subject and substance as engaged in a co-constitutive relation when it comes to how the
individual subject, and a society, considers and mediates its material embodiment, and the
particular example of patriarchy provides a helpful means by which to bridge the reading
of Hegel (by Žižek and a selection of feminist theorists) with care ethics.

The underlying premise of Gilligan and Snider's study is that there is a fundamental
"connection between the psychology of loss and the politics of patriarchy" [48] (p. 31).
The "psychology of loss" entails deeply engrained gendered modes of engaging in social

relations which serve as a means to avoid the trauma of loss, which, as mentioned a number of times in the text, appears most obviously in the fact of human mortality [48] (pp. 88–91). These gendered responses are united in the fact that they "denigrate and detach from those very relational capacities necessary for repairing the ruptures that patriarchy and all forms of hierarchy create" [48] (p. 90). Within this process, the "sacrifice of love" enables various social divisions and restrains our opposition in the sense that we become less able and/or willing to recognize ourselves as relational, vulnerable, and interdependent beings, and, as such, stop responding to others in light of that relational fact and the corresponding need for care that can arise [48] (p. 33).

More broadly, and following Gilligan and Snider, we could say that the denial of a relational social ontology premised upon shared vulnerability and interdependency becomes a constitutive function for forms of social, political, and economic hierarchy. This involves a mutually reinforcing relationship between social, material, and political structures premised upon an individualist social ontology, on the one hand, and a particular psychological reaction to vulnerability on the other hand. This psychological reaction adopts an atomistic view of society as a requirement for participating in contemporary social relations as organized around a series of hierarchies and divisions. In other words, once these hierarchies are established, individual participation in social relations requires a form of subjectivity which pre-emptively suppresses "our relational desires and capacities" [48] (p. 12). Gilligan and Snider thus identify a key "paradox: we give up relationship in order to have "relationships," meaning a place within the patriarchal order" [48] (p. 14) (see also (p. 18)). Snider, in reflecting on her reaction to the death of her father puts this well when she writes:

> Could it be, I began to wonder, that my psychological response to death—the sacrifice of love to avoid further loss—highlights the psychological dynamic of a culture which forces the sacrifice of connection for the sake of hierarchy? [48] (p. 31)

In light of feminist readings of Hegel, which see the subject as engaged in a dialectical relationship with other subjects and the broader cultural/natural context with which they are co-constitutive, we could say that the picture of the self that appears out of Gilligan and Snider's [48] study is one that experiences a relation of "non-identity" [19] between their subjective desire and capacity for relation, on the one hand, and an objective world which is premised instead on norms more appropriate to an individualist ontology; to use other Hegelian language, we could perhaps say that there is an apparent conflict between the "subjective will" of individual subjects and the social, political and economic institutions in which they find themselves, and that the "subjective will" of individuals and socio-political institutions, as the "objective" expression of "human will" and reason, is thus in need of some sort of reconciliation [13] (pp. 40–42). However, Gilligan and Snider do identify a source, or strategy, for an attempted reconciliation between our intuitive sense of self as vulnerable and relational beings and the structures which constitute the broader social world; they call this "healthy resistance." As Gilligan and Snider write:

> Put simply, our relational desires and capacities, which are present in rudimentary form from the very outset, keep opening a potential for love and democracy while the politics of patriarchy keep shutting it down. Knowing this, the solutions become twofold: one, joining the healthy resistance and by doing so encouraging the psychology that would transform the political; and two, naming and taking on the cultural and political forces that, by subverting the capacity to repair, drive a healthy resistance into despair and detachment, thus paving the way to oppression and injustice. [48] (pp. 120–121)

In this sense, Gilligan and Snider [48] are suggesting that subjects within the "culture of patriarchy" can be the source of *negativity*, engaged in a process whereby the individual self is transformed alongside the broader social and political environment in a dialectical relationship. This source of negativity, however, depends on the human subject's embodied desire for relationship and on a recognition of our inherent vulnerability and dependency on others. Hoppania and Vaittinen discuss a similar dynamic when it comes to the notion of *care*, in particular, when they write that: "when care is not adequately accounted for in some sphere of life, it starts to challenge that very sphere" [50] (p. 87). In both instances, there is a moment of contradiction, or "non-identity" to use Hutchings's [19] language, which then, through a process of negation, can lead to a form of transformation that seeks to reconcile the self-understanding of subjects—as ontologically vulnerable and relational—and what comes to be seen as an external (social) world that is structured around a denial of this fundamental reality. In a sense, *vulnerability*, and its denial by dominant cultural norms, is what can render those norms themselves as vulnerable [50] (pp. 87–88) and this very concrete example echoes the foregrounding of death in Hutchings's [20] reading of Hegel.

This continues a line of thought that seeks to make the Žižekian-Hegelian notion of the "negativity" of the subject a more embodied feature of human existence and was aided by the reflection on Hegelian themes in the work of Butler and Irigaray. With this present section, we see that via a mobilization of concepts such as *care, vulnerability,* and *relationality,* we can further render this notion of "negativity" as an embodied and relational "negativity" which can be implicated in how the relational ontology of the ethics of care can be an effective means to both understand contemporary social relations, but also constitute a contextualist moral and political theory by which to critique and potentially transform those relations—these concepts are both personal and political. The notion of relationality thus suggests a connection between the levels of philosophical reflection on the ontological and epistemological features of human life and also provides normative guidance for how to engage in relationship—both personal and political—considering our self-understanding as vulnerable and relational beings in need of care in particular and fluid contexts.

5. Conclusions

In this article, I have sought to uncover connections between Slavoj Žižek's influential reading of G.W.F Hegel and the feminist ethics of care. This adds to previous efforts to put thematic elements of care ethics in conversation with Žižek's philosophy [53]. I argue that this confrontation between care ethics and more 'mainstream' philosophy is valuable because it offers mutual contributions to both care ethics as a moral and political theory and the philosophy of Hegel and Žižek. For care ethics, a reading of Hegel through Žižek on the question of subjectivity contributes further resources to thinking about the nature and implications of vulnerability and interdependency in a way that does not require positing a 'thick' or essentialist description of either individual subjects or their relationships; on the other hand, uncovering some of the points of compatibility between Žižek's reading of Hegel, explicitly feminist mobilizations of Hegel, and care ethics, can contribute to Žižek's Hegelian reflections on the nature of the contemporary self and ideology within the social and political context of global capitalism in ways that better highlight the related themes of gender, embodiment, and relationality as potential examples and sources of the Hegelian negative reimagined according to a broadened understanding of vulnerability as a philosophical concept [2]. This was accomplished by first unpacking the theory of the self that is explored in Žižek's reading of Hegel, second, through a presentation of how Hegel has been interpreted and mobilized in critical feminist theory, and third, by demonstrating how the picture of the self and its relation to the world as presented by both Žižek and feminist thinkers can also be put into a philosophically interesting dialogue with the themes of vulnerability and relationality as presented by some contemporary theorists of care.

Funding: This research received no external funding.

Institutional Review Board Statement: Not applicable.

Informed Consent Statement: Not applicable.

Data Availability Statement: Not applicable.

Acknowledgments: I would like to thank Maggie FitzGerald and Maurice Hamington for their guidance and helpful suggestions.

Conflicts of Interest: The author declares no conflict of interest.

Notes:

1. The in-text citations for Hegel's *Phenomenology* in this article refer to the paragraph numbering used in the referenced version of the text rather than the text's page numbers.
2. The question of whether Žižek himself would agree with this reimagining of the negative as a form of vulnerability is of course a subject for further discussion.

References

1. Robinson, F. Resisting Hierarchies through Relationality in the Ethics of Care. *Int. J. Care Caring* **2020**, *4*, 11–23. [CrossRef]
2. Gilligan, C. *A Different Voice: Psychological Theory and Women's Development*; Harvard University Press: Cambridge, MA, USA, 1993.
3. Held, V. *The Ethics of Care: Personal, Political, and Global*; Oxford University Press: New York, NY, USA, 2006.
4. Tronto, J.C. *Moral Boundaries: A Political Argument for an Ethic of Care*; Routledge: New York, NY, USA, 1993.
5. Hegel, G.W.F. *Phenomenology of Spirit*; Miller, A.V., Translator; Oxford University Press: New York, NY, USA, 1977.
6. Taylor, C. Atomism. In *Philosophy and the Human Sciences: Philosophical Papers 2*; Cambridge University Press: New York, NY, USA, 1985; pp. 187–210.
7. Hekman, S. The Embodiment of the Subject: Feminism and the Communitarian Critique of Liberalism. *J. Politics* **1992**, *54*, 1098–1119. [CrossRef]
8. Taylor, C. The Politics of Recognition. In *Multiculturalism: Examining the Politics of Recognition*; Gutmann, A., Ed.; Princeton University Press: Princeton, NJ, USA, 1994; pp. 25–74.
9. Oliver, K. Witnessing, Recognition, and Response Ethics. *Philos. Rhetor.* **2015**, *48*, 473–493. [CrossRef]
10. Lacan, J. *On the Names-of-the-Father*; Fink, B., Translator; Polity Press: Malden, MA, USA, 2013.
11. Žižek, S. *Event*; Penguin Books: London, UK, 2014.
12. Žižek, S. *Less than Nothing: Hegel and the Shadow of Dialectical Materialism*; Verso: New York, NY, USA, 2012.
13. Hegel, G.W.F. *Introduction to the Philosophy of History*; Rauch, L., Translator; Hackett Publishing Company: Indianapolis, IN, USA, 1988.
14. Kojève, A. The Idea of Death in the Philosophy of Hegel. *Interpretation* **1973**, *3*, 114–156.
15. Oliver, K. Subjectivity as Responsivity: The Ethical Implications of Dependency. In *The Subject of Care: Feminist Perspectives on Dependency*; Kittay, E.F., Feder, E.K., Eds.; Rowman and Littlefield Publishers: Lanham, MD, USA, 2002; pp. 322–333.
16. Žižek, S. *For They Know Not What They Do: Enjoyment as a Political Factor*; Verso: New York, NY, USA, 1991.
17. McGowan, T. The Insubstantiality of Substance, or, Why We Should Read Hegel's *Philosophy of Nature*. *Int. J. Žižek Stud.* **2014**, *8*, 1–19.
18. Strauss, L. *On Tyranny*; Gourevitch, V., Roth, M.S., Eds.; University of Chicago Press: Chicago, IL, USA, 2013.
19. Hutchings, K. *Hegel and Feminist Philosophy*; Polity Press: Malden, MA, USA, 2003.
20. Hutchings, K. Knowing Thyself: Hegel, Feminism and an Ethics of Heteronomy. In *Hegel's Philosophy and Feminist Thought: Beyond Antigone?* Hutchings, K., Pulkkinen, T., Eds.; Palgrave Macmillan: New York, NY, USA, 2010; pp. 87–107.
21. Pulkkinen, T. Differing Spirits: Reflections on Hegelian Inspiration in Feminist Theory. In *Hegel's Philosophy and Feminist Thought: Beyond Antigone?* Hutchings, K., Pulkkinen, T., Eds.; Palgrave Macmillan: New York, NY, USA, 2010; pp. 19–37.
22. Butler, J. *The Psychic Life of Power: Theories in Subjection*; Stanford University Press: Stanford, CA, USA, 1997.
23. Butler, J. *Subjects of Desire: Hegelian Reflections in Twentieth-Century France*; Columbia University Press: New York, NY, USA, 1999.
24. Butler, J. Restaging the Universal: Hegemony and the Limits of Formalism. In *Contingency, Hegemony, Universality: Contemporary Dialogues on the Left*; Butler, J., Laclau, E., Žižek, S., Eds.; Verso: New York, NY, USA, 2000; pp. 11–43.
25. Butler, J. Competing Universalities. In *Contingency, Hegemony, Universality: Contemporary Dialogues on the Left*; Butler, J., Laclau, E., Žižek, S., Eds.; Verso: New York, NY, USA, 2000; pp. 136–181.
26. Irigaray, L. *An Ethics of Sexual Difference*; Burke, C.; Gill, G.C., Translators; Cornell University Press: Ithaca, NY, USA, 1993.
27. Irigaray, L. *I Love to You: Sketch of a Possible Felicity in History*; Martin, A., Translator; Routledge: New York, NY, USA, 1996.
28. Hutchings, K.; Pulkkinen, T. Introduction: Reading Hegel. In *Hegel's Philosophy and Feminist Thought: Beyond Antigone?* Hutchings, K., Pulkkinen, T., Eds.; Palgrave Macmillan: New York, NY, USA, 2010; pp. 1–15.
29. Changfoot, N. Feminist Standpoint Theory, Hegel and the Dialectical Self: Shifting the Foundations. *Philos. Soc. Crit.* **2004**, *30*, 477–502. [CrossRef]
30. Stone, A. Going beyond Oppositional Thinking? The Possibility of a Hegelian Feminist Philosophy. *Res. Publica* **2004**, *10*, 301–310. [CrossRef]
31. Hutchings, K. A Place of Greater Safety? Securing Judgment in International Ethics. In *The Vulnerable Subject: Beyond Rationalism in International Relations*; Beattie, A.R., Schick, K., Eds.; Palgrave Macmillan: Houndmills, UK, 2013; pp. 25–42.
32. Haraway, D. Situated Knowledges: The Science Question in Feminism and the Privilege of Partial Perspective. *Fem. Stud.* **1988**, *14*, 575–599. [CrossRef]

33. Hartsock, N.C.M. The Feminist Standpoint: Developing the Ground for a Specifically Feminist Historical Materialism. In *Discovering Reality*; Harding, S., Hintikka, M.B., Eds.; Reidel/Kluwer: Dordrecht, NL, USA, 1983; pp. 283–310.
34. Hekman, S. Truth and Method: Feminist Standpoint Theory Revisited. *Signs* **1997**, *22*, 341–365. [CrossRef]
35. Schick, K. Re-cognizing Recognition: Gillian Rose's "Radical Hegel" and Vulnerable Recognition. *Telos* **2015**, *173*, 87–105. [CrossRef]
36. Janicka, I. Hegel on a Carrousel: Universality and the Politics of Translation in the Work of Judith Butler. *Paragraph* **2013**, *36*, 361–375. [CrossRef]
37. Stark, H. Judith Butler's Post-Hegelian Ethics and the Problem with Recognition. *Fem. Theory* **2014**, *15*, 89–100. [CrossRef]
38. De Beauvoir, S. *The Second Sex*; Borde, C.; Malovany-Chevallier, S., Translators; Vintage Books: New York, NY, USA, 2011.
39. Chanter, T. *Ethics of Eros: Irigaray's Rewriting of the Philosophers*; Routledge: New York, NY, USA, 1995.
40. Winnubst, S. Exceeding Hegel and Lacan: Different Fields of Pleasure within Foucault and Irigaray. *Hypatia* **1999**, *14*, 13–37. [CrossRef]
41. Benhabib, S. On Hegel, Women, and Irony. In *Feminist Interpretations of Hegel*; Mills, P.J., Ed.; The Pennsylvania State University Press: University Park, PA, USA, 1996; pp. 25–43.
42. Werner, L. "That which is Different from Difference is Identity"—Hegel on Gender. *Nord. J. Women's Stud.* **2006**, *14*, 183–194. [CrossRef]
43. Perpich, D. Universality, Singularity, and Sexual Difference: Reflections on Political Community. *Philos. Soc. Crit.* **2005**, *31*, 445–460. [CrossRef]
44. Hegel, G.W.F. *Elements of the Philosophy of Right*; Nisbet, H.B., Translator; Wood, A.W., Ed.; Cambridge University Press: New York, NY, USA, 1991.
45. Jones, E.R. The Future of Sexuate Difference: Irigaray, Heidegger, Ontology, and Ethics. *L'Esprit Créateur* **2012**, *52*, 26–39. [CrossRef]
46. Zalloua, Z. Žižek with French Feminism: Enjoyment and the Feminine Logic of the "Not-All". *Intertexts* **2014**, *18*, 109–130. [CrossRef]
47. Robinson, F. *Globalizing Care: Ethics, Feminist Theory, and International Relations*; Westview Press: Boulder, CO, USA, 1999.
48. Gilligan, C.; Snider, N. *Why Does Patriarchy Persist?* Polity Press: Medford, MA, USA, 2018.
49. Petherbridge, D. What's Critical about Vulnerability? Rethinking Interdependence, Recognition, and Power. *Hypatia* **2016**, *31*, 589–604. [CrossRef]
50. Hoppania, H.-K.; Vaittinen, T. A Household Full of Bodies: Neoliberalism, Care and "the Political". *Glob. Soc.* **2015**, *29*, 70–88. [CrossRef]
51. Laugier, S. The Ethics of Care as a Politics of the Ordinary. *New Lit. Hist.* **2015**, *46*, 217–240. [CrossRef]
52. Ferrarese, E. Vulnerability: A Concept with Which to Undo the World As It Is. *Crit. Horiz.* **2016**, *17*, 149–159. [CrossRef]
53. Ghandeharian, S.; FitzGerald, M. COVID-19, the Trauma of the 'Real' and the Political Import of Vulnerability. *Int. J. Care Caring* **2022**, *6*, 33–47. [CrossRef]

 *philosophies*

Article

Care Ethics and the Feminist Personalism of Edith Stein

Petr Urban

Institute of Philosophy, Czech Academy of Sciences, Jilska 1, 110 00 Prague, Czech Republic; urban@flu.cas.cz

Abstract: The personalist ethics of Edith Stein and her feminist thought are intrinsically interrelated. This unique connection constitutes perhaps the main novelty of Stein's ethical thought that makes her a forerunner of some recent developments in feminist ethics, particularly ethics of care. A few scholars have noticed the resemblance between Stein's feminist personalism and care ethics, yet none of them have properly explored it. This paper offers an in-depth discussion of the overlaps and differences between Stein's ethical insights and the core ideas of care ethics. It argues that both Stein and care ethicists relocate a certain set of practices, values and attitudes from the periphery to the center of ethical reflection. This includes relationality, emotionality and care. The paper finally argues that it is plausible and fruitful to read Stein's advocacy of 'woman's values and attitudes' in a critical feminist way, rather than as an instance of essentialist difference feminism.

Keywords: Edith Stein; care ethics; personalism; feminism; empathy; emotions; caring; phenomenology

Citation: Urban, P. Care Ethics and the Feminist Personalism of Edith Stein. *Philosophies* **2022**, *7*, 60. https://doi.org/10.3390/philosophies7030060

Academic Editors: Maurice Hamington and Maggie FitzGerald

Received: 14 April 2022
Accepted: 28 May 2022
Published: 1 June 2022

Publisher's Note: MDPI stays neutral with regard to jurisdictional claims in published maps and institutional affiliations.

1. Introduction

Since its first formulation in the early 1980s, care ethics has developed into a burgeoning field of ethical inquiry that has spread worldwide as a viable alternative to the mainstream currents in moral and political philosophy. Care ethicists are rightly credited with refocusing ethical reflections on the relational nature of the human condition and revaluing care as a fundamental human practice that was historically marginalized and devalued as a matter of private life and family relationships. It has been widely acknowledged, though, that some seminal care ethical insights draw on the ideas of care ethics' predecessors or at least have striking parallels in the history of philosophy. A few scholars have noticed remarkable similarities between care ethics and the ideas of the German philosopher Edith Stein, who developed her phenomenological anthropology and social philosophy between the 1910s and the late 1930s [1–5]. Yet, most of the previous contributions to the question of the relationship between Stein's philosophy and care ethics are rather cursory or suffer from a disproportionate emphasis on one of the sides of the relationship, approaching it either from the perspective of Stein's thought or the one of care ethics. In this paper, I provide a more balanced and thorough account of this relationship. I start with a presentation of care ethics focusing mainly on its development in the work of Sara Ruddick, Carol Gilligan and Joan Tronto. In the next section, I explore the core ideas of Stein's phenomenological personalism and her feminist thought. Finally, I conclude by reflecting on the overlaps and differences between the two ethical approaches. This reflection can help us to arrive at a more adequate and relational understanding of the place of care ethics within the diverse landscape of traditional moral and political philosophy. It also provides an impulse to a more vivid dialogue between care ethicists and the current proponents of personalist ethics who often take Stein's philosophical work as a source of inspiration.

2. Care Ethics

In most general terms, care ethics can be described as an approach to ethics that foregrounds caring as a core human practice and conceives of the goals and values of this practice as fundamental for achieving good life at both individual and collective levels.

The idea of rehabilitating care as a critical human practice and value arose from the historical, intellectual and socio-political context of the 1970s in North America and Europe—in particular, from second-wave feminism and its critique of the dominant 'hegemonic masculinity' manifested in the image of the human person as an autonomous independent individual. In the early 1980s, the shift to the description and revaluation of care and caring relationality was reinforced by numerous works across many academic disciplines, such as epistemology [6], sociology [7–9], social policy [10–12], political economy [13], philosophy of education [14], social philosophy [15] and developmental psychology [16]. It was the field of developmental psychology where Carol Gilligan coined the term 'an ethic of care'. Yet, two years before the term 'an ethic of care' would appear in Gilligan's widely known book *In a Different Voice*, an American philosopher Sara Ruddick published an essay "Maternal Thinking" (1980) [17], in which she put forth several key ideas that have proven of central importance for the subsequent forty years of the development of care ethics. I start this section with a discussion of Ruddick's 1980 paper, followed by a brief introduction to the formation of care ethics in the 1980s and the early 1990s.

2.1. Revaluing Care as a Core Human Practice

Ruddick builds her reflections on the observation that our "working and caring with others" and the corresponding way of thinking plays a crucial role in human life. Her main point is that, although caring practices form the core of human existence, they have historically been marginalized, devalued and portrayed in a sentimental and romantic way. Ruddick seeks to provide an adequate philosophical description of this practice, point out its distinctiveness and explain its value as an important source of an alternative moral, social and political theory.

Though Ruddick links her central concept of 'maternal practice' primarily to the activity of taking care of and raising a child, she concedes that maternal thinking expresses itself "in various kinds of working and caring with others" [17] (p. 346). Maternal practice that gives rise to maternal thinking, Ruddick argues, is a response to three basic interests or demands of a child, namely for preservation, growth and acceptability. Ruddick defines 'maternal thinking' as a distinctive style of reflecting, judging and feeling that is guided by distinctive goals and interests of 'maternal practice'.

Ruddick distinguishes between degenerative and non-degenerative forms of maternal practice. The actor of the non-generative form of maternal practice would typically feature attention, love, humility, understanding, respect for the other, sense of complexity, the capacity to change (alongside with or in response to the changing reality), explore, create and insist upon one's own values and the ability to see and name existing forms of oppression and domination. In contrast, the actor of the degenerative form of maternal practice is characterized by rigid and excessive control over the other, self-refusal and uncritical acceptance of the values of the dominant culture or obedience—a sense of wanting to 'be good' in the 'eyes' of the dominant culture and society [17] (p. 354-55.).

For Ruddick "'maternal' is a social category" [17] (p. 346), which entails that her account focuses on the practice itself and by "concentrating on what mothers do" rather than on what they are suspends any question about the 'essence' of this practice. Ruddick rejects "the ideology of womanhood" and argues that it was invented by men and caused the oppression of women [17] (p. 345). Moreover, any identification of maternal practice with biological or adoptive motherhood is false, Ruddick argues, since it "obscures the many kinds of mothering performed by those who do not parent particular children in families" [17] (p. 363). Together with 'the ideology of womanhood', Ruddick rejects "all accounts of gender difference or maternal nature which would claim an essential and ineradicable difference between female and male parents" [17] (p. 346).[1] In sum, Ruddick describes maternal practice as a fundamental human practice that has been historically associated with women (and other marginalized groups), but in fact has no essential relation to any sex or gender identity.[2]

Ruddick finally borrows the notion of 'feminist consciousness' from Sandra Bartky [20] and concludes her essay by envisioning 'maternal thought transformed by feminist consciousness'. It is a task of 'feminist consciousness' to critique the current economic, social and political structures that perpetuate the marginalization and devaluation of the practice of 'working and caring with others' and that foster the dominant association of this practice with women and other oppressed groups. When shaped by 'feminist consciousness', maternal thinking reveals "the damaging effects of the prevailing sexual arrangements and social hierarchies on maternal lives" [17] (p. 356) and raises a voice "affirming its own criteria of acceptability, insisting that the dominant values are unacceptable and need not to be accepted" [17] (p. 357). In order to create a society based on the values and rationality of this practice, Ruddick argues we must "work to bring transformed maternal thought into the public realm" and to make it "a work of public conscience and legislation" [17] (p. 361). This would require, on the practical level, a transformation of politics and "moral reforms of economic life" [17] (p. 360) and, on the theoretical level, "articulating a theory of justice shaped by and incorporating maternal thinking" [17] (p. 361).

To summarize, the four key elements that are present in Ruddick's early essay and that prefigure what will constitute the core of the subsequent development of moral and political theory of care are: (1) The focus on caring as a human practice that, though fundamental to the human condition, was historically marginalized, devalued and kept outside the scope of the dominant Western moral, social and political thought; (2) the aim to provide an adequate analysis of this practice, which would replace the widespread sentimentalizing and romanticizing distortions that go often hand-in-hand with the sociocultural and political devaluating of the practice and the focus on the practice itself, which entails a rejection of its naturalistic and essentialist accounts;[3] (3) the emphasis on the transformative potential of such an analysis, which inspires a critique of the social, economic and political structures that hinder realization of the nondegenerative forms of the practice and (4) the insight that the relational values and ideals inherent in caring practice are connected with the values and ideals of justice and that promoting both requires a transformation of our social and political institutions.

2.2. Identifying a Different Voice in Ethics

As described above, the notion of an 'ethic of care' was coined by the American developmental psychologist Carol Gilligan in her widely acclaimed book *In a Different Voice* [16]. Gilligan famously characterizes an ethic of care as a distinctive style of moral judging and way of constructing moral problems which centers around the responsibility for human relationships, builds moral judgment on concrete knowledge of a particular situation and context, emphasizes the priority of connection and starts from the insight that there is no contradiction in acting responsibly towards oneself and others. As Gilligan puts it, "the ideal of care is thus an activity of relationship, of seeing and responding to need, taking care of the world by sustaining the web of connection that no one is left alone" [16] (p. 62).

It is noteworthy that Gilligan, in contrast to Ruddick, formulates the idea of an ethic of care within a fundamentally dualistic framework. In her view, an ethic of care is a 'different voice', which differs from the voice of an ethic of justice (or rights). In contrast to an ethic of care, an ethic of justice emphasizes the priority of the individual, derives moral judgement from formal and abstract rules, foregrounds the ideal of equality and impartiality and considers the struggle for individual rights as the fundamental dynamics of social relations. Despite the numerous harsh contrasts in her exposition of an ethic of care and an ethic of justice, Gilligan ultimately contends that the "two views of morality . . . are complementary rather than sequential or opposed" [16] (p. 33) and that "to understand how the tension between responsibilities and rights sustains the dialectic of human development is to see the integrity of two disparate modes of experiences that are in the end connected" [16] (p. 174). Yet, perhaps due to the fact that she expresses this view with restraint, or due to her failure to provide an account of how the two views of morality should be connected in the real life of individuals and communities, many of the critics as well as admirers of Gilligan's work have one-sidedly focused on the opposition of the "two different constructions of the moral domain" [16] (p. 69).[4]

Another notable duality that marks Gilligan's initial presentation of an ethic of care is the duality of the female and male 'voices', the female and male ways of telling the story of what it means to be oneself, to be an adult human being. The author of *In a Different Voice* contends that the male voice typically speaks "of the role of separation as it defines and empowers the self", whereas the female voice typically speaks "of the ongoing process of attachment that creates and sustains the human community" [16] (p. 156). Gilligan conceives of the dual way of defining the self and its relationships to other selves and the world as rooted in the difference between the psychology of men and "the psychology of women that has constantly been described as distinctive" [16] (p. 22). Yet, on the opening pages of her book, Gilligan assures her reader that an ethic of care "is characterized not by gender but theme". She maintains that "the contrasts between male and female voices are presented here to highlight a distinction between two modes of thought and to focus a problem of interpretation rather than to represent a generalization about either sex" [16] (p. 2). I agree with Tronto's remark that "the equation of Gilligan's work with women's morality is a cultural phenomenon, and not of Gilligan's making" [21] (p. 646).[5] However, I think that the conceptual ambiguity of Gilligan's early work opened the way for the formation of this cultural phenomenon, as well as the related misunderstanding as regards the nature of care ethics.

2.3. Care as a Political Concept

The American moral and political philosopher Joan Tronto was among the first care theorists who clearly showed that confusing care ethics with private life-oriented women's morality not only leads to an easy dismissal of the feminist 'different voice' in the context of dominant moral and political theories but also jeopardizes care ethics' feminists goals and may result in harmful consequences for women, such as sidestepping structural problems of domination, exploitation, oppression and marginalization (cf. [21]). To address this issue, Fisher and Tronto [23] took up the task of constructing a full moral and political theory of care by offering a broader definition and analysis of caring that enables the inclusion of the whole range of human activities and allows for taking into account the political dimensions of power and conflict entailed in all caring activities.[6]

Fisher and Tronto famously define caring as "a species activity that includes everything we do to maintain, continue, and repair our 'world' so that we can live in it as well as possible. That world includes our bodies, our selves, and our environment, all of which we seek to interweave in a complex, life-sustaining web" [23] (p. 40). This definition, which emphasizes the processual dimension of care and implies that the caring process may be directed not only toward people but also other living being and things, has been widely influential in further development of a moral and political theory of care and has served as a starting point for numerous applications of a care ethical perspective (which I discuss later on). The same holds true for Fisher and Tronto's related distinction and analysis of four intertwining phases or components of the caring process: (1) caring about—paying attention to something with a focus on continuity, maintenance and repair; (2) taking care of—taking responsibility for activities responding to the facts noticed in caring about; (3) care giving—the concrete tasks and the hands-on care work and (4) care-receiving—the responses of those toward whom caring is directed [23] (p. 40).

In a way similar to Ruddick's reflection on degenerative forms of 'maternal practice', Fisher and Tronto describe ineffective and destructive patterns in caring activities. They think of them as characterized by fragmentation and alienation in the caring process, as opposed to the integrity of caring where the four phases of the care process fit together into a whole. Such ineffective patterns in caring occur, for example, when caregivers suffer a shortage of time and/or other resources necessary for caring or when care-receivers have little control over how their needs are defined in the caring process. Against the background of the insight that how we think about care is deeply affected by existing social and political structures of power and inequality, Fisher and Tronto conclude that the patterns of fragmentation and imbalance of the caring process are mainly created by deficient social and political arrangements. Hence, a full-fledged moral theory of care needs to be developed hand in hand with a political theory of care that scrutinizes the workings of our social and political institutions (e.g., the household, the market and the state) from a critical perspective inspired by the ideal of good caring.

While an ethic of care envisions "a different world, one where the daily caring of people for each other is a valued premise of human existence, ... an alternative vison of life, one centred on human care and interdependence" [27] (p. x), a political theory of care reveals that "what this vision requires is that individuals and groups be frankly assessed in terms of the extent to which they are permitted to be care demanders and required to be care providers" [27] (p. 168). In her path-breaking book *Moral Boundaries* [27], Tronto lays ground for a full-fledged political theory of care that aims to explicate what "a just distribution of caring tasks and benefits" [27] (p. 169) entails and which social and political arrangements facilitate caring and contribute to creating "a more just world that embodies good caring" [27] (p. xii). A political theory of care sheds light on the close relationship between care and justice. On the one hand, to address the problems of care and to conceptualize the prerequisites of good caring requires concepts of justice, equality and democracy, since caring is always deeply affected by unequal power and access to material conditions and recourses necessary for caring. Thus, Tronto argues, "only in a just, pluralistic, democratic society can care flourish" [27] (p. 162). On the other hand, "care as a practice can inform the practices of democratic citizenship" [27] (p. 177), since it describes "the qualities necessary for democratic citizens to live together well in a pluralistic society" [27] (p. 161–62). Refection on the mutually enabling relationship, foregrounded by Tronto [27], between good caring and democratic citizenship in a just society is a thread that connects most subsequent developments in a political theory of care.

The exploration of a close relationship between caring, democracy, citizenship and equality inspired Tronto's more recent reflection on the practice of 'caring with' as constitutive for a 'caring democracy' [28]. To be a citizen in a democracy means, Tronto argues, "to care for citizens and to care for democracy itself" [28] (p. x). This requires that citizens take seriously the collective responsibility for 'caring with' each other and that democratic politics recognizes the centrality of "assigning responsibilities for care, and for ensuring that democratic citizens are as capable as possible of participating in this assignment of responsibilities" [28] (p. 30). Tronto expands the original distinction of the four phases of caring [23] by adding 'caring with' as the final fifth phase of the care process and identifying plurality, communication, trust, respect and solidarity as the key moral qualities that 'caring with' requires [28] (p. 35–36).

3. Edith Stein's Feminist Personalism

Edith Stein (1891–1942), a patron saint of Europe, was a German philosopher and religious thinker. She was a pupil and follower of the founder of phenomenological philosophy, Edmund Husserl. Stein was born in a German Jewish family, but she later converted to Catholicism and became a Carmelite nun. Nazis murdered her in Auschwitz in 1942. Stein has left an extensive philosophical and theological corpus of work. The following exposition of her thought focuses in particular on her personalist and feminist views in relation to ethics.

3.1. Stein's Personalist Ethics

In spite of the fact that Stein has never wrote any systematic work on ethics per se, it is plausible to argue that her entire philosophical corpus "implicitly entails a consciously developing ethical vision entering into conversation with ethical philosophy's major representatives" [29] (p. 73–74) and "ethical concern is deeply and thematically woven into the fabric of her studies in anthropology, community, and political existence" [29] (p. 86).

In the phenomenological phase of her philosophical work, from the late 1910s to the early 1930s, Stein's ethical views draw heavily on the personalist ethics developed by Edmund Husserl and Max Scheler.[7] Yet, as we will see in a moment, Stein enriches the personalist perspective of her phenomenological companions by a unique feminist tweak. Stein's latest thought shifts towards the Aristotelian-Thomist tradition, attempting to link the Christian perspective with the phenomenological position of her earlier works (see, e.g., [30]). For the purpose of this paper, I want to narrow my focus down to the phenomenological phase of Stein's ethical thought, which offers enough material for a comparison with care ethics.

Following the methods of her teacher, Husserl, Stein devotes a great deal of her philosophical project to answering the questions of what it means to be a self and how the self relates to the world. The question of how the world can be given to us as an objective world appears for Stein, as well as for Husserl, as inseparably connected with the question of how we can know and understand others as subjects who relate to the same shared world as we do. Stein conceives of the act of understanding or knowing the other subject as an act of empathy (Einfühlung) and characterizes it as a unique type of perception that differs from all other forms of perceiving. In empathy, Stein argues, I grasp the other person as a person who has her own perspective and experiences, whereas, in nonempathic perceptions, I perceive things and external objects in the world. Stein notes that there is a fundamental difference between the way in which I grasp my own inner life and the way in which I grasp the other's inner life: the content of empathy is never fully present for the empathizer, as long as it belongs to someone else, to a different person. Empathy is an other-oriented type of consciousness; the aspect of otherness is constitutive for empathy. However, this self-other distinction in empathy also entails that the empathizer is connected with her own experiences too. As Hamington puts it, for Stein, "empathy does not negate the self but actually strengthens self-concept" [31] (p. 80). Hence, in Stein's view, empathy plays an important double role in that it both constitutes the other self for me, and it constitutes my own self as different from the other.

Stein's investigations in the constitution of the human person led her to her study of the human person's relations and intersubjective links, which resulted in a fundamentally relational view of the human person. As Fuentes stresses, "it is through empathy that the individual, human person (psychophysical individual), becomes constructed as such—and I cannot be formed without a you—the possibility of knowing oneself in the other and knowing the other is inseparable—one cannot be oneself, build oneself as a self, or form one's own identity, without the reference of the other" [32] (p. 206). Human persons are referred from themselves to the other in order to be what they are and to become what they can be.

This deeply relational perspective on the human person becomes manifest in Stein's ethical reflections that start from the notion of the person. As she argues already in her dissertation, "one's own moral life and moral character is constituted alongside the moral encounter with the other and in one's own response to her moral character" [29] (p. 81). In the ethics derived from the relational perspective, "the other, the good of the other, is not only something tolerable or acceptable, but it is indispensable for the same comprehension and realization of one's own good—one's own good cannot be carried out without the other's own good and vice versa" [32] (p. 206).

Stein shares and further develops the view of other early phenomenologists that it is through emotions that a person grasps "the meaning of another being in relation to its own being, and then the significance of the inherent value of exterior things, of other persons, and impersonal things" [33] (p. 96). Emotions are the "essential organ for comprehension of the existent in its totality and its peculiarity" [33] (p. 96), and through emotions, we open ourselves to the world of values that Stein takes to be present in the world of persons. It is important to stress that by 'emotions' Stein does not mean fluctuating states of sentiment, although emotions may include sentiment. Stein relates the primordial recognition of others to the emotions as a peculiar spiritual capacity, present both in self-knowledge and empathy (Haney 458). What we ought to be and do shows itself to us through the feelings we develop in encountering the experiences and actions of other persons [34] (p. 757). In line with the emotional value realism of Scheler, Stein claims that the structure of personal depth and periphery is mapped out in response to a range of values and that the person ought to be affected in the deepest way by the highest values [29] (p. 74). On the top of the hierarchy of values resides the absolute value of the human person: "the human person is more precious than all objective values" [33] (p. 256—cited in [35]).

To be responsive to the highest value, to the absolute value of the person as person requires, in Stein's view, love. In love, the person opens herself to the value of other persons, as well as to the value of one's own person. Thus, love, Stein argues, is vital for the individual and community alike. By contrast, hate is a vital disvalue both for the individual and community: "love operates within the one who loves as an invigorating force that might even develop more powers within him than experiencing it costs him. And hate depletes his powers far more severely as a content than as an experiencing of hate. Thus, love and positive attitudes in general don't feed upon themselves; rather, they are a font from which I can nourish others without impoverishing myself" [36] (p. 212). Attitudes such as love, trust and gratitude have the effect of 'enlivening' the person who receives them, inasmuch as there is a real community among persons who are evaluatively 'affirmed' in and through them. The opposite acts of "distrust, aversion, hatred—in short, the whole set of 'rejecting' manner of behavior" [36] (p. 211) are devitalizing, because in them the person is evaluatively negated [29] (p. 78).

However, even the 'personal attitude' and 'love', in Stein's view, can become devitalizing and destructive if it takes on excessive forms, such an excess of interest in the other person, the urge to lose oneself completely in the other. Stein states that, in a passion of wanting to confiscate the other [33] (p. 257), one does justice neither to one's self nor to the humanity of another [33] (p. 257): "The woman who hovers anxiously over her children as if they were her own possessions will try to bind them to her in every way . . . She will try to curtail their freedom of development; she will check their development and destroy their happiness" [33] (p. 75). In contrast, the true capacity to love is the capacity to 'go out to the other' without losing oneself.

Along this trajectory, Stein arrives at a normative ideal of community (Gemeinschaft as opposed to Gesellschaft) as "the union of purely free persons who are united with their innermost 'personal' life, or the life of the soul, and each of whom feels for himself or herself and for the community" [36] (p. 273). This ideal of community—of love freely given and received—is oriented around the consciousness of collective and individual responsibility for one another [29] (pp. 81–82). The authentic community orders persons towards "not separated living but common living, fed from common sources and stirred by common motives" [36] (p. 215). In her Investigation Concerning the State, Stein conceives of political community as a major stage upon which social-ethical responsibility is born [29] (p. 82). Since real communities and polities deviate—to a greater or lesser extent—from the ideal patterns of the forms of freedom, love and co-responsibility, such deviations must be navigated ethically and addressed through a never completed process of moral reform and renovation (Erneuerung). Stein contends that, although the process of morality's reform must originate in the souls of those who are capable of intuiting the right order of values, the state can be utilized as the specific 'tool' of social reformation by transforming the prevailing morality through legal regulation, as well as through the development of institutions that facilitate desirable forms of moral and social life [29] (p. 84).

3.2. Stein's Feminism

As described above, Stein's ethical personalism has a unique character mainly due to a remarkable feminist element that is increasingly present in the development of her thought.[8] The 'question of woman' is one the of the questions that occupy Stein throughout her life and writing. Already as a young university student, Stein was a "radical fighter for women's rights" [37] (p. 185); she advocated women's suffrage and engaged in vocational counselling for female students. In the 1930s, after she had given a series of public lectures and radio addresses on women's issues in Germany, Austria and Switzerland, Stein gained a reputation as an international spokesperson for the catholic women's movement and a leading figure of the educational reform. Stein's theoretical reflections on the 'question of woman' appeared in the volume *Essays on Woman* [33], which serves as the primary source for our present interpretation.[9]

In her 1928 lecture "The significance of woman's intrinsic value in national life", Stein describes the situation of the European women's movement in the 1920s as follows: "We women have become aware once again of our peculiarity. […] And this 'self-awareness' could also develop the conviction that an intrinsic value resides in the peculiarity" [33] (p. 254). Even if Stein approaches the idea of the revaluation of 'woman's peculiarity' with caution—indeed, she resists painting a shining ideal of feminine nature with the hope that a realization of this ideal will be the cure for all contemporary problems—she defends the view that "the purely developed feminine nature does include a sublime vital value" as well as "ethical value" [33] (p. 46). Thus, in her lectures on woman, Stein aims not only to provide an account of 'woman's distinctive personality' but also to reveal the quality and significance of the value that is, in her view, inherent in woman's peculiar style of being a person.

The phenomenological method, which Stein uses to reveal the sense of 'woman's peculiarity' [33] (p. 255), requires her to focus on the form and structure of the intentional life as it is lived through by the person. It is on this experiential, phenomenal level that she finds the core differences between man and woman. Stein obviously does not think of 'woman's peculiarity' in terms of exclusive traits and faculties. The personal traits in question are primarily human ones, and all faculties that are present in woman's personality are also present in man's personality. Nonetheless, Stein argues, the human traits may generally appear in different degrees and relationships in man and woman [3] (p. 72). When it comes to the question of equality between the sexes, an attentive reading of Stein's lectures reveals that she insists on genuine equality between men and women. Thus, Stein consistently affirms her commitment to a distinctive feminine personality without thereby undermining the equality of the sexes [3] (p. 67f.).

Let us take a closer look at Stein's views of the peculiarity of woman's intentional life. With woman, Stein believes there is a more intense and complete unity of the living body and soul, which includes that women are more capable of being affected by that which they encounter as concrete persons living in and through the body. Stein also claims that "the strength of the woman lies in the emotional life" [33] (p. 96).[10] Due to the centrality of "understanding of the things of value" [33] (p. 73), a woman seems also more capable of feeling a "joy in creatures", which makes her "sensitive and attentive to all that lives, grows and strives for development" [33] (p. 73). Stein characterizes women's prevalent attitude as 'personal', which means several things: "in one instance she is happily involved with her total person in what she does; then, she has particular interest in the living, concrete person, and, indeed, as much for her own personal life as for other persons and their personal affairs" [33] (p. 255). Finally, Stein argues, "in woman, there lives a natural drive towards wholeness and completeness. And, again, this drive has a twofold direction: she herself would like to become a complete human being, one who is fully developed in every way; and she would like to help others to become so, and by all means, she would like to do justice to the complete human being whenever she has to deal with human beings" [33] (p. 255).

Women's personal attitude and tendency to completeness go, in Stein's view, hand in hand with two major existential tasks: being a mother and being a companion. Stein claims that "the innermost formative principle of woman's soul is the love" [33] (p. 57) and "the deepest feminine yearning is to achieve a loving union which, in its development, validates her maturation and simultaneously stimulates and furthers the desire for perfection in others" [33] (p. 94). Stein sees an intimate link between woman's task of being a mother and her yearning to embrace that what is living, personal and whole to cherish, guard, protect, nourish and advance growth [33] (p. 45). Women's peculiar orientation toward the personal, the concrete and living and toward the full development of each being comes to a special, intense expression in her motherhood. A similar set of values becomes manifest in a woman's task of being a companion: "where a human being is alone, especially one in bodily or psychological need, she stands lovingly participating and understanding, advising and helping; she is the companion of life who helps so that 'man is not alone'" [39] (p. 50).

A critical feminist reader may object that this view of Stein "reads as if she is trying to rehabilitate the patriarchy [40] (p. 214). Yet, it is Stein's firm contention that patriarchal society in its many destructive manifestations is abnormal and morally unacceptable and that "only subjective delusion could deny that women are capable of practicing vocations other than that of spouse and mother" [33] (p. 49). It needs to be stressed that what Stein means by motherhood and companionship is by no means mere physical motherhood and marital companionship. For Stein, to be a mother is to nourish and protect what is alive and bring it to development, to be a companion is to provide support and be a mainstay [33] (p. 256). Hence, any woman, regardless of her actual state in life, can take up the tasks of companionship and motherhood. Stein also emphasizes the possibility of spiritual companionship and motherhood that "extend to all people with whom woman comes into contact" [33] (p. 132), and stresses that the motherhood she has in mind "must be that which does not remain within the narrow circle of blood relations or of personal friends" [33] (p. 264)

Stein's idea of motherhood also has a deeper ontological meaning that reflects her fundamentally relational view of the human person. In her mature work, *Finite and Eternal Being*, Stein meditates our existence as something that is constantly given to us moment to moment anew. She describes human persisting in being as 'ontological security'. As Calcagno rightly notes, "the image she employs to give resonance to this insight is the image of a child being held in the arms of her mother, certain and comfortable that no danger will come to him or her while sleeping". This image, Calcagno concludes, "also shows how Stein conceives of being not as a solitary enterprise of an ego or a Dasein, but as a communal enterprise, the living of one in the security of the other" [5] (p. 74).

Drawing on her philosophy of the human person and authentic community, Stein contends that woman's peculiar attitudes and values can and should help us in transforming social and moral life of our communities [33] (p. 262). For example, Stein explains potential transformation of health care profession by stressing that in a still increasing medical specialization we should not forget that often it is not only the organ but the entire person who is sick along with the organ. Women, in Stein's view, have insight into diverse human situations and get to see clearly material and moral needs of others [33] (pp. 262–263). Counteracting abstract medical procedures, a woman's attitude is oriented towards the concrete and whole person. Stein recommends the healthcare professional to exercise courage in following her intuition and to liberate herself whenever necessary from methods learned and practiced according to formal rules. Yet, the intent must be to understand correctly the whole human situation, and to intervene helpfully not only by medical means but also as a mother or a sister [33] (pp. 111–112).

Finally, Stein stresses the significance of woman's unique attitudes and values in political life. In legislation, she observes, there is always danger that a resolution will be based on elaboration of the most perfect paragraphs without consideration of actual needs in practical life. Women, Stein argues, are suited to act in accordance with the concrete human needs, and so they are able to serve as redress here [33] (pp. 263–264). Stein refers to a particular historical example when in the deliberation of youth laws there was the danger that the project would end in failure by party opposition. At that time, the women of the differing parties worked together and reached an agreement [33] (p. 264). Women's attitudes and values can also work beneficially in the application of the law, provided it does not lead to abstract validation of the letter of the law but to the accomplishment of justice for humanity. Stein eventually does not restrict her account to the level of individual states and nations but maintains that "there is a connection between success and adversity in both private and national life; just so are the individual nations and states connected one with the other. . . . woman's sphere of action has been extended from the home to the world" [33] (p. 154).

4. Conclusions: Overlaps and Differences

There are obvious overlaps, as well as differences, between care ethics and Stein's feminist personalism. Let us first focus on some overlaps between the two approaches.

First, both care ethicists and Stein start from a fundamentally relational view of human beings. Human existence is inevitably marked by interdependence. Human persons are referred from themselves to the other in order to be what they are and to become what they can be. Thus, ethical reflections on what does it mean to live a good life and what we ought to do—both at individual and collective level—should refocus on the ways in which we relate to each other and examine the social and political structures that frame our relationships.

Second, both approaches relocate certain sets of practices, values and attitudes from the periphery to the center of ethics. The dominant currents of Western ethical and political thought typically devalue care and love as matters of intimate relationships and biological ties that belong to the narrow sphere of private life and, thus, do not constitute a proper subject of ethics and politics. In contrast, Stein and care ethicists share the view that the historically marginalized practices and attitudes of care and love build the core of our human existence and are of ultimate importance for any normative moral and political theory that aims to adequately respond to the true nature and complexity of human life. Yet, in both approaches, a key normative task concerns distinguishing the 'empowering' and 'enlivening' patterns of care and love from the 'degenerative' ones.

Third, the refocusing of ethics on the practices, values and attitudes of care and love goes hand-in-hand in both care ethics and Stein's philosophy, with revaluing the experiences of the members of certain marginalized groups, typically women, who historically bore the burden of excessive caring responsibilities. Both approaches emphasize the need for a more just attribution of these responsibilities and see a transformative potential of the realization of the corresponding values and attitudes in the everyday life of our communities and polities.

Finally, in contrast to the emphasis on abstract moral reasoning and rule following, both care ethics and Stein promote "concrete thinking" (in Sara Ruddick's phrase) based on practical experience of situated persons. They offer a counterbalance to the perspective which focuses one-sidedly on the cognitive and rational dimension of what is to be a human, by stressing that we are essentially embodied and emotional beings to who affects and emotions say important things about what is of value and how the life can be made better.

Let us turn to the differences between Stein's feminist personalism and care ethics. The most obvious difference seems to lie in Stein's embracing emotional value realism and the idea of the absolute value of the human person. Scheler's idea that there is an objective hierarchy of values that can be grasped in correct or incorrect ways by human persons in specific acts of value-feeling (emotions) is part and parcel of Stein's ethical personalism. Most care ethicists, however, emphasize the context-related and situated nature of all moral knowledge, as well as the importance, of particularity and singularity in the practice of caring. This is not to say that there is no place for particularity and singularity in Stein's ethical thought. Stein, as we have seen, conceives of empathy and personal attitude as inherently linked to the capacity of understanding the meaning and value of the concrete and particular. Yet, her insistence on the existence of a universal hierarchy of values and the distinction between rightness and wrongness of value-intuitions clearly restricts her appreciation of the relevance of context and situation in moral knowledge.

Stein's ethical personalism, as we could see, revolves around the idea that the value of the human person is the most precious of all values. Only a few care ethicists would embrace this view. Although some (mainly early) formulations of care ethics start with the image of a person-to-person relationship which implies the ethical centrality of the human being and her relationships, most recent developments in care ethics show a broader understanding of caring as a process which "includes everything we do to maintain, continue, and repair our 'world' so that we can live in it as well as possible." That world includes not only human persons, but also our environment "all of which we seek to interweave in a complex, life-sustaining web" [23] (p. 40). Drawing on this broader concept of caring, care ethicists have laid ground for non-anthropocentric environmental ethics which is hard to incorporate in the personalist perspective of Stein.

Finally, an important and tricky question concerns the difference between Stein's philosophy and care ethics as regards the feminist dimension of the two approaches. In the first section, I argued that the effort to dissociate care ethics from the idea of women's morality growing from 'woman's nature' was of a great importance for further development of care ethics towards a full-blown moral and political theory of care. Now, Stein's feminist personalism depends on an account of the sexual difference which seems to rely on some essentialist presuppositions. It is precisely an essentialist view of 'woman's capacities' that, for some scholars, provides the very grounds for calling Stein's ethics 'feminist': "Stein's ethics are correctly called feminist . . . because they include a capacity for which woman is especially well suited" [1] (p. 473). Other interpretations, on the contrary, take an issue with Stein's apparently essentialist view of womanhood and consider it as a weakness of her account. For example, Calcagno points out several ambiguities in Stein's description of 'the female essence' and argues: "one wonders whether the essence of woman as mother cannot also apply to men, especially to men who find themselves in situations where they are constrained to be both mother and father to a child. This brings to light the possibility that the female essence may be shared by both men and women, and need not be tied exclusively to the gender of the person" [5] (p. 73).

It is beyond any doubt that Stein adopts what Ales Bello [41] aptly calls 'dual anthropology', namely a view that woman and man differ in their specific natures and capacities. There is plenty of textual evidence for this claim across Stein's philosophical work. In her lectures on woman Stein makes a crystal-clear statement: "I am convinced that the species 'human' is actualized as a double species—'man' and 'woman'; that the essence of human being, whose features cannot be lacking in either one, becomes expressed in a binate way; that the entire essential structure demonstrates the specific stamp" [33] (p. 187f.). Yet, I do believe that it is possible and even correct not to interpret Stein as an advocate of a feminism characterized by essentialist difference. Elsewhere [42], I provided a detailed argument in favor of a phenomenological reading of Stein's 'dual anthropology' by stressing that Stein conceives of the sexual difference as a difference between two related styles of intentional life rather than a difference between two separate essences (regardless of if it is ontologically or biologically defined). From the phenomenological perspective it seems plausible to read Stein's descriptions of woman's specific capacities and attitudes as describing a particular life form that can be shared by women and men alike. Hence, the alternative options suggested by Calcagno in the quote above seem to me not only right but also compatible with Stein's own perspective.

This brings us back to the initial question concerning the difference between Stein's philosophy and care ethics. It is challenging to come up with a clear-cut answer. On the one hand, Stein's account of woman's specific capacities, attitudes and values and their importance for a renewal of the moral life of individuals and communities has some resemblances to the currents in care ethics that aim to promote a 'feminine approach to ethics' and advocate an essentialist difference feminism (e.g., [14]). This entails that Stein's feminist personalism is vulnerable when faced with some of the forms of criticisms that many raised against the 'feminine approach' in care ethics. On the other hand, Stein's feminist personalism, when detached from its essentialist interpretation—which is something that can and perhaps should be done—shares some seminal feminist insights with those care ethicists who reject gender essentialism and adopt a critical feminist perspective on various social and political issues.

The confrontation, or rather the encounter, between care ethics and Stein's philosophy that I explored in this article helps us better see and appreciate how several 'mainstream philosophers' anticipated some key care ethical insights. A deeper understanding of the alternative contexts of the birth of similar ethical insights can broaden the dominant self-concept of care ethics. A more relational understanding of the place of care ethics within the diverse landscape of traditional moral and political philosophy would certainly fit well in care ethics' relational perspective. Moreover, this encounter provides an impulse to a more vivid dialogue between care ethicists and the current proponents of personalist ethics who often take Stein's philosophical work as a source of inspiration. The awareness of shared core ideas can help the personalist ethicists to better appreciate the way in which care ethics decenters the human and allows for the relationality of all things, which makes a non-anthropocentric relational environmental ethics possible. The critical feminist emphasis on the analysis and normative assessment of our social and political arrangements of caring can also enrich the perspective of personalist ethics which tends to underestimate the salience of wider social and political contexts for the lives of individual persons and communities. Finally, the non-essentialist and non-differentialist understanding of feminism in contemporary care ethics can foster further development of non-essentialist variants of feminist personalism that follow some of the paths foreshadowed in Edith Stein's thought.

Funding: This research was funded by the Czech Science Foundation as a part of the project 'Towards a New Ontology of Social Cohesion', grant number GA19-20031S, realized at the Institute of Philosophy of the Czech Academy of Sciences.

Institutional Review Board Statement: Not applicable.

Informed Consent Statement: Not applicable.

Data Availability Statement: Not applicable.

Acknowledgments: The second section of this paper draws on a book chapter "Introducing the Contexts of Moral and Political Theory of Care" which I co-authored with Lizzie Ward in 2020. I am deeply indebted to Lizzie for her contribution to the ideas that I further develop here.

Conflicts of Interest: The author declares no conflict of interest. The funder had no role in the design of the study; in the collection, analyses or interpretation of data; in the writing of the manuscript or in the decision to publish the results.

Notes

[1]   Ruddick quotes in an affirmative tone Chodorow's claim that "we cannot know what children would make of their bodies in a nongender or nonsexually organized social world. . . . It is not obvious that there would be major significance to biological sex differences, to gender difference or to different sexualities" [18] (p. 66, cited in [17] p. 364).

[2]   Robinson [19] similarly concludes that "contrary to the arguments of some critics, Ruddick's work neither upholds gender roles nor idealizes the values and activities of mothering. On the contrary, Ruddick's philosophy politicizes motherhood and draws our attention to the ambivalent relationship that mothers have with the societies in which they live" [19] (p. 106).

[3]   I certainly do not suggest that care ethicists in general refused to view care as an *essentially* feminine practice. There can be no doubt that several care theorists have proposed accounts of caring built on an essentialist account of sexual difference and defined care ethics as a distinctive "feminine approach to ethics" [14]. Yet, most of the feminist care theorists, including those of the first generation, have opposed such a view and sought to dissociate care ethics from any essentialism, cf. [21–23]. I think it's plausible to argue that the latter approach has been decisive for further formation and development of feminist ethics of care.

[4]   It is fair to note that in her later work Gilligan did elaborate on how care and justice may be connected in the real life of individuals and communities. See in particular her studies on patriarchy and democracy [24] and African-American young women [25].

[5]   Soon after she had published her 1982 book, Gilligan herself made it clear that this was a very limited interpretation of her research—cf. [26].

[6]   Virginia Held's valuable work represents a parallel attempt to construct a full-blown feminist moral theory as an alternative to dominant moral and social theories. Held's approach, in contrast to Tronto and Fisher [23], foregrounds mothering as the paradigm caring practice.

[7]   Personalism emphasizes the centrality of the person as the primary locus of inquiry for philosophical, theological, and humanistic investigation. Humans are considered the ultimate explanatory epistemological and ontological principle of reality. Personalism has a variety of manifestations but phenomenology is closely associated with it.

[8]   Haney speaks of Stein's "gradual identification of her early personalism with feminism" [1] (p. 451).

9   In this paper, I mostly offer a thorough modification of the available English translation. For an apt comment on the inaccuracy of Oben's translation of *Die Frau*, see [38] (p. 326; 335, fn. 16 and 17).

10   Cf. "her [woman's] strength lies in her intuitive grasp of the concrete and the living, especially of the personal. She has the gift of adapting herself to the inner life of others, to their goal orientation and working methods. Feelings are central to her as the faculty which grasps concrete being in its unique nature and specific value; and it is through feeling that she expresses her attitude. She desires to bring humanity in its specific and individual character in herself and in others to the most perfect development possible" [33] (p. 188).

References

1. Haney, K. Edith Stein: Woman and Essence. In *Feminist Phenomenology*; Fisher, L., Embree, L., Eds.; Springer: Dordrecht, The Netherlands, 2000; pp. 213–235.
2. Brenner, R.F. Edith Stein: A Reading of Her Feminist Thought. In *Contemplating Edith Stein*; Berkman, J.A., Ed.; University of Notre Dame Press: Notre Dame, IN, USA, 2006; pp. 212–225.
3. Borden, S. *Edith Stein*; Continuum: New York, NY, USA, 2003.
4. Gachenga, E. Stein's Ethic of Care: An Alternative Perspective to Reflections on Women Lawyering. In *Alternative Perspectives on Lawyers and Legal Ethics: Reimagining the Profession*; Mortensen, R., Bartlett, F., Tranter, K., Eds.; Routledge: London, UK, 2010; pp. 151–168.
5. Calcagno, A. *The Philosophy of Edith Stein*; Duquesne University Press: Pittsburgh, PA, USA, 2007.
6. Rose, H. Hand, Brain, and Heart: A Feminist Epistemology for the Natural Sciences. *Sings J. Women Cult. Soc.* **1983**, *1*, 73–90. [CrossRef]
7. Ungerson, C. Why Do Women Care? In *A Labour of Love: Women, Work and Caring*; Finch, J., Groves, D., Eds.; Routledge and Kegan Paul: London, UK, 1983; pp. 31–49.
8. Graham, H. Caring: A Labour of Love. In *A Labour of Love: Women, Work and Caring*; Finch, J., Groves, D., Eds.; Routledge and Kegan Paul: London, UK, 1983; pp. 13–30.
9. Waerness, K. Caring as Women's Work in the Welfare State. In *Patriarchy in a Welfare Society*; Holter, H., Ed.; Universitetsforlaget: Oslo, Norway, 1984; pp. 67–87.
10. Finch, J.; Groves, D. (Eds.) *A Labour of Love: Women, Work and Caring*; Routledge and Kegan Paul: London, UK, 1983.
11. Finch, J. Community Care: Developing Non-sexist Alternatives. *Crit. Soc. Policy* **1984**, *9*, 6–18. [CrossRef]
12. Parker, G. *With Due Care and Attention: A Review of Research on Informal Care*; Family Policy Studies Centre: London, UK, 1985.
13. Hartsock, N. The Feminist Standpoint: Developing the Ground for a Specifically Feminist Historical Materialism. In *Discovering Reality: Feminist Perspectives on Epistemology, Metaphysics, Methodology, and Philosophy of Science*; Harding, S., Hintikka, M.B., Eds.; Kluwer Academic Publishers: New York, NY, USA, 1984; pp. 283–310.
14. Noddings, N. *Caring: A Feminine Approach to Ethics and Moral Education*; University of California Press: Berkeley, CA, USA, 1984.
15. Held, V. The Obligations of Mothers and Fathers. In *Mothering: Essays in Feminist Theory*; Treblicot, J., Ed.; Rowman and Allanheld: Totowa, NJ, USA, 1983; pp. 9–20.
16. Gilligan, C. *In a Different Voice: Psychological Theory and Women's Development*, 38th ed.; Harvard University Press: Cambridge, MA, USA, 2003.
17. Ruddick, S. Maternal Thinking. *Fem. Stud.* **1980**, *2*, 342–367. [CrossRef]
18. Chodorow, N. Feminism and Difference: Gender, Relation, and Difference in Psychoanalytic Perspective. *Soc. Rev.* **1979**, *46*, 42–64.
19. Robinson, F. Discourses of Motherhood and Women's Health: Maternal Thinking as Feminist Politics. *J. Int. Political Theory* **2014**, *1*, 94–108. [CrossRef]
20. Bartky, S.L. Toward a Phenomenology of Feminist Consciousness. *Soc. Theory Pract.* **1975**, *3*, 425–439. [CrossRef]
21. Tronto, C.J. Beyond Gender Difference to a Theory of Care. *Signs J. Women Cult. Soc.* **1987**, *4*, 644–663. [CrossRef]
22. Ruddick, S. *Maternal Thinking: Toward a Politics of Peace*; Beacon Press: Boston, MA, USA, 1989.
23. Fisher, B.; Tronto, C.J. Toward a Feminist Theory of Caring. In *Circles of Care: Work and Identity in Women's Lives*; Abel, E.K., Nelson, M.K., Eds.; SUNY Press: Albany, NY, USA, 1990; pp. 35–62.
24. Gilligan, C.; Richards, D.A.J. *The Deepening Darkness: Patriarchy, Resistance, and Democracy's Future*; Cambridge University Press: Cambridge, UK, 2009.
25. McLean Taylor, J.; Gilligan, C.; Sullivan, A.M. *Between Voice and Silence: Women and Girls, Race and Relationship*; Harvard University Press: Cambridge, MA, USA, 1995.
26. Gilligan, C. Reply (to the Critics). *Signs J. Women Cult. Soc.* **1986**, *2*, 324–333. [CrossRef]
27. Tronto, C.J. *Moral Boundaries: A Political Argument for an Ethic of Care*; Routledge: London, UK, 1993.
28. Tronto, C.J. *Caring Democracy: Markets, Equality, and Justice*; New York University Press: New York, NY, USA, 2013.
29. Tullius, W.E. Person in Community, Repentance, and Historical Meaning: From an Individual to a Social Ethics in Stein's Early Phenomenological Treatises. In *Ethics and Metaphysics in the Philosophy of Edith Stein: Applications and Implications*; Andrews, M.F., Calcagno, A., Eds.; Springer: Cham, Switzerland, 2022; pp. 73–87.

30. Kueter Petersen, M. Love Divined: Discerning a Contemplative Ethic in the Philosophy of Edith Stein. In *Ethics and Metaphysics in the Philosophy of Edith Stein: Applications and Implications*; Andrews, M.F., Calcagno, A., Eds.; Springer: Cham, Switzerland, 2022; pp. 55–71.
31. Hamington, M. *Embodied Care: Jane Addams, Maurice Merleau-Ponty, and Feminist Ethics*; University of Illinois Press: Urbana, IL, USA, 2004.
32. Monjaraz Fuentes, P. Ontology and Relational Ethics in Edith Stein's Thought. In *Ethics and Metaphysics in the Philosophy of Edith Stein: Applications and Implications*; Andrews, M.F., Calcagno, A., Eds.; Springer: Cham, Switzerland, 2022; pp. 201–209.
33. Stein, E. *Essays on Woman*, 2nd ed.; Oben, F.M., Translator; ICS Publications: Washington, DC, USA, 1996.
34. Svenaeus, F. Edith Stein's Phenomenology of Sensual and Emotional Empathy. *Phenomenol. Cogn. Sci.* **2018**, *17*, 741–760. [CrossRef]
35. Andrews, M.F. Edith Stein's Understanding of the Personal Attitude: Applications and Implications for a New Ethics. In *Ethics and Metaphysics in the Philosophy of Edith Stein: Applications and Implications*; Andrews, M.F., Calcagno, A., Eds.; Springer: Cham, Switzerland, 2022; pp. 225–239.
36. Stein, E. *Philosophy of Psychology and the Humanities*; Baseheart, M.C., Sawicki, M., Eds.; ICS Publications: Washington, DC, USA, 2000.
37. Stein, E. *Selbstbildnis in Briefen I (1916–1933)*, 3rd ed.; Verlag Herder: Freiburg im Breisgau, Germany, 2000.
38. Borden, S. Review of Literature in English on Edith Stein. In *Contemplating Edith Stein*; Berkman, J.A., Ed.; University of Notre Dame Press: Notre Dame, IN, USA, 2006.
39. Stein, E. *Die Frau*; Verlag Herder: Freiburg im Breisgau, Germany, 2000.
40. Haney, K.; Valiquette, J. Edith Stein: Woman as Ethical Type. In *Phenomenological Approaches to Moral Philosophy: A Handbook*; Drummond, J.J., Embree, L., Eds.; Springer: Dordrecht, The Netherlands, 2002; pp. 451–473.
41. Ales Bello, A. Dual Anthropology as the Imago Dei in Edith Stein. *Open Theol.* **2019**, *5*, 95–106. [CrossRef]
42. Urban, P. Edith Stein's Phenomenology of Woman's Personality and Value. In *'Alles Wesentliche Lässt Sich Nicht Schreiben': Leben und Denken Edith Steins im Spiegel Ihres Gesamtwerks*; Regh, S., Speer, A., Eds.; Verlag Herder: Freiburg im Breisgau, Germany, 2016; pp. 538–555.

 *philosophies*

Article

Jacques Rancière and Care Ethics: Four Lessons in (Feminist) Emancipation

Sophie Bourgault

School of Political Studies, University of Ottawa, Ottawa, ON K1N 6N5, Canada; sbourgau@uottawa.ca

Abstract: This paper proposes a conversation between Jacques Rancière and feminist care ethicists. It argues that there are important resonances between these two bodies of scholarship, thanks to their similar indictments of Western hierarchies and binaries, their shared invitation to "blur boundaries" and embrace a politics of "impropriety", and their views on the significance of storytelling/narratives and of the ordinary. Drawing largely on *Disagreement, Proletarian Nights,* and *The Ignorant Schoolmaster: Five Lessons in Intellectual Emancipation,* I also indicate that Rancière's work offers crucial and timely insights for care ethicists on the importance of attending to desire and hope in research, the inevitability of conflict in social transformation, and the need to think *together* the transformation of care work/practices and of dominant social norms.

Keywords: care ethics; feminism; Jacques Rancière; Carol Gilligan; politics

Citation: Bourgault, S. Jacques Rancière and Care Ethics: Four Lessons in (Feminist) Emancipation. *Philosophies* **2022**, *7*, 62. https://doi.org/10.3390/philosophies7030062

Academic Editors: Maurice Hamington and Maggie FitzGerald

Received: 14 April 2022
Accepted: 2 June 2022
Published: 8 June 2022

Publisher's Note: MDPI stays neutral with regard to jurisdictional claims in published maps and institutional affiliations.

1. Introduction

In *Joining the Resistance* [1], Carol Gilligan reflects on the troubled reception history of her first book *In a Different Voice* [2]—a work of moral psychology that largely founded feminist care ethics.[1] Looking back to both the enthusiasm and the stir it caused, she suggests that her book was misunderstood by some because it was read through a patriarchal lens. Indeed, several of the claims she made to describe and rehabilitate a marginalized "different" moral voice were wrongly read as essentialist claims about women: "these misinterpretations reflect an assimilation of my work to the very gender norms and values I was contesting." [1] (p. 19) Indeed, the sad irony here is that the book's point was precisely to show that the dominant paradigm in moral psychology—which considered disincarnated rationality, impartiality, and universality in moral reasoning as the clearest signs of ethical maturity—was a patriarchal paradigm that pre-orients us to *mishear* or dismiss the "different", more marginalized way of approaching moral dilemmas.[2] Gilligan's reception history is, in a sense, a fine illustration of what contemporary political theorist Jacques Rancière calls a "més-entente". An important term of art in Rancière's work, "més-entente" has no good equivalent in English;[3] it plays on the double-meaning of the French *"entendre"* as hearing and understanding. A *"més-entente"* thus refers at once to a *sensory* failure in listening/hearing, and a *cognitive* failure in understanding. Gilligan's book was the object of this twofold failure on the part of some.

This paper's main purpose is not to use Rancierian concepts to revisit the reception history of Gilligan's ground-breaking text. I rather wish to orchestrate, more widely, a conversation between Rancière's work and care ethics. This is largely unchartered territory for care ethics, if one sets aside three notable exceptions: Gradon Diprose's [3] appeal to Rancière's account of radical equality to analyze alternative ways of exchanging care labour, Ella Myers' recourse to Rancière to theorize a caring democratic ethos [4], and Jorma Heier's [5–7] illuminating resort to Rancière's writings to theorize repair and the political more generally. Apart from Diprose, Myers, and Heier, scholars have yet to explore at length the affinities between Rancière and care ethics. This paper proposes a modest first step in this direction.

Naturally, care ethics cannot be boiled down to a single, unified perspective; it is a multidisciplinary and polyphonic scholarly field (e.g., Engster and Hamington [8]; Barnes et al. [9]; Urban and Ward [10]; Bourgault and Vosman [11]). My characterization of care ethics will hence be incomplete, and some of the parallels drawn between Rancière and care ethics might do violence to some articulations of care ethics. Nevertheless, I show that Rancière's work speaks most clearly (*but not exclusively*) to that one "strand" of care ethics that sees this ethics as a critical theory devoted to undermining patriarchal dichotomies and hierarchies (e.g., Gilligan [1,2,12]; Laugier (in [11]); Paperman and Molinier in [11]; Robinson [13]; FitzGerald [14]). The second strand of care ethics (as identified by Fiona Robinson and Maggie FitzGerald) is that which envisions care ethics chiefly as something to be *applied* to specific policies, occupations, or institutions (health care, home care, etc.).[4] I shall return to this typology later in the paper, in order to complicate it.

More specifically, this paper's chief objective is to show that Rancière's thought is particularly relevant for care ethics in light of: (1) its indictment of Western philosophical binaries and hierarchies; (2) its account of politics as a matter of challenging "miscounts" (i.e., socio-political exclusions) and of blurring boundaries; (3) its emphasis on story-telling/narrative and listening as crucial for 'un-learning' inequality; and (4) its call to attend to *ordinary* heroes and to *desiring*. I also indicate, throughout, some differences between Rancière's work and care ethics scholarship—differences that, I suggest, serve as useful correctives or ways to illuminate blind spots and tensions in their respective works.

The paper will argue that reading Rancière is quite instructive for care ethicists because his work resolutely underscores the conflictual nature of politics. As I will explain below, this is an insight I consider of utmost importance for care ethics scholars as they reflect not only on how to (re)distribute care responsibilities and challenge privileged positions, but also on what some of the short-term *effects* of this redistribution might be (namely, a fair amount of discomfort, grumbling, and tensions). What my conversation between Rancière and care ethics also indicates is the desirability of interrogating the separation of care ethics research into two separate strands. My intention here is not to deny that there are numerous and quite distinctive types of work done (and methodologies used) in care ethics. But what I suggest, with Rancière, is that the most fruitful way to carry out social transformation might be to attend to both "strands" *simultaneously*. And finally, I argue that another uptake of reading Rancière for care ethicists lies in his compelling invitation to always emphasize hope and desire in research on oppression and inequality.

Some might wonder whether Rancière has ever read (or cited) care ethics scholarship. As far as I know, the answer is no. But in a sense, the question is moot. In the spirit of Rancière's own method, my intention is not to demonstrate intellectual debts or trace direct lines of influence. It is, more modestly, to seek out a few *resonances* and offer a prolegomenon to more sustained conversations between Rancière scholarship and care ethics. As we shall see, Rancière's methodological motto (see [15])—*find resonances*[5]—is one he invites scholars to take up not only when studying old philosophical *texts*, but also when engaging with all real-life experiences and contemporary stories. For him, it is in part through the creation of these resonances that concerned scholars, teachers, parents, civil servants, or activists might contribute (if modestly) to an emancipatory politics.

2. Putting the Western Philosophical Tradition on Trial: Of Hierarchies and Binaries

> "I've never imagined my work developing *from* politics *to* aesthetics, especially since *it has always sought to blur boundaries*." Rancière [16] (p. 203; my italics)

Rancière's oeuvre is vast and multidisciplinary, dealing with subjects as diverse as literary theory, education, cinema, workers' history, aesthetic theory, Marxism, etc. This is hardly surprising in light of Rancière's own self-description as someone who has always been committed to challenging disciplinary boundaries—if not, in fact, *all* boundaries. But there is still one crucial idea or concept that traverses all his writings: namely, that of equality—the equality of intelligence of all human beings. Indeed, Rancière is of the view that all individuals are capable of thinking, learning, and communicating with others, and this thesis informs everything he says about socio-political life[6] (see [17]), art, and education. For Rancière, equality is not an ontological principle, a ground for politics or a (future-oriented) *goal* to be pursued via various social reforms. It is a *presupposition* that ought to inform how people think, feel, and act *right now*. For him, equality is an opinion that, when upheld, can have radical effects—particularly when it is upheld by those on the margins of socio-political life. For this reason, he writes: "our problem isn't proving that all intelligence is equal. *It's seeing what can be done under that supposition.* And for this, it's enough for us that the opinion be possible" [18] (p. 46; my italics).

It is this radical egalitarianism that also informs Rancière's critique of the Western tradition. Indeed, his writings propose an indictment of the way Western philosophers have, ever since antiquity, divided the world into two types of people (superior vs. inferior, 'those who think' vs. 'those who don't')—a division closely tied to other binaries such as reason/affects, mind/body, public/private. From Aristotle to Arendt and Bourdieu, Rancière sees in intellectuals (including the most well-meaning ones) a regrettable hierarchical "division into two" of individuals, to which various expectations and norms are attached (whether the ranked division is *fought against* or *celebrated* by a philosopher is unimportant for Rancière; the *effects* are ultimately the same). First, there are those individuals whose possession of rational speech (*logos*) makes them fit for political participation, leisurely activities, and detached reflection on justice. Second, there are others (workers, slaves, women) whose laboring bodies are unceasingly preoccupied with need satisfaction and who are unfit for politics and noble leisure; they lack *logos*, energy, and time, and they only seem capable of communicating emotions and "noise"—a "mere" expression of suffering and pain. Throughout his career, Rancière tirelessly sought to contest this "division into two"—whether it concerns that between necessity/freedom, emotion/reason, manual/intellectual work. Rancière's first major book, *Proletarian Nights* [19], sought to show how poorly these divisions captured the reality of working-class experience and abilities. The book chronicles, based on a lengthy study of 19th century workers' archives (diaries, letters, plays, poetry), everyday blurrings of the dichotomies mentioned above: countless examples of metalworkers, shoemakers, and seamstresses using their evenings and nights *not* to rest their tired bodies or sedate their suffering with "light" entertainment (as Bourdieusian intellectuals tend to think, according to Rancière), but rather, to engage in leisurely pursuits devoted to many other things than simply expressing suffering. Through these pursuits, workers effectively *demonstrated* their equality: they took up artistic and intellectual activities that did not match the expectations of the dominant classes. Note here that Rancière does not deny that many of these (exploited) workers were indeed *suffering*—he does acknowledge harsh work conditions and low pay—but his work is chiefly devoted to showing that: (1) workers' bodies and minds are about *so much more* than suffering or the dire pull of necessity; (2) workers do not need to have their suffering *explained* to them (they already understand it).

The inegalitarian division of the world into (ranked) types is part and parcel of what Rancière refers to as "the police": that is, a hierarchical way of ordering or "partitioning" the socio-symbolic order, a hierarchical way to share power and authority. In Rancière's view, all societies are shaped by implicit laws and norms that govern their *sensory order*, their modes of perception (i.e., what gets heard and seen vs. what does not). The police "distributes bodies within the space of their visibility/invisibility, and aligns ways of being, ways of doing and ways of saying *appropriate to each*" [20] (p. 28; my italics). The police's norms, processes and implicit laws "organize" bodies, names, and places not only according to a principle of hierarchy, but also according to a principle of "saturation"; the police *claims* that everyone has been "counted"/acknowledged, and thus that "all is well" [21] (thesis 8). But all is not well; there *are* always un-counted or excluded individuals in the police order (*les sans-parts*, those "without a part"). There are always *wrongs*, miscounts in "the count of community parts"; all societies are based on exclusions and miscounts, and processing (or contesting) those wrongs is what politics is about [20].

We shall return below to what politics accomplishes. For now, let us focus on the binaries that organize our world and briefly discuss the Rancierian concept of the "partition of the sensible" (*le partage du sensible*). With this term of art, Rancière plays with the double meaning of the French word *partager*, referring at once to "that which separates and excludes", and "that which allows participation" or inclusion [21] (thesis 7). A partition of the sensory thus always produces both exclusions and inclusion; and as noted above, an (inegalitarian) partition such as the police's will shape how we perceive and relate to others. We come to "imbibe" the prevalent *opinion* of inequality, which affects how we acquire and *share* knowledge and how we live our socio-political lives more generally. An *opinion* for Rancière [22] (p. 26) "is the framework within which we learn and know, within which the work of our mind is linked with that of all the other minds". This explains why emancipation requires *un-learning* the opinion of inequality.

For now, what I wish to suggest is that we can describe patriarchy (as defined by care ethics scholars) as an inegalitarian "distribution of the sensible". Indeed, one finds in Gilligan, Robinson, Laugier, Molinier, Heier[7], and FitzGerald's respective works an indictment of the Western philosophical tradition that is quite similar to Rancière's: they fault our tradition for having split humans into two (superior vs. inferior), for having posited dubious binaries between men and women, mind and body, thinking and feeling. In *Why does patriarchy persist?*, Gilligan and Snider define patriarchy as a framework that profoundly shapes both our affective/sensory and our cognitive/intellectual skills (like Rancière, the authors regard the two sets of skills as inseparable). "Patriarchy exists as a set of rules and values, codes and scripts that specify how men and women should act and be in the world", the authors write [23] (p. 6). "Patriarchy also exists internally, shaping how we think and feel how we perceive and judge ourselves, our desires, our relationships and the world we live in." Patriarchy is hence akin to a Rancièrian *police* distribution of the sensory, operating based on an inegalitarian logic. Gilligan and Snider [23] (p. 33) write: "Patriarchy is an order of living that privileges some men over other men (straight over gay, rich over poor, white over black, fathers over sons [. . .]) and all men over women. The politics of patriarchy is the politics of domination." The sensory and epistemic framework that is patriarchy is thus responsible for multiple forms of oppression; and it is on the basis of passages such as these that Robinson [13] argues that care ethics *can* respond to the charge that care ethics is almost exclusively concerned with *gender* (or differently put, that it is not intersectional enough). For Robinson [13] (p. 17), the criticism is misplaced, since care ethics challenges *multiple* hierarchical divides at once—not simply gender divides.

For Robinson [13] (p. 20), one can effectively answer the critique if one regards care ethics not chiefly as one to be "applied" to policies or institutions (one strand of care ethics, for Robinson), but rather as an epistemic, critical framework that seeks to challenge patriarchy's binary logic (the second main strand of care ethics; see [13]). This is also the way Maggie FitzGerald defines care ethics—namely, as a "critique of existing hierarchies [. . . that] has the potential to transform governing norms, the institutions shaped by these norms and values, and most radically, the current configuration of 'the political' itself." [14] (p. 261) While FitzGerald does not draw on Rancière, much of what she has to say about "the political" clearly speaks to a Rancierian account of political life and reiterates his invitation to "blur boundaries"; for FitzGerald, critical care ethics has the power to "erase the boundaries that define dualistic categories" ([14], (p. 252)). For Robinson, the two most crucial binaries challenged by care ethics are those between reason and emotion, and between mind and body. Robinson is absolutely correct to observe that "care ethics brings back the body into view" [13] (p. 17), as the work of Pascale Molinier [11,24] and Maurice Hamington—to mention only those two—clearly illustrates. The latter's writings have certainly emphasized care's radically *embodied* nature. For Hamington, care is regarded as a "mind-body activity", an "embodied performance that can disrupt knowledge" [8] (p. 279). As we will see, Rancière also underscores—albeit from a different angle—this ability of bodies (particularly bodies *out of place*) to disrupt epistemic frameworks.

But if both Rancière and care ethics challenge the mind/body divide, I want to argue that care ethics' treatment of the dichotomy is more robust and radical because it is *consistently* tied to a challenge of the distinction between ('low') manual labor vs. ('high') intellectual labor. (This is one of the many reasons why a dialogue between Rancierian scholarship and care ethics is valuable: namely, for uncovering some of the *limits* of Rancière's treatment of (manual) labor—limits that have been too rarely noted by researchers.) Rancière understood his oeuvre as devoted to challenging the "very old hierarchy that subordinates those who are 'destined' to work with their hands to those who have received the privilege of thought,"[8] but it seems to me that this dichotomy eventually sneaks in through the back door in his writings. Rancière is so intent on underscoring that workers *can* "think" and do art/philosophy, and that they are equal, that he ends up reiterating at times the old (Aristotelian) assumption that some pursuits (e.g., poetry, painting, theater) are *higher* than those involving the body (e.g., carpentry, farming, or knitting). Indeed, despite his best intention, I think that *Proletarian Nights* ultimately presents "bourgeois" leisurely activities as highly desirable *escapes* from physical work. By contrast, I wish to suggest that care ethicists have much more *consistently* sought to underscore the *nobility* and indispensability of more manual work, in its own right (e.g., Molinier [24]).

One other difference between care ethics and Rancière is that the latter does not seem to be as preoccupied with explaining *why* our epistemic/sensory frameworks are so informed by binaries. Much of Gilligan's recent work is about figuring out *why* humans come up with and hold onto binaries if they are so toxic. Gilligan's answer is multi-layered [1,23,25]. She first acknowledges the evident fact that patriarchy greatly *benefits* some individuals (an obvious obstacle to challenging the privileged irresponsibility of many). But she also proposes another, more psychologically-rooted explanation for patriarchal binaries: they exist because human beings (incorrectly) believe patriarchy will shield them from suffering. Individuals sacrifice love and equality for hierarchy because they hope—after having experienced some (traumatic) losses of love—that this will protect them from *further* losses and vulnerability.

That Rancière may not seem interested in the emotional or psychological forces that could explain *why* individuals embrace hierarchies will not surprise readers. After all, if care ethics is commonly associated with an interrogation of the emotion/reason split and with celebrations of emotions such as empathy (e.g., Slote [26]; Brugère [27]; Pulcini [28]; cf. Paperman [29])—few associate Rancière's name with a deep concern for emotions (many rather associate him with the opposite (e.g., Davis [30])). But Brigitte Bargetz has shown that while *explicit* references to emotions are rare in Rancière, one may still label him a "political philosopher of affect." [31] (p. 588) It is possible to do so, she insists, if one considers *le partage du sensible* as a partitioning that concerns affects and emotions (two terms Bargetz uses interchangeably). In Bargetz's view, the distribution of the sensible not only entails a separation between those who (are said to) think/speak and those who don't; it also "marks a distinction between those whose feelings *constitute the existing distribution* of the sensible and *those whose feelings are excluded.*" [31] (p. 589) Rancière's work could thus enrich feminist theory in light of its fine grasp of the power of "feeling scripts" (i.e., what should (or should not) be felt or expressed, by whom, and in what circumstances). Following Bargetz's lead, one might indeed try using a Rancierian lens on affects to take up Gilligan's depiction [1] of the (patriarchal) feeling-script tied to anger—i.e., anger is an "appropriate" emotion in boisterous boys and "real men", and an "improper" one in "good" girls. A Rancierian perspective on anger and gender roles could in fact bring new energy, more widely, to the existing feminist literature on the value of anger for social transformation (e.g., see the writings of Sara Ahmed, Audre Lorde, and bell hooks). What Rancière would bring to the subject is not only a rich account of the "distribution of the sensible", but also singular reflections on what *education* and *everyday/ordinary* affective gestures can do to trouble existing gendered and racialized feeling scripts. (There is one additional way to nuance the prevalent view that Rancière has almost nothing to say about affects, and it is to consider the significance he ascribes to the *passion* for equality, which I will do below.)

To recapitulate, what I have claimed is that there are good grounds to embrace Bargetz' reading of Rancière as a theorist of emotions. But what I am *not* claiming is that Rancière is *as* concerned by emotions (or by the emotion/reason split) as care ethicists are. For one thing, the remarks on the subject are too rare. Although this is speculative on my part, I think that what partially explains his reticence to discuss emotions at length might be his often-expressed fear (e.g., [32]: p. 112) of making any strong ontological or anthropological claim about humans (whether he *can* in fact avoid these altogether is, of course, is a complex issue that would be worthy of an entirely different paper).

3. Laboring Bodies, Women, and Dirty Diapers "Out of Place": Politics as the Processing of Wrongs and Blurring of Boundaries

As noted above, if "the police" is an inegalitarian logic or series of processes that organize people according to identities, functions and places, "politics" for Rancière is an egalitarian logic that illuminates and *disturbs* the police. In line with his idiosyncratic way of (re)defining standard concepts, Rancière uses the term "politics" in a very particular manner. Much of what we usually understand as "political" (policymaking, parliamentary life, elections, parties) Rancière [18,20] subsumes under the heading of the "police"; this becomes a gigantic concept that engulfs norms, values and implicit laws, *and* "concrete" processes and institutions (courts, political parties, media, police force, etc.). In turn, politics takes on a specific, narrow meaning, referring to fairly rare moments when the police order is disturbed by the appearance of an excluded group, a "part of those without a part".

At the heart of the Rancierian portrait of the police and of politics (see [33–35]),[9] one finds a radically categorical dismissal of institutions: Rancière repeatedly insists that *all institutions* inevitably rest on an inegalitarian explicatory logic and that, as such, no institution "will ever emancipate a single person" [18] (p. 102). While these statements are unsurprising given Rancière's characterization of his work as *polemical* interventions [17] (p. 116), they do preclude any extensive discussion of what *some* state-funded programs could do (and what they *already do*) to participate in an emancipatory feminist politics (e.g., universal, accessible daycare services). This is a significant gap, and the fact that this neglect of institutions might be the "mere" result of his lack of interest in them (as he suggests in an interview[10]) does not make this any less of a problem. Care ethics offers here a healthy corrective, since many of its scholars have argued that a truly egalitarian democratic politics can very well coexist with (caring) institutions and social programs (e.g., Tronto [36]). For feminist care ethicists, we cannot have social transformation without attending to *both* individuals and institutions—and the latter need not *necessarily* be an embodiment of an inegalitarian logic.

So far, I have noted that Rancière sees politics as what can modify a *partage du sensible*. Politics accomplishes this partially by revealing that there are individuals who are not "counted", and it can do so in several different ways. One of these is simply by having bodies show up in a place or at a time they are *not* expected to, or where they take up tasks in a manner they were not expected to be able to. For him, "Politics begins exactly when those who 'cannot' do something show that in fact they can" [16] (p. 202). Rancierian politics is thus, once again, about impropriety or, to use his own term, "dis-identification": it is a stepping aside from the categories, norms, and expectations assigned by the police. And it is precisely in these moments of *dis*-identification that a political subject is created and that a distribution of the sensible gets illuminated *and changed*. Dominant norms get probed as bodies move "out of place", as politics takes place; what was accepted as self-evident becomes—momentarily—less so.

A politics worthy of the name for Rancière is thus about the blurring of various boundaries (a blurring that obviously echoes much feminist theory, not only care ethics). As he explains:

> Political action consists in viewing as political what was viewed as 'social', 'economic', or 'domestic'. It consists in blurring the boundaries. It is what happens whenever 'domestic' agents—workers or women, for instance—reconfigure their

> quarrel as a quarrel concerning the common […] *there is politics when there is a disagreement about what is politics, when the boundary separating the political from the social or the public from the domestic is put into question.* Politics is a way of re-partitioning the political from the non-political. This is why *it generally occurs 'out of place'.* [35] (p. 4; my italics)

What Rancière attacks here, then, are the binaries of ancient philosophers such as Aristotle, but also, insistently so, those of Hannah Arendt. While Rancière acknowledges his debts towards her, he nevertheless understands his own work as a *reaction* to her oeuvre and particularly her distinctions between "mere life"/"political life", "the social"/politics, needs/freedom, noise/logos [16] (p. 202). He is convinced that these distinctions regrettably reiterate old inegalitarian distributions of the sensible and that Arendt failed to appreciate what politics is about: a contest over these very distinctions. For him, "Political activity is whatever *shifts a body from the place assigned to it* [… .] It makes visible *what had no business being seen*, and makes heard a discourse where once there was only place for noise" [20] (p. 30; my italics).

One vivid example used by Rancière [32] (p. 120) to capture the way boundaries can be disputed by a mere body "out of place" is that of Rosa Parks: Parks' body sitting on a bus where it was *not supposed to*, a black body causing through this impropriety significant political effects. Another story of a woman "out of place" often used by Rancière [20,32] is that of French revolutionary Jeanne Deroin. When Deroin ran as a *candidate for office* in 1849 (at a time when women could not even vote), she effectively exposed the "misfit" or gap between the egalitarian rhetoric of universal rights and suffrage, and the (inegalitarian) logic of women's relegation to the domestic sphere. Women are both included and excluded in the police order of Republican France, and Deroin's act exposed that paradox. For Rancière, Olympe de Gouges grasped the paradoxical inclusion/exclusion of women quite well when she reflected on her tragic fate as a female revolutionary: if one can be guillotined for political reasons, she wondered, shouldn't one be allowed to climb up on the tribune [37] (p. 35)? In brief, what political action accomplishes is to expose the various paradoxes and "miscounts" in the dominant social order. It makes visible "what had no business being seen" in public (manual workers, pregnant bodies,[11] dirty diapers, female revolutionary heroes), and hereby blurs boundaries. Rancierian politics is, in short, a politics of *impropriety* (as Davide Panagia has convincingly shown).

While their overall theories are far from equivalent, there is an important resonance between Rancière's commitment to impropriety and the "blurring of boundaries" and Tronto's critique of several binaries. Recall that in her ground-breaking *Moral Boundaries*, Tronto sought to denounce several regrettable dichotomies that have led to care's disregard (not all of these are central to Rancière's work, but most certainly the private/public one is). Many feminist scholars have heeded Tronto's call to blur boundaries and to address one grave "miscount": the non-recognition of the significance of care work and the invisibilisation of all those individuals who 'maintain and repair' our world. Countless care ethicists have busied themselves with revealing this miscount, making visible what most privileged individuals would "prefer not to see" (Molinier [24]). In Gilligan and Brugère as well, democratic politics gets framed largely in terms of addressing "miscounts" and achieving *equality of voice.* Indeed, for Brugère, politics largely hangs unto what is heard/not heard. Hers is a "sensate" theory of democracy: a caring democracy worthy of the name (equality of voice) will not come without a *simultaneous* reform of our socio-political normative and sensory order—a far-reaching disturbing of the "established order of places" (Brugère [38] (p. 145).[12]

Note here that for Brugère, Robinson, and Gilligan, as much as for Rancière, the concept of "voice" is a radically *relational* one. Speaking matters only if it is perceived, i.e., meaningfully *listened* to. What politics calls into question, then, is not only the private/public divide, but also the divide between what is *said to* constitute "discourse'/rational speech" and what is *said to* constitute "noise". Politics entails changing our *perception*, reconfiguring *"the way we share out* or divide places and times, *speech and silence,* the visible and the invisible" [16] (p. 203). Rancière thus complicates the old binary between speech and silence posited by our logocentric tradition. Since Aristotle, most have associated rational speech (*logos*) with intelligence, freedom, and activity, and silence/listening with stupidity, servility, or passivity, and many have *used* this logocentric binary to *justify* the distinction between those who (should) have citizenship and those who should not. Rancière challenges this dichotomy—most notably in *Disagreement*, but also in an essay on "un-learning" inequality [22], where he invites us to reconsider the view that politics necessarily requires *audible* claims. He does so in part by underscoring the significance of *silent* bodies assembling (see his discussion of the quiet resistance of the "silent standing man" in Turkey's Gezi Park ([22]: p. 45).

The interrogation of the old binary "speech" vs. "silence/listening" is something I consider central to care ethics; a few scholars have critically re-evaluated the dichotomy or highlighted the crucial role of (silent) attention and listening (e.g., Brugère [27]; Casalini [39]; Laugier [11]; Robinson [40]). (see also [41])[13] For several care ethics scholars and Rancière, listening (and non-listening) plays a key role in the mechanics of domination *and* of politics/emancipation. After all, as Rancière [21] (thesis 8) correctly observes: "If there is someone you do not wish to recognize as a political being, you begin by not seeing them as the bearers of politicalness [. . .] by not hearing that it is an utterance coming out of their mouths." Naturally, neither care ethicists nor Rancière dismiss the desirability of having a "voice" altogether, but they insist that without *genuine* listening (e.g., attentive, caring silence), speech means little. As Gilligan puts it: "speaking depends on listening and *being heard*, it is an intensely relational act" [2] (p. xvi) (see also [42])[14] In her piece "Stop Talking and Listen" [40], Robinson also offers a crucial corrective to mainstream discourse ethics, one that speaks to Gilligan's insights.[15] Robinson argues that merely including in dialogue previously excluded parties is inadequate without a genuine commitment to *listen* on the part of dominant groups—a commitment that ought to be informed by a clear awareness of various relationships of dependency, and of the norms and power inequalities that have shaped (and continue to shape) the dominants' relationship to the excluded (e.g., female migrant care workers [40]). Robinson's depiction [13] (p. 20) of what care ethics calls for in order to remedy "miscounts" is something Rancière would be sympathetic to: "The task for care ethicists . . . is not simply to 'bring in' marginalised perspectives [. . .] it is to make more visible the way in which care ethics can offer a critique of the psychological and structural forces that actively repress and militate against our ability to think and respond to others through relational lenses".

In short, I am suggesting that the Rancierian account of politics and of the reconfiguration of the sensory is quite relevant for care ethicists' theorization of social transformation (and here I follow in the footsteps of Heier [7]). To capture succinctly this relevance, allow me to juxtapose briefly a "care ethics" voice to Rancière's voice, in this passage from *Disagreement*:

> The police is thus first an order of bodies that defines the allocation of *ways of doing, ways of being, and ways of saying* . . . an order of the visible and the sayable that sees that a *particular activity* is visible [e.g., production or protection] and another is not [care and reproductive work], that this speech [about markets or war] is understood as discourse and another [about child care or decent housing] as noise . . . I propose to reserve the term politics for an extremely determined activity *antagonistic* to policing: [. . .] political activity is always a mode of expression that undoes the perceptible [patriarchal] divisions of the police order. [20] (pp. 29–30; my italics)

Two important things are worth noting here in my view: first, the extent to which, according to Rancière, particular activities and occupations are *inseparable* from the epistemico-sensory order (e.g., what we see and hear/what we don't). That a particular activity/work might be deemed "invisible" (washing floors, changing diapers) and thus poorly paid and funded is not separable from the symbolic order, from the ways we allocate "ways of doing, being and seeing". Stated differently, the "miscounting" of essential care workers and the lack of decent funding and attention given to certain care services and policies are very closely tied according to a Rancierian lens to the norms of the "police" order and to our field of perception. My point here is that Rancière's work may force us to consider anew whether the gulf between care ethics' two "strands" is as wide as we think (and if it is so, whether we wish *to keep it* that way). As mentioned above, several scholars have suggested that there are two main strands in care ethics (FitzGerald, Robinson, Conradi): one that uses care ethics as a framework to assess concrete care *work*/practices/policies, and another that is said to consist in a *critical* theory that seeks to trouble existing norms and dominant epistemologies (e.g., Robinson [13]; FitzGerald [14]). What I suggest here is that Rancière's oeuvre interrogates the view that it might be possible—or desirable from a feminist activist point of view—to separate these two strands.

The second thing I wish to underscore from that passage from *Disagreement* is the *necessarily conflictual* nature of politics—something that Rancière asserts loud and clear throughout his writings, but that care ethicists have been a lot less vocal about. While care ethicists have not completely overlooked conflict (e.g., Tronto [43]: p. 109), they have overall given it a fairly modest place in their account of what social transformation might imply (notable exceptions here are Mayer [44], Heier [7], Vosman [45]). Rancière's thesis might thus be worth considering at greater length by care ethicists; after all, denouncing and rectifying miscounts (including miscounts pertaining to care responsibilities) is bound to cause discomfort and conflict. The processing of wrongs (i.e., meaningful public debates about who does what/who does not, and in what condition, etc.) is no easy deliberative affair. For one, people who benefit greatly from a hierarchical way of "distributing places" will not easily accept a reconfiguration of names, responsibilities and roles, especially if this leads to a significant *loss* in privilege and power. As we saw above, Rancierian politics is what *disturbs* existing norms and existing ways of doing/saying/sharing [20] [21]; it is hence what causes (at least in the short term) a fair amount of unease and trouble. After all, politics shows what had "no business being seen" and it voices *wrongs* (past *and* present); it brings attention to the least admirable periods of our histories, to the margins, and to what many (privileged) individuals would prefer *not* to see, *not* to know. It is, as such, bound to generate conflict and unease. Indeed, (Rancierian) politics is no picnic. And what I wish to argue here is that it might be of utmost importance for care ethicists to keep this (apparently) simple thesis in mind as they reflect on how to challenge unjust care distribution and hierarchical structures.

4. Un-Learning Hierarchies, Making Voices Resonate: The Significance of Narratives

"The teacher is first of all a person who speaks to another, who tells stories . . . " Rancière [18] (p. 6)

"My research began with questions about voice: who is speaking, and to whom? In what body? Telling what stories about relationships? In what societal and cultural frameworks?" Gilligan [1], (p. 5)

I now wish to turn to a book of utmost importance for Rancière's intellectual trajectory, *The Ignorant Schoolmaster. Five Lessons in Emancipation*. This is no standard work of philosophy: it is, rather, storytelling. The (true) story in question concerns Joseph Jacotot, a French revolutionary forced into exile in Holland in the 1820s who has to teach literature to Flemish students who know no French (and he knows no Flemish). Thanks to these odd circumstances, Jacotot ends up "discovering" a controversial method of teaching, based around the following principles: that humans possess equal intelligence, that students need no detailed explanation to learn, and teachers need not know about a subject to "teach". Indeed, the "Jacotot method" makes it possible for an illiterate mother to teach her children how to read.

If students do not need detailed explanations to learn, why are these so central to our education system? Part of Jacotot/Rancière's answer is that "it is the *explicator* who needs the incapable". After all, "to explain something to someone is first of all to show him he cannot understand it by himself. [. . .] explication is [. . .] the parable of a world divided into knowing minds and ignorant ones." [18] (p. 6) The "explanatory logic" is thus a stultifying way of interacting with others that subordinates, and that seeks to justify the division of the world into two. This is not only a logic found in schools, but in all institutions for Rancière: "explication is not only the stultifying weapon of pedagogues but the very bond of the social order. Whoever says order says distribution into ranks. Putting into ranks presupposes explication, the distributory, *justificatory* fiction of *an inequality that has no other reason for being*" [18] (p. 117; my italics).

Now, one cannot conclude from this that Rancière wants to get rid of "teachers" altogether (a tempting conclusion to draw given Rancière's numerous anarchist quips): teachers *can* play a role in emancipatory practice (a role that can be taken up by *all* concerned social scientists, activists, or parents in his view).[16] Part of this role consists in working on students' will to learn and their attentiveness, which can be accomplished in part by making them embrace the *opinion* of equality. For Rancière/Jacotot, an emancipator is someone who can offer the "consciousness of what an intelligence can do when it considers itself equal to any other and considers any other equal to itself […] Emancipation is the consciousness of that reciprocity" [18] (p. 39). Moreover, an "ignorant"[17] teacher can also ask questions [18] (p. 30); through this interactive questioning, *both* parties learn *and* teach something to the other. Finally, and most significantly for our purposes, a good teacher will *tell stories*, as Rancière thinks he is doing in *The Ignorant Schoolmaster* (a book in which he merges his voice with that of Jacotot). What ought to replace the dominant "explanatory logic" of institutions is a narrative[18] one: storytelling is a way of *sharing* knowledge that does not subordinate one's interlocutor(s). It is a communicative mode that *presumes* a basic equality; when one tells a story, one rarely pauses to *explain* what is going on or to append lengthy theoretical addendums.

Rancière also employed this narrative mode in *Proletarian Nights* [19], doing his best to resist the temptation social scientists regularly face to offer, first, a "pure" rendition of a working-class archive (via the recounting of a letter or poem), followed by a theoretical analysis to *explain* the worker's experience. As noted above, explanations are undesirable (they have stultifying effects); they are also, most often,[19] completely unnecessary. For Rancière, workers (and most marginalized groups) are aware of their domination; they rarely benefit from a well-meaning sociologist such as Bourdieu explaining it *to them*. There are thus close ties between Rancière's celebration of storytelling, and his polemics against intellectuals such as Bourdieu. The latter, according to Rancière, problematically claimed to know best workers' domination and thought, as such, that he could speak *for* them.

Rancierian storytelling is a way to speak *with* rather than *for* people—something Rancière wants social scientists to do to help "erase the hierarchical privilege of the comment (of the scholar) whose words explain the words that are its 'object'" [22] (p.38). For Rancière, there is a way to present stories—via a deliberate yet subtle intermeshing of voices—that won't reinforce existing epistemic power hierarchies. For him, this is an intermeshing that ultimately reflects the narrative ambiguity of *all* human experience and that can inform an emancipatory politics, by purposefully blurring the line between *past* stories of emancipation and *present* possibilities. What is produced via this intermeshing of different voices, temporalities, and levels of meaning are what Rancière calls "resonances", which help create an *intensification* of stories (they allow us to grasp and feel better a particular situation). Whether one is entirely convinced by Rancière's clear-cut distinction between *his* narrative method and more "standard" ways of presenting archival material is an open question (I am of the view that his polemical attacks against Bourdieu have clouded his judgment on this). Also open is the question of whether there is a contradiction between Rancière's egalitarian call to dismantle (epistemic) hierarchies, and the baffling impenetrability of some of his writings. (Tackling these complicated questions would require another paper.)

Seeking out resonances and ascribing significance to the narrative mode are also crucial for Gilligan's method and project, as she herself acknowledges repeatedly. In *Joining the Resistance*, for instance, she writes: "I am a woman who listens. My research began with questions about voice: who is speaking, and to whom? In what body? Telling what stories about relationships?" [1] (p. 5). Like Rancière, Gilligan thinks narratives can powerfully capture both affective experiences and *reflection* on these experiences: "My interest lies in the interaction of experience and thought [. . .] in the stories we tell about our lives." [2] (p. 2) Moreover, what Gilligan [46] (p. 121) believes she accomplished in *In a Different Voice* is somewhat akin to what Rancière thought he did: namely, she made voices resonate. Gilligan sought to interlace discreetly her voice *and* that of other women to that of Amy (one of the young girls she interviewed for her 1982 book), so that the latter's voice could be "intensified". Through these "relational resonances" [46] (p. 125), Gilligan hoped that Amy's (different) perspective on what matters for moral reasoning could be taken more seriously by unreceptive academic institutions and by a patriarchal culture. I would like to argue that behind Gilligan's embrace of the desirability of "making resonate" rests the (Rancierian) view that what concerned scholars should do is not chiefly *speak for* or *speak of*, but rather, what I have referred to as *speak with*. Moreover, like Rancière, Gilligan considers the narrative mode helpful for communicating knowledge in an accessible, *egalitarian* manner and for seizing the polyphonic nature of human experience (her "Listening Guide" is there to help scholars "tune their ear to the multiplicity of voices" present in every human being; see [47] (p. 76); [2] (p. 2).

Several care theorists have appealed to the work of Gilligan and that of Iris Marion Young [48] and Margaret Walker [49] to emphasize the importance of narratives (but cf. Tronto [36]: p. 63). Jorma Heier [5], for instance, has shown the significance of Young's discussion of alternative forms of communication (e.g., greeting, storytelling) for addressing historical injustices and improving democratic participation more generally. Care ethicists such as Alain Loute and Patrick Schuchter also propose fine reflections on "storytelling" in their work on the relevance of narrative ethics within clinical settings. Inspired by the work of Rita Charron (an authority on narrative ethics), Loute [50] argues that such ethics can help democratize healthcare institutions and participate in the cultivation of better listening and attention skills in health care professionals. (Note that Gilligan's "listening guide" [47] begins with a brief passage from Charron).

Schuchter and Heller's [51] analysis of "care dialog" echoes partially Rancière in that it underscores the powerful *effects* of sharing stories and of *thinking with* stories. Also similarly to Rancière, these authors suggest that storytelling can blur the line between who teaches and who learns; it can help flatten (epistemic) hierarchies in ethical decision-making. Based on many narrative workshops held in nursing homes, they came to one of the following conclusions:

> The only chance to equalize asymmetries and create understanding beyond social roles is to give priority to the elementary storytelling of the individuals concerned. By giving narratives (stories) the status of the central language game in ethical deliberation, the dominance of expert knowledge is annulled in favor of a democratization of the opportunities to speak and a consequent participation. [51] (p. 59)

Like other care ethicists interested in narratives, Schuchter and Heller do not, of course, completely dismiss the relevance of medical knowledge/expertise, but they think patients ought to better participate in the decision-making that concerns them. (Needless to say, this widely embraced acknowledgement on care ethicists' part that narrative modes may *not* be appropriate in *all* settings and situations is in harmony with care ethics' commitment to always attend to particulars.) I wish to suggest that care ethics' and Rancière's insights on narratives are quite pertinent for the communities that seek to answer their citizens' desire to have more deliberative, *shared* decision-making in health care and to finally put to rest the old paternalistic (and hierarchical) relationships between doctors and patients. For example, the Canadian Royal College of Physicians [52] has explicitly committed to an approach to physician-patient relations that is more dialogical and participatory, and that gives a pride of place to the knowledge and situatedness of patients—to their values, preferences, and experiences (all of which are often communicated in a *narrative mode*). While Rancière himself might have suspicions about the possibility of enacting these learning models chiefly in traditional, hierarchical medical education, I would like to suggest that a very practical application of his insights (and of care ethics') might entail medical and nursing faculties adding more training in narrative ethics to their curricula. After all, as some researchers have indicated [53], it is precisely those groups most marginalized (e.g., racialized groups, the elderly) that have so much to gain from a flattening of epistemic hierarchies and from a better appreciation of the powers of story-telling.

5. The Importance of "Ordinary Heroes" and of Desiring Equality

"Love is the enemy of patriarchy, crossing its boundaries, dissolving its hierarchies." Gilligan [1] (p. 42)

The previous section indicated that Rancière's analysis of narratives echoes with care ethics scholars', but we have yet to consider what "type" of stories we are talking about here. What I would like to suggest below is that what both Rancière and care ethics seek are stories about the *ordinary*. One of the more implicit intentions in what follows is to nuance the prevalent interpretation of Rancière as a theorist of radical exceptions or ruptures. (If that prevalent reading were entirely correct, this would obviously make him a poor friend of care ethics in my view.) I wish to argue that if Rancière is partially a theorist of ruptures—he *is* fascinated by singular "events" and the heroic, as indicated in his book *Disagreement*—he is *also* an inspiring theorist of the ordinary. For both care ethicists and Rancière, the word "ordinary hero" is no oxymoron.

The care ethics scholar who has been most central for giving center stage to the ordinary is Sandra Laugier, whose writings have heavily drawn on Ordinary Language Philosophy' (e.g., Stanley Cavell and Cora Diamond). In Laugier's view, care ethics—just like Ordinary Language Philosophy—redefines ethics as a matter of "attention to ordinary life", to "what we are unable to see, to what is right before our eyes" (Laugier in [11]) (p. 33). Contrary to dominant moral theories that begin *from* universal principles, care ethics for Laugier *begins* with *experiences of the everyday* and the "moral problems of real people in their ordinary lives." As such, care ethics has contributed to a revolution in philosophy by giving "a voice to the ordinary" (Laugier in [11]) (p. 31)—precisely what Gilligan accomplished in her 1982 ground-breaking work according to Laugier. For Pascale Molinier as well (Molinier in [11]) (p. 88), care ethics' study of the "ordinary" and everyday life radically changed philosophy and disciplines like sociology of work.[20] One can see Molinier's strong interest in ordinary men and women in much of her scholarship, where she gives central stage to janitors, personal support workers, and numerous ordinary care *receivers* caught up in various challenges and (all-too-human) desires, joys, and frustrations. Consider the (real) story of long-term care home resident "Monsieur Georges", thoughtfully recounted by Molinier to capture the messy complexity of care receiver/giver dynamics, and the subtle heroism of the staff that attends to the disagreeable Monsieur Georges [24] (pp. 237–241).

If the claim that care ethics is inseparable from attentiveness to the ordinary is hardly a controversial one, the same cannot be said for Rancière, whose name tends to be commonly associated with the "exceptional". But perhaps there is another, more nuanced and appropriate way of reading Rancière. Above, we discussed various aspects of his method of equality, which informs his approach to knowledge (co)production and sharing, and his approach to social transformation. Recall that Rancière wants us to appreciate the *radical effects* the presupposition of equality can have—but he seeks to unearth these in *everyday* practices. Jacotot's method can be set to "work everywhere at any time" [32] (p. 155); and dissensus, Rancière insists, "can start from an imperceptible modification of the forms of *everyday experience*" [32] (p. 140). Recall that this is what *Proletarian Nights* depicted: the (heroic) power of "simple" nightly pursuits taken up by workers.

Jason Frank [54] (p. 259) is thus correct to suggest that there is a rich "politics of the ordinary" in Rancière's early work. We also see this "politics of the ordinary" cohabitate with a "politics of the extraordinary" in a brief text [37] where Rancière comments on the publication of a massive history of women. Rancière celebrates the project of bringing women out of history's margins, but he faults the editors of *L'Histoire des femmes* for having paid insufficient attention to women's work—effectively making "the quotidian disappear" [37]—and for having paid insufficient attention to singular or "atypical", "heroic cases". While most of us would see an evident tension in these reproaches (there seems to be at once insufficient and excessive attention paid to the "ordinary"), Rancière sees none. In harmony with his account of politics as a "crossing of boundaries", Rancière is convinced that many historical cases and stories can cross the border "between the event and non-event, the ordinary and non-ordinary" [22] (p. 39). Differently put: the singular can live *in* the ordinary, the heroic in the everyday.

Now, if attending to "small dramas of sensory redistribution" [54] (p. 259) in our *scholarly* work can have radical transformative effects, so can the actual performance of seemingly "ordinary" political acts according to Rancière (recall the example of the silent man protesting in Gezi Park). Rancière likes to underscore the power of certain forms of (non-spectacular) occupation: for instance, an "unexpected crowd of anonymous persons" gathering without "specific claims, just to affirm their refusal of the way in which the spaces and times or (their) lives are managed by the alliance of state powers and financial oligarchies" [22] (p. 43). While these (modest) gatherings may be fleeting and small-scale, he argues that they can chip away at dominant norms and institutions.

We noted above that Rancière is radically dismissive of institutions' ability to participate directly in emancipation. And yet, his is not a theory of socio-political life that entirely discards hope in institutional change (even if ultimately Rancière refuses to see institutional life and change as worthy of the term *politics*). Rancière might be (personally) *uninterested* in institutions, but he does recognize their ability to be improved. And significantly for our purposes, Rancière on occasions notes that this change might come via small-scale contesting of norms and rules. What he terms a "slow subversion" through "ordinary" gestures and claims [22] (p. 43) can have immense significance. After all, "there is no group strength independent of the strength with which individuals tear themselves out of the ether world of inarticulate sounds […] emancipation [makes] its way forward through a multitude of individual experiences" [55] (p. 50).

Hence, in spite of his polemical attack against everything that the police represents and does, Rancière still acknowledges, in rare but significant passages, that "there is a worse and a better police" [20] (pp. 30–31). While this is not really spelled out, he says that one way to identify a *better* police is on basis of whether it has regularly faced (and been receptive to) various mobilizations that affirm equality [20] (p. 31). In brief, there seem to be 'police processes' and institutions that are more amenable than others to the expressions of our passion for equality. What this suggests in my view is that the line between hierarchical/"bad" institutions and egalitarian/"good" politics might be much finer than Rancière is willing to acknowledge. Consider the following passage, where he locates (modest) emancipatory practices *within* institutions:

> There is no un-explicative institution. But there are a multiplicity of practices, *inside* or outside the dominant institutions, which extend the community of equal speaking beings and open new paths, […] new forms of access to research and knowledge. [22] (p. 43; my italics)

Quite significantly for our purposes, this multiplicity of practices includes moments of subjectivation that are *not* located on the front line of spectacular revolutionary political battles. He confides in an interview that he never wanted to claim that "equality exists only on the barricades"[21]; he is rather convinced that politics and equality are also found in the actions and speech claims of *ordinary* men and women.

This section has, thus far, sought to indicate that the "ordinary" is given a pride of place in Rancière and Gilligan's account of what stories *matter*. What also brings the two authors together is that both seek out (ordinary) stories about something we *all already* have within ourselves: a desire. For Gilligan, this is the desire for love; for Rancière it is the desire for equality. We do not need to 'learn' this desire or have it explained to us: both authors insist we *already know this*. But our desire for equality is buried under something that needs *un*-learning: the *opinion* of inequality.

One can see the significance of the desire for equality in the exchange Rancière had with Axel Honneth. Here, appealing to the stories of Jeanne Deroin and Rosa Parks, Rancière faults Honnethian theory for analyzing politics chiefly in terms of pathologies/suffering that individuals mobilize to address. "When Jeanne Deroin made this claim to be a candidate, she didn't need to run as a candidate in order to respond to [pathologies]". Rather, Rancière insists, "she did it in order to construct another world, another relation between the domestic and the political space" [32] (pp. 119–120). For Rancière, what both Parks and Deroin were driven by was a strong *desire* for another world, for "other ways of being" [32] (p. 126).

In Rancière and Gilligan, there is in my view a similar oscillation between exceedingly somber "diagnostics" regarding the hierarchies that have structured our lives for times immemorial, and a disconcertingly naive optimism regarding what is required to overcome these hierarchies. Illustrative of this optimism is the way they both present emancipation/democratic politics as "easy" to achieve: we "just" need to listen to what we *already know* and *feel*. "This is what continues to fuel my optimism", candidly observes Gilligan, "that we have within ourselves the potential to free our humanity from a false story" [1] (p. 178). Jacotot/Rancière echoes back: surely "you already have all it takes to emancipate yourself"; all you need is the *will* to embrace equality. "What more is needed?" asks Rancière [18] (p. 23).[22] The answer: nothing more than *attending* to this desire; it is *that* simple for him.

The feminist materialists among us would undoubtedly want to ask: … *but is it?* There is indeed a problematic bootstrap kind of argument in Rancière, an argument that is sometimes expressed in individualist overtones that do violence to care ethics (an ethics that ascribes much significance to interdependence and relationality). (Rancière's bootstrap argument is largely the result of his wariness about institutions and his concern that the latter tend to "stultify" individuals [18].) As noted above, it is here that care ethics can offer a healthy corrective to Rancière, by stressing the *indispensability* and desirability of welfare programs (such as child care) and social distributive measures for (individual) well-being, autonomy, and gender equality. Indeed, care ethics' radical commitment to a *relational* account of autonomy makes it more fit at assessing what advancing equality and freedom requires (e.g., Doucet [56] on parental leaves and social justice; see also [8–10]). Rancière polemically affirms in several places that just like education, liberty is won and lost "solely by each person's effort" [18] (p. 62). While Rancière recognizes that differences in wealth, race or linguistic capital will affect one's life chances, he is so insistent on polemically proposing a theory that starts *from* emancipation (rather than suffering) and that starts *from* the opinion of equality (rather than inequality) that I believe he sometimes completely loses touch with the "Weight of the World"—that weight powerfully described by Bourdieu and his colleagues in a book bearing this very title.[23]

6. Conclusions

This paper has sought to initiate a dialogue between care ethics scholarship and Rancière's work—most notably around their shared critiques of hierarchy and dichotomies, their call to "blur boundaries" and their reflections on the significance of narratives and of the ordinary. My list of affinities between Rancière's work and care ethics is obviously not exhaustive; much could have been said, for instance, about their shared conviction that *attention* is crucial for ethical life, or about their rejection of the theory/practice split. Throughout, I have also underscored differences in their perspectives, which I argued can serve as friendly mutual correctives. In particular, I noted that feminist care ethics can enrich Rancièrian theory in light of its more nuanced perspective on institutions' role, and in light of its more radical and *consistent* challenge to the hierarchy between physical and intellectual labor. (Rancière's "failure" here might partially be the result of his gloomy assessment of the extent to which *permanent* (as opposed to fleeting) social transformation was ultimately possible).

I have also made the claim that reading Rancière is timely and valuable for care ethicists because of Rancière's insight on the *necessarily* conflictual nature of politics (politics here understood in the specific Rancierian sense). This insight, I argued, is something care scholars might wish to consider a bit more closely as they pursue their reflections on what it would *require* to take away the "free passes" from the hands of the privileged and, more significantly, what the *effects* of this might be. Using Rancière's work, I have also called into question the desirability of separating too neatly the so-called two 'strands' in care ethics research.

In the remaining space at my disposal, I would like to underscore another 'uptake' offered by a dialogue between Rancière and care ethics—one rooted in the issue of desire and hope. What we saw above is that for Rancière, it is critical that the stories told about people not only be about suffering, but also about their desire (and hope) to create a different order (this is what he stressed in his account of the stories of Rosa Parks and Jeanne Deroin, and in those concerning workers (in *Proletarian Nights*)). For him, a Bourdieusian lexicon of "suffering" cannot capture what largely leads to a politics: "acting politically, very often, comes because some forms of ruptures appear *possible* […] It is very rare that suffering produces politics by itself" [32] (p. 126; my italics). Rancière thus invites scholars to exercise caution with a grammar of suffering, which is not equivalent to putting aside the language of injustice: "we cannot break with the logic of the reproduction of suffering if we don't also break with the very language of suffering in approaching society and individuals. Taking injustice as a starting point is not the same as starting from suffering" [32] (p. 127).

What Rancière is claiming here is reminiscent of Indigenous scholar Eve Tuck's characterization of the two main types of research done on marginalized groups: desire-based and damage-based research. For Tuck [57], the later refers to scholarly work that chiefly seeks to chronicle people's stories of pain/brokenness, and while doing so, utterly disregards stories of hope or desire (the latter are deemed less useful for provoking indignation and rectifying oppression). For Tuck, while the strategic goal may be laudable, it is important that scholars be more cognizant of the *impact* this research can have on their research "objects" (here, Indigenous communities). There are consequences for these "objects" to be described *in terms of* suffering or to come to define themselves *chiefly as damaged* [57] (p. 415). Hence Tuck's call for more heavily desire-based research, which documents "not only the painful elements of social realities but also the wisdom and hope" [57] (p. 416). But to be absolutely clear: the goal is definitely *not* to brush aside past (wrongs) or present suffering; it is rather to de-pathologize "the experiences of dispossessed and disenfranchised communities so that people are seen *as more than broken* and conquered. This is to say that even when communities *are* broken and conquered, *they are so much more than that*" [57] (p. 416; my italics). This is why Tuck regards desire-based research *not* as the opposite of damage-based inquiry, but as a type of research that *troubles* the binary between the endless reproduction (of suffering) and resistance/change.

Rancière's invitation to *begin* with equality and desire in research, combined with his view that suffering workers (and marginalized women) are about *so much more* than tired, "broken" bodies speaks—if imperfectly—to what Tuck is calling for. As we saw, *Proletarian Nights* is desire-based research on 19th century manual workers, whose nightly activities allowed to express pain *and* hopeful desire, and whose accomplishments showed that they could *"already* live" the impossible [19] (p. 8; my italics). This emphasis on hope, desire and present possibilities represents another crucial Rancierian insight that care ethicists may wish to pay attention to. Indeed, as they make their most powerful case against patriarchy-induced suffering and colonial injustices, and as they seek to ameliorate the situation of exploited essential workers, care ethicists might wish to be wary of succumbing to the temptation of conducting (excessively) "damage-based" research. Not only might that bolster the dignity and agency of the "objects" of their research, it might also help them finally lay to rest the old—and dubious—charge that care ethics is tainted by paternalism.

Funding: This research received no external funding.

Acknowledgments: I wish to thank Maggie FitzGerald, Maurice Hamington and two anonymous reviewers for their helpful feedback on this paper. All errors and limitations that remain are, naturally, mine. I also want to thank Stéphanie Mayer for many rich conversations we have had about some of the issues raised in the paper.

Conflicts of Interest: The author declares no conflict of interest.

Notes

1   I say "largely" because one cannot overstate the importance of Sara Ruddick's "Maternal Thinking" (published before Gilligan's work). On Ruddick's significance for care ethics, see Urban and Ward 2020 (introduction).

2   "Within a patriarchal framework, care is a feminine ethic. Within a *democratic* framework, care is a human ethic" [1], p. 22.

3   The term 'disagreement' is typically used by translators, but most acknowledge the inadequacy of the translation.

4   See Robinson (2020) and FitzGerald's (2020) characterization.

5   He describes his approach to the study of 19th century texts as follows: "My idea was that there are resonances, and *things you can feel and understand based on those resonances*; there's no need to know if the workers read Jacotot or if Marx read this particular Saint-Simonian pamphlet or whatever. There are signifiers that circulate and crystallize historic experiences, situations, movements, projects." [15], p.35; my italics.

6   E.g., democratic action is "the affirmation of the equal capacity of anybody" [17], p. 120.

7   Heier [7] (p. 9) argues that Rancière's account of binaries is highly instructive because it illuminates the problems with too neat a dichotomy between care receivers vs. care givers—a binary that still unfortunately ends to pop up in care ethics scholarship according to her. Hence, for Heier, care ethicists could go *even further* in their "Rancierian" embrace of challenging dichotomies.

8   Rancière [19], p. 8 (my translation).

9   Much ink has been spilled over whether this peculiar, quite "narrow" conception of politics is fruitful or problematic. Some have suggested that there is a kind "politics of purity" here that ends up reiterating the binaries Rancière sought to interrogate in the first place (e.g., Slavoj Zizěk & Jodi Dean); others have contested this view. For two important texts on this issue, see Samuel Chambers [33] and Todd May [34]. In my view, Chambers convincingly shows us that politics is always interlaced with the police—it is always 'impure'. See how Rancière himself answers the charge of 'purity' at Rancière [35], pp. 2–3.

10   Rancière [16], pp. 199–200.

11   Here I am alluding to the certain "horror" with which Arendt describes, in the *Human Condition* (ch.2), the growth of the "social" and the "irruption of pregnant bodies" in the public sphere.

12   Brugère [38], p. 145 (my translation); also Hamrouni 2020, 157.

13   As I argued elsewhere [41], this has important implications for how we theorize *responsibility* for redressing problems of democratic exclusion: the burden gets shifted onto *listeners* instead of resting chiefly on (marginalized) speakers.

14   Gilligan strikingly sums up her entire output as one "grounded in listening" [10], p. xiii. In some of the interviews she conducted with 'at risk' young Black women in the USA (e.g., *Between Voice and Silence* [42]), Gilligan showed quite powerfully that these women had what some people would call a "voice" (many of them were in fact criticized by teachers and by members of dominant institutions for speaking *too loudly*), but they were clearly not *listened* to. Hence, to put it most succinctly: these young women had no voice worthy of the name.

15   I cannot delve into an exploration of Rancière's correctives to Habermasian discourse ethics; I merely note that some of these echo Robinson's (e.g., Rancière [18] (ch.5); [20]).

16   We can move between spheres because as Rancière/Jacotot tells us, *The Ignorant Schoolmaster* is not about education—it is about emancipation and about how the "explanatory logic" affects *all* institutions ([20], p. 33).

17   With this term 'ignorant' he emphasizes *ignoring* as discounting or overlooking (of the opinion of inequality).

18   I use the term even though Rancière [38] (p. 62) expressed hesitation about it (he preferred 'storytelling'). But his qualms only concern a *particular* type of narrative—i.e., "grand narratives" that subsume in a single story a "multiplicity of voices, identities, and language games".

19   I say most often because ultimately, Rancière does resort to *some* explanations or guidance (so did his 'model' Jacotot).

20   See also Hamrouni's insightful account of *ordinary* vulnerability (2020).

21   This is from an interview (2012), where he also said: "I am not a thinker of the event, of the upsurge, but rather of emancipation as something with its own tradition, with a history that isn't just made up of great striking deeds but also of the ongoing effort to create forms of the common different from the ones on offer from the state" (cited in [54] 259).

22   "Don't say you can't. You know how to see, how to speak, you know how to show, you can remember. What more is needed?" [18], p. 23.

23   I am obviously referring here to Bourdieu's edited volume on social suffering, which bears in English the title *The Weight of the World: Social Suffering in Contemporary Society.*

References

1. Gilligan, C. *Joining the Resistance*; Polity Press: Cambridge, UK, 2017.
2. Gilligan, C. *In a Different Voice*, 2nd ed.; Harvard University Press: Cambridge, MA, USA, 1993.
3. Diprose, G. Radical Equality, Care and Labour in a Community Economy. *Gend. Place Cult.* **2017**, *24*, 834–850. [CrossRef]
4. Myers, E. *Worldly Ethics. Democratic Politics and Care for the World*; Duke University Press: Durham, UK, 2013.
5. Heier, J. Disabling Constraints on Democratic Participation. *Ethics Soc. Welf.* **2015**, *9*, 201–208. [CrossRef]
6. Heier, J. Democratic Inclusion through Caring Together with Others. In *Care Ethics, Democratic Citizenship and the State*; Urban, P., Ward, L., Eds.; Palgrave Macmillan: London, UK, 2020.
7. Heier, J. Political Repair in Relation to Tronto's Political Ethics of Care. 2018. Available online: ethicsofcare.org (accessed on 10 October 2021).
8. Engster, D.; Hamington, M. (Eds.) *Care Ethics and Political Theory*; Oxford University Press: Oxford, UK, 2015.
9. Barnes, M.; Brannelly, T. (Eds.) *Ethics of Care. Critical Advances in International Perspective*; Polity: Bristol, UK, 2015.
10. Urban, P.; Ward, L. (Eds.) *Care Ethics, Democratic Citizenship and the State*; Palgrave: Cham, Switzerland, 2020.
11. Bourgault, S.; Vosman, F. (Eds.) *Care Ethics in Yet a Different Voice*; Peeters: Leuven, Belgium, 2020.
12. Gilligan, C. Strong democracy and a different voice: What stands in the way? In *Strong Democracy in Crisis*; Norris, T., Ed.; Lexington: Lanham, MD, USA, 2016.
13. Robinson, F. Resisting Hierarchies through Relationality in the Ethics of Care. *Int. J. Care Caring* **2020**, *4*, 11–23. [CrossRef]
14. FitzGerald, M. Reimagining Government with the Ethics of Care: A Department of Care. *Ethics Soc. Welf.* **2020**, *14*, 248–265. [CrossRef]
15. Rancière, J. *The Method of Equality*; Polity: Cambridge, UK, 2016.
16. Rancière, J. Politics and Aesthetics. *Angelaki* **2003**, *8*, 191–211. [CrossRef]
17. Rancière, J. A Few Remarks on the Method of Jacques Rancière. *Parallax* **2009**, *15*, 114–123. [CrossRef]
18. Rancière, J. *The Ignorant Schoolmaster: Five Lessons in Intellectual Emancipation*; Ross, K., Translator; Stanford University Press: Stanford, CA, USA, 1991.
19. Rancière, J. *La Nuit des Prolétaires*; Fayard: Paris, France, 2012.
20. Rancière, J. *Disagreement: Politics and Philosophy*; Rose, J., Translator; University of Minnesota Press: Minneapolis, MN, USA, 1999.
21. Rancière, J. Ten Theses on Politics. *Theory Event* **2001**, *5*. [CrossRef]
22. Rancière, J. Un-What? In *The Pedagogics of Unlearning*; Seery, A., Dunne, É., Eds.; Punctum Books: Santa Barbara, CA, USA, 2016.
23. Gilligan, C.; Snider, M. *Why Does Patriarchy Persist?* Polity Press: Cambridge, UK, 2018.
24. Molinier, P. Quel est le bon témoin du care. In *Qu'est-ce Que Le Care?* Molinier, P., Laugier, S., Paperman, P., Eds.; Payot: Paris, France, 2009.
25. Gilligan, C. Moral Orientation and moral development. In *Women and Moral Theory*; Kittay, E.F., Meyers, D.T., Eds.; Rowman and Littlefield: Lanham, MD, USA, 1987.
26. Slote, M. *The Ethics of Care and Empathy*; Routledge: London, UK, 2007.
27. Brugère, F. *Le Sexe de la Sollicitude*; Seuil: Paris, France, 2008.
28. Pulcini, E. What Emotions Motivate Care? *Emot. Rev.* **2016**, *9*, 64–71. [CrossRef]
29. Paperman, P. *Care et Sentiments*; PUF: Paris, France, 2013.
30. Davis, O. *Jacques Rancière*; Polity: Boston, USA, 2010.
31. Bargetz, B. The Distribution of Emotions: Affective Politics of Emancipation. *Hypatia* **2015**, *30*, 580–596. [CrossRef]
32. Rancière, J.; Honneth, A. *Recognition or Disagreement*; Genel, K., Deranty, J.P., Eds.; Columbia University Press: New York, NY, USA, 2017.
33. Chambers, S.A. Jacques Rancière and the problem of pure politics. *Eur. J. Political Theory* **2011**, *10*, 303–326. [CrossRef]
34. May, T. *The Political Thought of Rancière*; Penn State University Press: University Park, PA, USA, 2008.
35. Rancière, J. The thinking of dissensus: Politics and aesthetics. In *Reading Rancière*; Bowman, P., Stamp, R., Eds.; Continuum: New York, NY, USA, 2011.
36. Tronto, J. *Caring Democracy: Markets, Equality, and Justice*; New York University Press: New York, NY, USA, 2013.
37. Rancière, J. Sur le Devenir-Sujet Politique 'des' Femmes. *Cah. du Genre* **1994**, *11*, 23–40.
38. Brugère, F. Quelle politique du 'care' dans un monde néolibéral? In *Politiser le Care*; Le Bord de L'Eau: Lormont, France, 2012.
39. Casalini, B. (Ed.) *Il Peso del Corpo e la Bilancia Della Guistizia*; IF Press: Roma, Italy, 2017.
40. Robinson, F. Stop Talking and Listen: Discourse Ethics and Feminist Care Ethics in International Political Theory. *Millenium* **2011**, *39*, 845–860. [CrossRef]
41. Bourgault, S. Attentive listening and care in a neoliberal era: Weilian insights for hurried times. In *Etica & Politica/Ethics & Politics XVIII/3"*; EUT Edizioni Università di Trieste: Trieste, Italy, 2016; pp. 311–337.
42. Taylor, J.M.; Gilligan, C.; Sullivan, A.M. *Between Voice and Silence. Women and Girls, Race and Relationship*; Harvard University Press: Cambridge, MA, USA, 1997.
43. Tronto, J. *Moral Boundaries*; Routledge: New York, NY, USA, 1994.

44. Mayer, S. Irresponsabilités Relationnelles, Asymétries de Pouvoir et Injustices Ordinaires du Quotidien. *Can. J. Political Sci.* **2021**, *54*, 791–808. [CrossRef]
45. Vosman, F. The disenchantment of care ethics: A critical cartography. In *The Ethics of Care: The State of the Art*; Vosman, F., Baart, A., Hoffman, J., Eds.; Peeters: Louvain, Belgium, 2020.
46. Gilligan, C. Hearing the Difference: Theorizing Connection. *Hypatia* **1995**, *10*, 120–127. [CrossRef]
47. Gilligan, C.; Eddy, J. Listening as a Path to Psychological Discovery. *Perspect. Med. Educ.* **2017**, *6*, 76–81. [CrossRef] [PubMed]
48. Young, M. *Inclusion and Democracy*; Oxford University Press: Oxford, UK, 2010.
49. Walker, M. *Moral Understandings. A Feminist Study in Ethics*, 2nd ed.; Oxford University Press: Oxford, UK, 2007.
50. Loute, A. L'éthique clinique narrative porte-t-elle attention aux subjectivités? In *Valeurs de L'Attention*; Grandjean, N., Loute, A., Eds.; Septentrion: Lille, France, 2019.
51. Schuchter, P.; Heller, A. The Care Dialogue: The 'Ethics of Care' Approach and its Importance for Clinic Ethics Consultation. *Med. Health Care Philos.* **2018**, *21*, 51–62. [CrossRef] [PubMed]
52. Royal College of Physicians and Surgeons of Canada. Can-MEDS. Available online: https://www.royalcollege.ca/rcsite/canmeds/framework/canmeds-role-communicator-e (accessed on 13 April 2022).
53. Haesebaert, J.; Adekpedjou, R.; Croteau, J.; Robitaille, H.; Légaré, F. Shared decision-making experienced by Canadians facing health care decisions: A Web-based survey. *CMAJ* **2019**, *7*, E210–E216. [CrossRef] [PubMed]
54. Frank, J. Logical Revolts: Jacques Rancière and Political Subjectivization. *Political Theory* **2015**, *43*, 249–261. [CrossRef]
55. Rancière, J. *On the Shores of Politics*; Verso: London, UK, 2007.
56. Doucet, A. The ethics of care and the radical potential of fathers 'home alone on leave': Care as practice, relational ontology, and social justice. In *Comparative Perspectives on Work-Life Balance and Gender Equality. Life Course Research and Social Policies*; O'Brien, M., Wall, K., Eds.; Springer: Cham, Switzerland, 2017; Volume 6.
57. Tuck, E. Suspending Damage: A Letter to Communities. *Harv. Educ. Rev.* **2009**, *79*, 409–427. [CrossRef]

philosophies

MDPI

Article

Violence and Care: Fanon and the Ethics of Care on Harm, Trauma, and Repair

Maggie FitzGerald 

Department of Political Studies, University of Saskatchewan, Saskatoon, SK S7N 5A2, Canada; maggie.fitzgerald@usask.ca

Abstract: According to Frantz Fanon, the psychological and social-political are deeply intertwined in the colonial context. Psychologically, the colonizers perceive the colonized as inferior and the colonized internalize this in an inferiority complex. This psychological reality is co-constitutive of and by material relations of power—the imaginary of inferiority both creates and is created by colonial relations of power. It is also in this context that violence takes on significant political import: violence deployed by the colonized to rebel against these colonial relations and enact a different world will also be violent in its fundamental disruption of this imaginary. The ethics of care, on the other hand, does not seem to sit well with violence, and thus Fanon's political theory more generally. Care ethics is concerned with everything we do to maintain and repair our worlds as well as reasonably possible. Violence, which ruptures our psycho-affective, material, and social-political realities, seems antithetical to this task. This article seeks to reconsider this apparent antinomy between violence and care via a dialogue between Fanon and the ethics of care. In so doing, this article mobilizes a relational conceptualization of violence that allows for the possibility that certain violences may, in fact, be justifiable from a care ethics perspective. At the same time, I contend that violence in any form will also eventually demand a caring response. Ultimately, this productive reading of Fanon's political theory and the ethics of care encourages both postcolonial philosophers and care ethicists alike to examine critically the relation between violence and care, and the ways in which we cannot a priori draw lines between the two.

Keywords: care ethics; Fanon; colonialism; violence; care; harm; relationality

Citation: FitzGerald, M. Violence and Care: Fanon and the Ethics of Care on Harm, Trauma, and Repair. *Philosophies* **2022**, *7*, 64. https://doi.org/10.3390/philosophies7030064

Academic Editors: Marcin J. Schroeder and Maurice Hamington

Received: 16 March 2022
Accepted: 6 June 2022
Published: 8 June 2022

Publisher's Note: MDPI stays neutral with regard to jurisdictional claims in published maps and institutional affiliations.

1. Introduction

Frantz Fanon is an important figure in contemporary post-colonial political thought and Black Atlantic theory, known for his theoretical reflections on and participation in anti-colonial liberation struggles (perhaps most notably, the Algerian revolutionary struggle in the 1950s). Fanon's political theory [1–3] is structured by two recurring themes. First, for Fanon, the psycho-affective realm and the social-political are deeply intertwined in the colonial context. Psychologically, the colonizers perceive the colonized as inferior, while the colonized internalize this in an inferiority complex. This psychological reality is co-constitutive of and by material relations of power—the imaginary of inferiority both creates and is created by colonial relations of power. It is also in this context that violence takes on significant political import: violence deployed by the colonized to rebel against these colonial relations and enact a different world will be violent in its fundamental disruption of this imaginary. Violence, therefore, is a second key theme in Fanon's work. Violence runs through the colonial project, the psyches of both colonizers and colonized, and if properly directed, violence has the potential to liberate colonized subjects from their chains.

At first glance, the ethics of care does not seem to sit well with violence, and thus Fanon's political theory more generally. Care ethics is concerned with everything we do to maintain and repair ourselves and our worlds as well as reasonably possible [4]. Violence, which ruptures our psycho-affective, material, and social-political realities, seems antithetical to this task. This article seeks to reconsider this apparent antinomy between violence and care. Specifically, via a dialogue between Fanon's political theory and the ethics of care, this article mobilizes a relational conceptualization of violence that allows for the possibility that certain forms or instances of violence may, in fact, be justifiable from a care ethics perspective. At the same time, I also contend that violence in any form will eventually demand a response of care. In this way, violence and care, I suggest, are not necessarily diametrically opposed. Instead, certain violences may be necessary for (or at the least, not always and already incompatible with) establishing care (relations and practices that allow us to live well). And, certainly, I assert, the aftermath of violence will always demand care.

In demonstrating these lines of connection between/across violence and care, I also lay groundwork for a continued dialogue between Fanon's political theory and a critical and political ethics of care. Fanon's political theory, concerned with undoing the oppressive and dominating relations that produce and sustain colonialism, has much to offer care ethicists who also seek to "undo the world as it is" [5]. The focus on violence—and how violence operates in and across different registers—can provide strategic vantage points for thinking through when, where, and in what ways the care ethics project of building a more caring world may enact or even require violence. At the same time, a care ethics lens amends Fanon's political theory, as it foregrounds the fact that even if violence is justifiable when it is directed at the destruction of unequal relations of power, violence never leaves anyone unmarked. As a result, I argue that a care-oriented intervention will always eventually be needed to interrupt cycles of violence and tend to the trauma and harm that arise in its wake. More simply, my hope is that this productive reading of Fanon's political theory and the ethics of care encourages both postcolonial philosophers and care ethicists alike to examine critically the relation between violence and care, and the ways in which we cannot a priori draw lines between the two. Neither violence nor care can be meaningfully understood, defined, or even critiqued when abstracted from context and the experiences of those living in and enacting it.

To make this argument, this article proceeds as follows. First, I review in detail Fanon's political theory, focusing on the two themes noted above: the intimate relation between the social-political and the psycho-affective realms, and violence. Following Elizabeth Frazer and Kimberly Hutchings [6], I argue that these two themes together allow Fanon to assert violence as both a doing (an instrumental tool that can be wielded towards anti-colonial ends) and a being (a unique libidinal energy that profoundly shapes the colonial subject). As Fanon argues, when the being and doing of violence are productively directed, they hold the potential to undo colonial relations and decolonize the mind. Next, I turn to a brief discussion of Fanon's final substantive chapter in *The Wretched of the Earth* [2], where Fanon presents several cases of mental disorders that have arisen in liberatory war. Here, I argue, Fanon undermines much of his claims about the liberatory potential of violence: in contrast, those who have enacted violence are presented as traumatized and deeply harmed. Violence does not, in the end, seem to deliver on the promise of fostering the conditions under which liberated persons can create themselves anew. To help address this dissonance in Fanon's theory, I introduce a critical and political ethics of care, and argue that the relational social ontology and normative trajectory of care provide a theoretical vantage point from which to rethink violence along relational lines. Through such a rethinking, I maintain space for Fanon's claims about the liberating possibilities of violence by demonstrating that violence cannot be abstractly pre-defined or determined (see also [7]); by extension, violence cannot abstractly be deemed always and already unjustifiable. At the same time, however, I also contend that even when justifiable, one cannot end with violence—a caring response is inevitably required. Care, I conclude, can respond directly to violence and help move us from trauma and toward repair and the reconstructive task of building worlds anew.

2. Fanon's Political Theory in Two Themes

2.1. Theme I: The Social-Political and the Psycho-Affective Realms

As just noted, two interrelated themes weave together Frantz Fanon's political theory as presented in *Black Skin, White Masks* [3], *The Wretched of the Earth* [2], and *A Dying Colonialism* [1] [1]. First, Fanon's analysis of the colonial condition demonstrates repeatedly that individuals' psychological infrastructure and the social-political sphere—particularly the social-political structures of colonialism—are intimately and inescapably intertwined. Colonialism, as Fanon illustrates, is a project that marks all parts of the human person, and neither colonizer nor colonized is untouched. Second, the key to understanding and ultimately liberating all from the colonial relation is violence. The violence of colonial relations, the internalization of this violence, and the immanent productive capacities of this violence run through the contours of colonialism's totalizing project—and perhaps its end.

To trace these themes, which are central to this article's broader argument which seeks to foster a meaningful dialogue between a critical and political ethics of care and Fanon's political theory, it is most fruitful to begin with a brief presentation of *Black Skin, White Masks* [3]. As Fanon [3] (p. xvi) writes, the thesis of his first book is that colonialism and the corresponding "juxtaposition of the black and white races has resulted in a massive psycho-existential complex":

> *The white man is locked in his whiteness.*
>
> *The black man in his blackness.*
>
> *We shall endeavour to determine the tendencies of this double narcissism and the motivations behind it.*

Specifically, Fanon argues that the colonial relation and anti-Black racism fundamentally structures the social-political *and* internal psycho-affective landscapes of all those involved. On the one hand, the very notion of 'black people' and 'white people,' and the hierarchy between them, is a consequence of colonial relations of power, as Fanon attests repeatedly throughout the text [2]. Further, Fanon argues that the psychological outcome of this reproduction of a hierarchy of black and white is an internalized superiority complex by the colonizers and a corresponding internalized inferiority complex by the colonized. The integral relation between these 'psycho-existential complexes' and the colonial context are perhaps no better illustrated than in the first substantive chapter of the text, in which Fanon [3] (pp. 2–3) discusses the relationship between colonialism, racism, and language:

> *The problem we shall tackle in this chapter is as follows: the more the Black Antillean assimilates the French language, the whiter he gets—i.e., the closer he comes to becoming a true human being. […] A man who possesses a language possesses as an indirect consequence the world expressed and implied by this language.*
>
> *All colonized people—in other words, people in whom an inferiority complex has taken root, whose local cultural originality has been committed to the grave—position themselves in relation to the civilizing language: i.e., the metropolitan culture. The more the colonized has assimilated the cultural values of the metropolis, the more he will have escaped the bush. The more he rejects the bush, the whiter he will become.*

In this passage, one can see Fanon's argument clearly: language is a key part of world-making and world-sustaining. In a world shaped by colonial relations—by a racist hierarchy of colonizers versus colonized—to participate in the world-making practices of the language of the colonizers is to sustain and reproduce these relations. Subsequently, colonized peoples who adopt the language of the colonizers sustain and reproduce their own oppression. For the Black Antillean, speaking the French language is to "reflect the very structures of your alienation in everything from vocabulary to syntax to intonation" [8] (np). Furthermore, this type of reproduction of the colonial world by the colonized plays out, psychologically, in complex ways for both colonized and colonizer. For instance, as

Fanon notes [3] (pp. 18–9), when the Black Antillean speaks the French of France, he is met by the colonizer with surprise:

> *The fact is that the European has a set idea of the black man, and there is nothing more exasperating than to hear: 'How long have you lived in France? You speak such good French?' [. . .] There is nothing more sensational than a black man speaking correctly, for he is appropriating the white world.*

This brief passage is telling. The fact that the European is 'surprised' to hear such 'good French' reveals their superiority complex: in a colonial world, they have internalized their superiority, and language is one of the markers of this superiority. To hear, then, "a black man speaking correctly" is surprising in that it ruptures this psychic structuring principle (itself co-constituted by the colonial social-political context) of superiority. Simultaneously, the exasperation experienced by the black man upon encountering this surprise is tied to the ways in which the exchange foregrounds his 'inferiority'. "To speak a language is to appropriate its world and culture" [3] (p. 21). When one encounters surprise at the sheer fact that one *could* appropriate a world and culture that is demarcated as superior to one's own, one is reminded of said demarcation, of the reality that *within such a system, one is constructed and positioned as inferior.*

This theme recurs time and time again through *Black Skin, White Masks* [3], as Fanon demonstrates how the colonial division of superior/inferior is "embodied in desire and discontent, in neurosis" [6] (p. 95). The colonial context both traps the black man as a slave to his 'inferiority,' and the white man as a slave to his 'superiority' (both of which are, of course, constituted in and by colonial relations of power), and this can be evidenced by neurotic behaviours in both.

> *The problem of colonization, therefore, comprises not only the intersection of historical and objective conditions but also man's attitudes toward these conditions.*

[3] (p. 65)

The problem of colonialism involves, more precisely, ongoing attempts "to decipher the changing scale (measure, judgment) of a problem, event, identity, or action as it comes to be represented or framed in the shifting ratios and relations that exist between the realms of political and psycho-affective experience" [9] (p. xxxvii). Colonialism can only be assessed, and the behaviours of both colonizers and colonized can only be grasped, by foregrounding the ways in which the social-political realm and the psycho-affective realm are inextricably intertwined [3].

2.2. Theme II: Violence

With the interrelation of social-political and psycho-affective colonialism established, Fanon's well-known focus on violence, the second theme explored here, can be contextualized. Violence, in Fanon's work, operates from, in, and across a multitude of interconnected registers. First, there is the physical violence of the colonial relation. As Fanon writes in *The Wretched of the Earth* [2] (p. 23), "colonialism is not a machine capable of thinking, a body endowed with reason. It is naked violence". This violence, largely rooted in physical harm, i.e., "the infliction of physical injury and damage without benefit to the victim" [7] (p. 112), is key to the production and reproduction of the colonial relation, and thus the social-political colonial world. It is also directly implicated in a 'second' violence—the violent reordering of the psycho-affective realm that results in and from colonialism. The relation between these first two registers of violence brings us back to the theme just discussed. "The initial move, colonialism, inserts violence into the world" [6] (p. 95); this violence, and the social-political order it founds, reorganizes the psyches of both colonizer and colonized (albeit in different ways). This reorganization constitutes its own violence, especially for the colonized who experience the trauma of internalized inferiority complexes. The linking of these two violences, then, is a crucial pin that ties together Fanon's reflections in *Black Skin, White Masks* [3] and *The Wretched of the Earth* [2]. While his earlier writing in *Black Skin, White Masks* focuses on the effects of the internalized inferiority/superiority complexes that are produced in the colonial social-political landscape, these consequences are more acutely understood as forms of violence in *The Wretched of the Earth* [6] (p. 95): the colonial world is violent in every way for the colonized. As Fanon describes, for example, in *A Dying Colonialism* [1] (p. 128):

> *There is, first of all, the fact that the colonized person, […] perceives life not as a flowering or a development of an essential productiveness, but as a permanent struggle against an omnipresent death. This ever-menacing death is experienced as endemic famine, unemployment, a high death rate, an inferiority complex and the absence of any hope for the future.*
>
> *All this gnawing at the existence of the colonized tends to make of life something resembling an incomplete death.*

More simply, as Fanon [2] (p. 46) writes, "It is understandable how in such an atmosphere [i.e., colonial relations of domination] everyday life becomes impossible".

It is also the combination of these two interrelated violences that brings us to a third register of violence, pertaining to a libidinal energy which manifests in the colonial subject. Violence in this sense refers to a force or energy which stems from the ways in which "political and economic structures of exploitation and oppression are embodied in rage and resentment, and finally in the pathologies of body and mind" [6] (p. 95). The violence of colonialism generates a sui generis force [6] (p. 97) that is tied to the "struggle for psycho-affective survival and a search for human agency in the midst of the agony of oppression" [9] (p. xxxvi). Colonialism festers a rage and resentment, an anxiety, a libidinal energy that must ultimately, Fanon contends, be expressed. "In every society, in every community, there exists, or must exist, a channel, an outlet whereby the energy accumulated in the form of aggressiveness is released" [3] (p. 124).

Releasing this libidinal energy—itself the product of living in and internalizing conditions of violence—is necessary, and, for Fanon, requires physical violence [2] (p. 50–52]. The redemptive part of this violence, and therefore of violence *as such* for Fanon, is that if this libidinal energy is harnessed and released through violence that is directed at destroying the shackles of the unjust social-political order of the present, it can be embodied in a creative and productive way, with the ultimate hope of building the world anew. In this way, a fourth register of Fanonian violence is liberating: "What we are striving for is to liberate the black man from the arsenal of complexes that germinated in a colonial situation" [3] (p. 14). This libidinal energy, this "violence rippling under the skin" [2] (p. 31) becomes generative when "the colonized subject discovers reality and transforms through his praxis, his deployment of violence and his agenda for liberation" [2] (p. 21; see also pp. 89–96).

Reflexivity and consciousness are, notably, key for the productive release and enactment of this libidinal energy. That is, liberatory violence is not violence for violence's sake. This is perhaps best exemplified in the final chapter of *Black Skin, White Masks*, where Fanon [3] (p. 206) famously writes, "O my body, always make me a man who questions!":

> *It is through self-consciousness and renunciation, through a permanent tension of his freedom, that man can create the ideal conditions of existence for a human world.*

While the violence of the colonized is necessarily a violence that responds to the violence of colonialism, and as such, it is inherently a reactive violence, it also must be a violence that recognizes "itself as the source of a new world, a new order" [6] (p. 96). Reflection, questioning, and understanding the conditions which give rise to the necessity of this violent release is of crucial import for Fanon's vision of how this violence can lead to liberation. In their awareness of how things are—of the violence inflicted upon them, of the unjustness of their circumstance—the colonized are able to "skim over this absurd drama that others have staged" [3] (p. 174) and build a different world. The reflexive colonized subject will be able to direct meaningfully their violence toward the disruption and destruction of the social-political structure. Another passage from *Black Skin, White Masks* [3] (p. 80, emphasis in original) helps to demonstrate this point:

> *In other words, the black man should no longer have to be faced with the dilemma 'whiten or perish,' but must become aware of the possibility of existence; in still other words, if society creates difficulties for him because of his colour, if I see in his dreams the expression of an unconscious desire to change colour, my objective will not be to dissuade him by advising him to 'keep his distance'; on the contrary, once his motives have been identified, my objective will be to enable him to choose action (or passivity) with respect to the real source of the conflict, i.e., the social structure.*

Through understanding the context of colonialism and 'becoming aware of the possibility of existence', the colonized subject can choose to act in response to the real source of the conflict, the real source of the violence: the social-political structure of colonialism. And, to be sure, the colonized subject is, as Fanon highlights, ready for such reflection: "their entire recent history has prepared them to 'understand' the situation" [2] (p. 40). "At the very moment when they discover their humanity", when they realize that they are not inferior, as the colonial system has suggested, the colonized "begin to sharpen their weapons to secure its victory" [2] (p. 8). Questioning, critique, and awareness are crucially important in Fanon's political theory, as it is such consciousness that has the potential to harness violence (which is inevitable within the colonial context, given the violence of colonialism itself) and target it at its source, i.e., the social-political order. "This violent praxis is totalizing since each individual represents a violent link in the great chain, in the almighty body of violence rearing up in reaction to the primary violence of the colonizer" [2] (p. 50).

This brings us to the final form or register of violence in Fanon, again intertwined with the violences discussed above, which is the violence of decolonization. This violence is the most productive and creative of all. As Fanon [2] (pp. 1–2) boldly claims at the beginning of *The Wretched of the Earth*, "decolonization is always a violent event":

> *Decolonization, which sets out to change the order of the world, is clearly an agenda for total disorder.*

The violence of shattering a social-political structure, like colonialism, is undeniable. It will be a reordering of the world, and the opening up of productive possibilities for new worlds. When revolutionary violence is "purposeful, intentional, and oriented toward world-making" [8], "violence can be embodied in a creative way" [6] (p. 96) and it becomes associated with world making and creative genius that cannot yet be known [10]. Through this final phase of violence comes decolonization and horizons of possibility for a world without colonial oppression. This world is a future-oriented world, a world of freedom.

2.3. Themes I and II: The Productive Harmony of Violence, Context, and Subject

In this discussion of Fanonian violence, I have distilled five registers of violence which are, clearly, deeply interrelated. Indeed, Fanon's theory is "shot through with violence" and "connections are specified at every level—political and historical setting, interpersonal encounters at the level of the social network, down to the intra-personal level of emotional reaction and motivation" [6] (p. 95). Of these connections, I focus on one for the remainder of this section: the link between violence, context, and subject formation. Notably, this connection is best understood, as I endeavour to demonstrate now, by joining the two themes described above (that is, the relation between the social-political realm and the psycho-affective realm, and the theme of violence).

As Frazer and Hutchings [6] (p. 93) summarize, and as detailed above, violence, for Fanon, is "to begin with, immanent in political structures of power; second, it is embodied and libidinal". Violence permeates both the social-political context and the formation of subjectivities in these contexts. This point is captured by the interrelation between the first three registers of violence explicated above. It is for this same reason that violence, for Fanon, is also assigned a dual task [8], captured in the last two registers of violence explicated in the previous section. On the one hand, physical violence enacted by the colonial subject against those who dominate and oppress them can eliminate the violence of the material social-political reality of colonialism. It can violently rupture the social-political system, and clear the way for new social-political relations, new worlds. For the colonized, "shattering their chains" [2] (p. 34), which have themselves been forcibly and violently applied and maintained, can only come about through violence; the "naked violence" of colonialism "only gives in when confronted with greater violence" [2] (p. 23). On the other hand, violence is simultaneously tasked with eliminating the colonial system at the level of the psycho-affective realm. As colonized subjects direct and release their libidinal energy through violence, they release themselves from the colonial imaginary which positions them as inferior (discussed most fully in *Black Skin, White Masks*) and in so doing, open the possibility of constructing themselves anew. As Fanon [2] (p. 51) writes,

> At the individual level, violence is a cleansing force. It rids the colonized of their inferiority complex, of their passive and despairing attitude. It emboldens them and restores their self-confidence.

This dual task—in which violence both 'cleanses' the individual and releases their psyche from the colonial imaginary *and* ruptures the relations of power that sustain the colonial order—can also be thought of as violence as a doing and a being [6]. As Frazer and Hutchings [6] (p. 98) argue, violence as a doing captures violence "understood as a tool that can be used and then abandoned, which is of course why it is possible that violent revolution can nevertheless be the way to a new and peaceful world". This is the violence that will unmake colonial relations in the material social-political sense, and pave the way for a liberated world. Violence as being, as a libidinal energy, "is a force that is inherent in colonial structures of oppression, in everyday colonial life, in the psyche of the native-turned-citizen-solider" [6] (p. 98). This is the violence that, when directed in certain productive directions, will unmake the colonial subject and pave the way for new liberated subjectivities. In other words, when properly harnessed, a potential harmony between violence, context, and subjects can be found, one which can make "a new start, develop a new way of thinking, and endeavour to create a new man" [2] (p. 239).

2.4. An Accidental: 'Colonial War and Mental Disorders'

In music composition, an accidental is a note of a pitch (or pitch class) that is not a member of the scale or mode indicated by the most recently applied key signature. In this section, I suggest that the chapter 'Colonial War and Mental Disorders' from *The Wretched of the Earth* [2] serves as an accidental in Fanon's political theory as presented above. In particular, I argue that this chapter, as an accidental, is not 'in pitch' with the rest of Fanon's theory—it sits in dissonance with Fanon's claim that violence can serve as a mechanism to escape or end violence. As I argue subsequently, something more than violence may be (I suspect, frequently will be) necessary to build a new world and create a new humanity—this something else, I contend, is care.

Before developing this line of argument, however, it is necessary to explore the dissonance that is the chapter 'Colonial War and Mental Disorders'. In this chapter, Fanon deals with the problem of mental disorders that arise out of the national war of liberation waged by the Algerian people. Despite Fanon's earlier claims that violence, when properly aimed toward the social-political structure, will liberate the colonized subject, allow them to reassert their agency, and plant the seeds for the formation of new identities and subjectivities, he here tells a different story regarding the consequence of violence:

> Today the all-out national war of liberation waged by the Algerian people for seven years has become a breeding ground for mental disorders. […]

> These disorders last for months, wage a massive attack on the ego, and almost invariably leave behind a vulnerability virtually visible to the naked eye. In all evidence the future of these patients is compromised.

> [2] (pp. 182–4)

Examining a series of psychiatric cases, Fanon shows how those impacted and complicit in violence—including those inflicting violence in the name of the colonial project, as well as those committing violence as part of the liberation struggle—are deeply affected, rendered vulnerable, and often disturbed. This trend, Fanon notes, aligns with what he has witnessed when treating patients from other African countries which have successfully waged a war for their independence [2] (p. 184). An example is worth quoting at length:

> *In a certain African country, independent for some years now, we had the opportunity of treating a patriot and former resistance fighter. The man, in his thirties, would come and ask us for advice and help, since he was afflicted with insomnia together with anxiety attacks and obsession with suicide around a certain date of the year. The critical date corresponded to the day he had been ordered to place a bomb somewhere. Ten people had perished during the attack.*

> *The circumstances surrounding the symptoms are interesting for several reasons. Several months after his country had gained independence he had made the acquaintance of nationals from the former colonizing nation. They became friends. These men and women welcomed the newly acquired independence and unhesitatingly paid tribute to the courage of the patriots in the national liberation struggle. The militant was then overcome by a kind of vertigo. He anxiously asked himself whether among the victims of his bomb there might have been individuals similar to his new acquaintances. It was true the bombed café was known to be the haunt of notorious racists, but nothing could stop any passerby from entering and having a drink. From that day on the man tried to avoid thinking of past events. But paradoxically a few days before the critical date the first symptoms would break out. They have been a regular occurrence ever since.* [2] (pp. 184–5)

What these cases suggest is that the use of violence did not liberate those who fought against colonial relations, like this resistance fighter. Instead, violence ensnared these fighters in a cycle of violence, trauma, and harm. This chapter, I suggest, thereby disrupts many of Fanon's political theoretical claims: "The idea that using violence may be a way to escape being in violence is countered by case after case in which people remain trapped in the violence they have inflicted and suffered" [6] (p. 98). Or, as Fanon [2] (p. 185) writes, "Our actions never cease to haunt us".

Thus, while Jean-Paul Sartre [11] (p. lxii) boldly claims in his Preface to *The Wretched of the Earth* that "Violence, like Achilles' spear, can heal the wounds it has inflicted", the reader, by the end of the text, is left wondering whether this can be true. In this chapter on mental disorders, it seems that further violence has not broken the cycle of violence initiated and sustained by the colonial project—it has not 'healed'. Those who have inflicted violence, even towards liberatory objectives, are 'haunted' by their actions, traumatized, and unable to build and enact a flourishing life—the very goal of the liberatory violence in the first place. As Frazer and Hutchings [6] (p. 106) argue, while Fanon clearly foregrounds how violence in the social-political context is deeply implicated in the formation of subjectivities, and comes to be embodied by those who inflict it and suffer from it, "he fails to explain how the vicious circle between the doing and 'being' of violence can be broken through the doing of further violence".

To put this point differently, earlier in the text, Fanon [2] (p. 21) writes, "The challenge now is to seize this violence as it realigns itself". However, he offers little explanation of how this violence can be fully contained and controlled; the 'realignment' of violence, while perhaps offering an instrumental tool for liberation from the colonial context, appears to also realign the psyches of those who inflict it. The cycle of violence, while perhaps redirected, continues.

3. The Ethics of Care: Violence, Trauma, and Repair

For the remainder of this article, I assert that a critical and political ethics of care can serve as an important additive to Fanon's political theory, most notably in terms of this problem of the cycle of violence. While there is a rich literature describing the tenets of an ethics of care, for this article, I focus on two key characteristics of care ethics that are useful for thinking through the issue of violence: care ethics' relational social ontology and the normative criteria of care. These two tenets, I suggest, orient us towards violence, trauma, and repair in a way that does not wholesale dismiss Fanon's political theoretical claims, but rather foregrounds that contexts of violence and rupture eventually demand a different intervention, one which is focused on repair.

To make this argument, I first provide an overview of a critical, political ethics of care. I then attend to the seeming incompatibility between a political theory that valorises violence, such as Fanon's, and a political theory that prioritizes care. In so doing, I demonstrate that instead of being antithetical, Fanonian violence and an ethics of care can mutually add to each perspective. Fanon's theory can remind care ethicists that the line between care and violence can never be drawn abstracted from context and, moreover, that care is especially significant in the context of violence. Care ethics likewise amends Fanon's political theory by emphasizing that cycles of violence will eventually demand a response other than violence, a response steeped in and from care.

3.1. A Critical, Political Ethics of Care

While the ethics of care is a diverse literature, there is broad agreement that care ethics is, first and foremost, premised upon a relational social ontology, which asserts that "we are the product of our relationships and cannot shed our social existence" [12] (p. 310). This means that we do not just 'have' relations; we are processual selves [13] who emerge, are changed by, and are therefore vulnerable to, relations that constitute us. From this vantage point, ethical action and thinking also stems in and from our unique relations and broader context, and it must inherently be plural [14–16]. A multitude of moral voices will emerge through unique sets of relations, and each of these moral voices carries the weight of the context in and from which it comes.

Given this, the ethics of care orients us towards a different moral task:

> This is not an abstract ethics about the application of rules, but a phenomenology of moral
> life which recognizes that addressing moral problems involves, first, an understanding
> of identities, relationships, and contexts, and second, a degree of social coordination and
> co-operation in order to try to answer questions and disputes about who cares for whom,
> and about how responsibilities will be discharged.

[17] (p. 31)

For this reason, the ethics of care must be a critical and political theory [14]. If moral subjects and moral knowledges are constituted in and by unique webs of social relations, a care ethical orientation to morality means that we must pay attention to how and why subjects and knowledges emerge in and through these unique relations, and especially to how power permeates all of this, shaping our needs, desires, and visions for living well. We must interrogate these relations critically, and seek to understand how they come to constitute what counts as morality *as such* and how they shape our moral understandings, our "moral forms of life" [18] (p. 105). Further, this is a collective task of mutual exchange, in which we listen and attempt to respond well to others so as to better meet their needs.

It is by extension of it being a relational approach to moral philosophizing that the ethics of care is also a thin "organizing trajectory around attentive/responsive living" [19] (p. 287). As I [14] (p. 143) describe elsewhere:

> This 'organizing trajectory' is, to be sure, only thinly normative; how to be attentive and
> respond to a plurality of situated needs and heterogenous vulnerable moral selves cannot
> be predetermined or prescribed, only gleaned through tentative, and sometimes agonistic,
> practices of care which are continually assessed and revised. Nonetheless, what is of moral
> value from an ethics of care perspective is care, and the substance of moral practices is
> found in the ways in which we attempt, often through many iterations, to live well, to
> respond to needs, and to minimize harm and suffering.

Again, given our relational and therefore vulnerable being (in that we are vulnerable as finite, material, and embodied beings, and in that we are vulnerable to the social relations that constitute our subjectivities), care ethics orients us to the ongoing task of fostering relations that support life, that allow for repair, and that respond to needs and desires. At the same time, care ethics' recognition of the multiplicity of moral selves and moral forms of life (a recognition which stems from its relational social ontology) means that what, exactly, is a need, and correspondingly, the content of 'care', cannot be taken for granted. As I have argued, alongside scholars like Kirsten Cloyes [20], care must be agonized. Different versions of care, heterogeneous understandings of needs, and divergent desires and interests must be deliberated on, and they will sometimes be at odds. A key moral task involves attempting to understand needs and forms of care that may be very different from our own, so that we may humbly and reflexively consider and adjudicate amongst competing visions of care. As Margaret Urban Walker [18] (p. 7) writes, "the justification of the moral understandings that are woven through a particular lifeway rests on the goods to be found in living it". Care ethics, as an approach to moral thinking which prioritizes the good to be found in living any form of life, does not rely on universal principles to make moral claims and justifications. Instead, it "recasts moral deliberation as the difficult and messy task of attempting to decentre one's own judgement (never fully possible) in an attempt to know the other (never fully possible) in an ongoing and iterative way" [14] (p. 143) so as to undertake the often "tough" [21] (p. 99) and pain-staking work of building and repairing relations that allow all to live as well as is reasonably possible [4] (p. 40) given our vulnerable embodiment and relational subjectivity.

3.2. Reconceiving Violence and Care: A Relational Approach

Given the understanding of care ethics just outlined, it might appear that care ethics is inherently and diametrically opposed to violence. Care ethics is concerned with everything we do to maintain and repair ourselves and our worlds as well as reasonably possible [4] (p. 40); violence, as that which ruptures our psychological and material reality, seems antithetical to this task. If care ethics is defined by a thorough-going pacifism, then staging this dialogue between care ethics and Fanonian violence is perhaps pointless. From such a vantage point, care ethics would condemn both colonial violence and the violence that Fanon asserts is key to the liberation of the colonized subject—further discussion would not be necessary. This is a complicated subject, and one which I perhaps cannot attend to fully here. However, following Frazer and Hutchings' [7] reading of Sara Ruddick's *Maternal Thinking: Towards a Politics of Peace* [22], I argue that care ethics, while highly critical of violence, does not inherently assert that violence is always and necessarily unjustifiable. In so doing, I claim that Fanon's political theory and the ethics of care can be productively read together, and I call for a more nuanced and sustained engagement with issues of political violence by care ethicists.

To illustrate, let us consider Ruddick's approach to ethics based on maternal thinking. While Ruddick did not use the language of care ethics, she has been linked to the care ethics literature in that her approach to ethics is relational and steeped in maternal thinking, a type of thinking which has as its general organizing principle the key aim of nurturing and socializing (i.e., caring for) the child. For instance, as Ruddick [22] (pp. 18–9) notes,

> *Maternal practice begins with a double vision—seeing the fact of biological vulnerability as socially significant and as demanding care. [. . .] To be committed to meeting children's demand for preservation does not require enthusiasm or even love; it simply means to see vulnerability and to respond to it with care rather than abuse, indifference, or flight.*

This maternal practice is very much in line with the ethics of care, which, as outlined above, is premised on a relational social ontology (and thus foregrounds our inescapable vulnerability) and a commitment to responding to this vulnerability with care. It is also because of this orientation that Ruddick has largely been read as a pacifist. Nurturing life, again, seems to be fundamentally at odds with violence. Yet, Frazer and Hutchings [7] (p. 116) suggest that Ruddick's work is better conceived of along the lines of "non-violent peacemaking". This difference may seem small, but it is salient. Like care ethics, Ruddick's relational ethics suggests that all ethical dilemmas and moral quandaries must be examined critically in the context in which they emerge. This includes issues related to violence, and even the meaning of violence itself. As Frazer and Hutchings [7] (p. 119) write,

> *From the standpoint of maternal thinking [and, as I argue here, the ethics of care], a reliance on violence contradicts the immanent meaning of the practice of mothering [or care more broadly] and reproduces subjectivities for whom others are 'killable'. Violence figures in [Ruddick's] work as something that cuts, breaks or freezes the possibility of constructive relationships between those in conflict, both interpersonally and collectively. Non-violence by contrast preserves, maintains and creates constructive relationships between those in conflict, interpersonally and collectively.*
>
> *What follows from this is that a feminist peacemaker cannot make conclusive judgements relating to whether a particular set of strategies and tactics are either violent or non-violent outside of a holistic understanding of the conflict at issue. Further, an appreciation of the importance of position in relation to the means by which political struggles are being fought is needed.*

In defining violence from a relational perspective, we can conceptually think about violence as that which ends relationships, and non-violence as that which preserves them (and of special concern for Ruddick and a critical and political ethics of care that focuses on relations of domination, oppression, and exploitation, non-violence is that which allows the *less* powerful to live well in their relations). From this vantage point, however, categorizing certain actions outside of the context in and through which they unfold as belonging to either of these categories (that is, either violence or non-violence) does not make sense. Instead, we must understand which relations the action will preserve and which they will end. Moreover, we must be attuned to the fact that certainly some relationships, such as relations of domination, should be ended if the goal is to preserve and maintain life.

This is an important point, as some readings of the ethics of care posit that care ethics requires moral subjects, and particularly women, to maintain certain relations at all costs—even when these relationships are harmful. Such readings rely on an understanding of care as a symptom or negative outcome of patriarchy, as opposed to a critical and political challenge to relations of power (like those that constitute patriarchy). From the standpoint of the former, patriarchy has shaped care and women's subjectivities so fundamentally that care is seen as (strictly or largely) oppressive for women. An example of this type of thinking would argue that because of patriarchy, women "have not learned to discriminate well between good [relationships] and bad ones but have learned instead to assume responsibility for maintaining whatever relationships 'fate' seems to throw [their] way" [23] (p. 86). This suggests that women lack autonomy, and the ability to judge whether a relationship is harmful or not, as they have been positioned to always 'self-sacrifice' for the good of the relationship. Put differently, this reading of the ethics of care as 'self-sacrifice' [13] (p. 58) suggests that because care ethics is premised on a relational ontology, and because the ethics of care has been associated with women's voices, it strips women of autonomy and agency. However, as scholars like Grace Clement [24] point out, the ethics of care actually argues that we are all constituted in and through our relations and, by extension, it is only through our relations, and "only as a result of the care of others", that one can become autonomous [24] (p. 32). This, when combined with the fact that a critical, political ethics of care requires us to agonize care, to interrogate when relations are 'caring' or not, and to reflect on the messiness of relations, means that sometimes 'good' care may actually mean ending a relation. Following a similar logic, I believe that it is possible to conceive of 'violence' which seeks to undo relations of exploitation, oppression, and domination (for instance, colonial relations) as perhaps somewhat aligned with the goals of care: depending on the context and the relations of power at play, fighting to end relations that are not caring relations could be seen in some sense as care, or, more accurately, as a movement towards more caring relations.

At the same time, a care (and maternal) perspective is "very powerfully weighted against the use of violence" [7] (p. 119). Violence as an instrument is to be viewed suspiciously [7] (p. 119). Yet, Ruddick's argument about the importance of context—an argument espoused by a critical and political care ethics perspective more broadly—suggests that "it is not until one has grasped the specific context of a particular conflict, and taken on board the perspectives of those who are/will be fighting, that a proper judgement can be made as to what actions may be in accord with an orientation to making rather than breaking relationships in which preservation, nurture and mutual accommodation are possible" [7] (p. 119). Violence cannot be abstractly pre-defined or determined; by extension, violence cannot abstractly be deemed always and already unjustifiable.

This point is worth dwelling on in the context of this dialogue between the ethics of care and Fanon's political theory. As Fanon reminds us, there are many registers (as I have called them above) of violence. Some, like brute physical violence, might seem more obviously condemnable (although again, I maintain that such determinations cannot be made without critically examining the contexts in and through which such violence occurs). However, other forms of violence—such as the violence of radically restructuring a social-political order—are perhaps more difficult to dismiss. Indeed, when care ethicists call for a reorganization of the social-political order—for a more caring world, for new social-political formations structured around care (see, for example, [14,25–27])—it would be erroneous to assume that this is not, in some sense, a call for violence (as it has been defined above). Rupturing a social-political order is violent and can have violent effects on people's psycho-affective realm (a lesson that Fanon has taught us so well).

For instance, elsewhere I have argued with my colleague Sacha Ghandeharian [27] that the COVID-19 global pandemic was traumatic in part because it demanded that our society confront our inherent ontological vulnerability. This was deeply traumatizing for many people as their structuring principles (based on a liberal myth of self-sufficiency and independence) were shaken to their very core. As people continue to navigate this trauma in a variety of ways—with responses ranging from denying outright the threat posed by COVID-19, to extreme forms of self-isolation—we can see that this event has had serious effects on the psyches of many. Investing in building a world that foregrounds our vulnerability—and therefore 'undoes' the liberal world [5,28]—could have, and likely will have, the same violent and traumatic affects. And yet, this is precisely what a critical and political ethics of care calls for. Again, if we redefine violence along relational lines to refer to that which "freezes the possibility of constructive relationships between those in conflict", and correspondingly, if we redefine non-violence as that which preserves, maintains and creates the conditions of possibility for constructive relationships [7] (p. 119), then a care ethics perspective cannot a priori conclude that all 'violence' is unjustifiable.

At the same time, however, we can adjudicate between and across different forms of 'violence'—that is emphatically the point. Perhaps building a caring world, in which we must confront our inherent vulnerability, will be violent to those subjects constituted in/through the liberal imaginary. But, in considering the broader context, and understanding that such a world would allow us to foster political dialogue on how to meet caring needs, how to rectify systemic injustices, and how to enhance the lives of all, we see that such violence is, perhaps, non-violent (as it is defined above), or at the very least, it may not be totally unjustified. It is from such a vantage point that I believe we can productively put care ethics and Fanon's political theory in conversation: Fanonian violence need not be condemned a priori from a care ethics perspective, despite the perceived and real tension between care and violence. Rather, an ethics of care approach to interrogating conflicts understands that the meaning and experience of violence (including what counts as violence) is contextual and can only be considered by examining the social-political context in and through which it unfolds. Such an approach

> stresses how the meaning of violence is not given simply by the infliction of pain and injury but also by the relations of power into which inflictions of pain and injury are introduced. [. . . This] opens up the possibility that some contexts that are apparently free of pain and injury may actually be violent, and some contexts in which pain and injury are inflicted may be non-violent, or at least not incompatible with non-violence.

[7] (p. 120)

An example of the former situation would be where systems of power, like colonial relations, make it difficult for some to meet their needs—often without the use of any direct force or violence. An example of the latter might very well be the type of violent revolution that Fanon envisions—a form of inflicting pain and injury so as to create 'non-violence,' meaning relations in which all, and particularly the least powerful, can flourish. From this point of view, care ethics, I suggest, is not necessarily opposed to Fanon's violence; moreover, care ethics could perhaps help us assess Fanon's claim about the efficacy of violence *in specific contexts*.

In some ways, then, this dialogue between Fanon's political theory and the ethics of care reminds care ethicists of the importance of attending to violence. This may seem like an odd statement, for surely care ethics is concerned with violence, with that which harms and prevents care. Yet, surprisingly little work confronts violence in the care ethics literature [4] [29,30]. Fanon's theory, in linking violence to both harm and to the possibility of liberated futures for all, reminds us of the pressing need to understand how violence itself is only meaningful, and can only be judged, in the context through which it emerges and in dialogue with those who are or will be fighting [6]. It reminds us of the importance of investigating how our own care practices, including our calls for new caring worlds, can, and perhaps even inevitably will, be violent. And lastly, it also reminds us that care carries extraordinary significance in the context of violence: care can respond directly to violence and help move us from trauma and towards repair. I conclude this argument by turning to this point now.

3.3. From Trauma and Towards Repair

Having demonstrated that violence and care are not inherently antithetical, and thereby cleared the way to continue building this dialogue between a critical, political ethics of care and Fanon's political theory, I return to the issue of the inescapability of the cycle of violence. As argued above, while Fanon correlates certain violences with an assertion of agency, and thus a creative building of the self and a new world, his accounts of mental disorders in liberation wars suggest another outcome: violence, including committing violence so as to dismantle colonial relations, marks those who enact it with trauma. It is unclear how those with such traumas will escape this cycle of violence—how "the new man" will "triumph" [2] (p. 233)—even if we accept the possibility that the violence itself is justified in so far as it undoes colonial relations.

It is here that care ethics, I believe, can productively amend or reorient Fanon's political theory. The ethics of care is both a critical and productive project. On the one hand, care ethics is attuned to the need for critique, to interrogating relations of power that impede well-being, and to disrupting those relations of power. For this reason, as demonstrated in the section above, care ethics holds space for the type of violence Fanon supports. Yet, care ethics, as a thin normative trajectory organized around attentive and responsive living, is also a "reconstructive" [19] (p. 286) theory. Care demands that we do more than critique and disrupt; we must actively respond and repair. "Care claims a tangible moral idea" [19] (p. 286): responding to need, minimizing harm, and fostering nurturing relations for all, orients us continually to traumas and how we can repair them, or at least lessen their effects. It is a moral orientation which "forces us to think concretely about people's real needs and to evaluate how those needs will be met" [17], (p. 31). Importantly, thinking concretely about people's needs must be an iterative and ongoing project. Needs shift and change, and our responses to harms may not be sufficient—in such cases, an ethics of care would demand we respond again, reflectively and through attentive listening to the other. From a care ethics perspective, then, while violence, as in inflicting pain and harm, might be non-violence (or, at least, it might be compatible with non-violence, as outlined in the previous section), we also cannot assume that it would ever be sufficient to rectify whatever harm it is disrupting.

Instead, the messy moral life of care requires persistent address. It requires careful examination of how our responses to injustices and harms—whether those responses involve violence or not—are productive and enabling of relations that support meaningful lives. The vulnerability of our material bodies and relational subjectivities demand ceaseless response, and care ethics orients us towards that response ceaselessly. While care is not, of course, a panacea, it is a thoroughgoing commitment to responding to our vulnerability, our traumas, and thereby striving for a 'good enough' life. Thus, while in the above I have suggested that care might in some instances be aligned with Fanon's violence, I conclude this argument by contending that care is, undoubtedly, that which must come after. An ethics of care holds us to the *unending process* of responding to need and trauma. Furthermore, this unending process requires innovative work and research. Indeed, a nascent literature on Fanon's recently collected and translated psychiatric writings ([31,32], see also [33]) is illuminating exciting insights related to psychiatric practices that might be a fruitful resource for enacting the care that must follow violence. An avenue for future research for care ethics concerned with violence and trauma would be to engage seriously with this emerging work and other therapeutic methods.

In short, both Fanon's political theory, and a critical political ethics of care, I believe, are concerned with building new worlds, new horizons of possibility which are freed from relations of domination, oppression, and exploitation, and new social-political realities in which all can flourish. While Fanon emphasizes the potential productive capacity of violence towards such a goal, the ethics of care emphasizes the productive capacity of care. When taken together, these theories help remind us of the importance of both. As Fanon shows us, what seems to be 'violent' might be compatible with care, particularly if and when the violence is done so as to rupture unequal relations of power. As care ethics reminds us, even in such cases, rupture is not enough. Something must be built in the aftermath of the "total disorder" [2] (p. 2) inherent to undoing social-political structures that dominate and oppress. Striving to live well is a productive and reconstructive task, and something more will be needed to repair and rebuild the world. The ethics of care is a normative commitment to endeavouring continually to enact that 'something more'.

Funding: This research received no external funding.

Acknowledgments: Thank you to Maurice Hamington for reading and providing thoughtful feedback on an earlier version of this piece. Thank you also to Silem Didier and Oleksa Wasylow for many generative conversations regarding Fanonian violence. Lastly, special thanks to Christina McRorie for invaluable editorial support during the final stages of preparing this manuscript.

Conflicts of Interest: The author declares no conflict of interest.

Notes

[1] Please note that I am relying on English translations of Fanon's work in this article.

[2] For instance, Fanon [3] (p. xviii) writes, "what is called the black soul is a construction by white folk". Elsewhere, he [3] (p. 128) notes, "The black man is unaware of [his skin] as long as he lives among his own people; but at the first white gaze, he feels the weight of his melanin".

[3] It is worth noting that Fanon is not reductionist in this sense; he also gives weight to the ways in which familial relations and interpersonal relations (which are, undeniably, also shaped by the social-political realm) can have serious effects on the individual's psyche. In other words, Fanon does not locate all neurosis as coming from the colonial system, although his goal is to trace the many ways in which this system does fundamentally shape people's psycho-affective landscapes. See, in particular, chapter 3 of *Black Skin, White Masks* [3].

[4] There are notable exceptions. Virginia Held [29] tackles the question of care ethics and violence directly, although in a way that is distinct from what has been argued here. For instance, while I have argued that a care ethics approach to understanding violence would entail investigating what practices constitute violence or non-violence within specific contexts, Held [29] (p. 126) spends less time interrogating these nuances and instead asserts that "violence damages and destroys what care labours to create". Fiona Robinson's [30] exploration of care ethics and human security is another important exception. Robinson expands notions of human security from a care ethics lens to reconsider what should count as 'human security;' in so doing, she implicitly expands on notions of violence, which are often conceived of narrowly and in terms of direct bodily harm within the international relations literature.

References

1. Fanon, F. *A Dying Colonialism*; Chevalier, H., Translator; Grove Press: New York, NY, USA, 1965.
2. Fanon, F. *The Wretched of the Earth*; Philcox, R., Translator; Grove Press: New York, NY, USA, 2004.
3. Fanon, F. *Black Skins, White Masks*; Philcox, R., Translator; Grove Press: New York, NY, USA, 2008.
4. Fisher, B.; Tronto, J. Toward a Feminist Theory of Caring. In *Circles of Care: Work and Identity in Women's Lives*; Abel, E.K., Nelson, M.K., Eds.; State University of New York Press: Albany, NY, USA, 1990; pp. 35–62.
5. Ferrarese, E. Vulnerability: A Concept with Which to Undo the World as It Is? *Crit. Horiz. A J. Philos. Soc. Theory* **2016**, *17*, 149–159. [CrossRef]
6. Frazer, E.; Hutchings, K. Revisiting Ruddick: Feminism, Pacifism and Non-violence. *J. Int. Political Theory* **2013**, *10*, 109–124. [CrossRef]
7. Frazer, E. On Politics and Violence: Arendt Contra Fanon. *Contemp. Political Theory* **2008**, *7*, 90–108. [CrossRef]
8. Drabinski, J. Frantz Fanon. In *The Stanford Encyclopedia of Philosophy*; Zalta, E.N., Ed.; The Metaphysics Research Lab: Stanford, CA, USA, 2019; Available online: https://plato.stanford.edu/archives/spr2019/entries/frantz-fanon (accessed on 15 March 2022).
9. Bhaba, H. Foreword: Framing Fanon. In *The Wretched of the Earth*; Fanon, F., Author; Philcox, R., Translator; Grove Press: New York, NY, USA, 2004; pp. VII–XLI.
10. Frazer, E.; Hutchings, K. Argument and Rhetoric in the Justification of Political Violence. *Eur. J. Political Theory* **2007**, *6*, 180–199. [CrossRef]
11. Sartre, J.-P. Preface. In *The Wretched of the Earth*; Fanon, F., Author; Philcox, R., Translator; Grove Press: New York, NY, USA, 2004; pp. VII–XLI.
12. Hamington, M. The Care Ethics Moment: International Innovations. *Int. J. Care Caring* **2018**, *2*, 309–318. [CrossRef]
13. Sevenhujsen, S. *Citizenship and the Ethics of Care: Feminist Considerations on Justice, Morality, and Politics*; Routledge: London, UK, 1998.
14. FitzGerald, M. *Care and the Pluriverse: Rethinking Global Ethics*; Bristol University Press: Bristol, UK, 2022.
15. Hekman, S. *Moral Voices, Moral Selves: Carol Gilligan and Feminist Moral Theory*; The Pennsylvania State University Press: University Park, PA, USA, 1995.
16. Robinson, F. Resisting Hierarchies through Relationality in the Ethics of Care. *Int. J. Care Caring* **2019**, *4*, 11–23. [CrossRef]
17. Robinson, F. *Globalizing Care: Ethics, Feminist Theory, and International Relations*; Westview Press: Boulder, CO, USA, 1999.
18. Walker, M.U. *Moral Understandings: A Feminist Study in Ethics*, 2nd ed.; Routledge: New York, NY, USA, 2007.
19. Hamington, M. Politics is Not a Game: The Radical Potential of Care. In *Care Ethics and Political Theory*; Engster, D., Hamington, M., Eds.; Oxford University Press: Oxford, UK, 2015; pp. 272–292.
20. Cloyes, K. Agonizing Care: Care Ethics, Agonistic Feminism and a Political Theory of Care. *Nurs. Inq.* **2002**, *9*, 203–214. [CrossRef] [PubMed]
21. Noddings, N. *Caring: A Feminine Approach to Ethics and Moral Education*, 2nd ed.; University of California Press: Berkeley, CA, USA, 2003.
22. Ruddick, S. *Maternal Thinking: Towards a Politics of Peace*; Beacon Press: Boston, MA, USA, 1989.
23. Card, C. Gender and Moral Luck. In *Justice and Care: Essential Readings in Feminist Ethics*; Held, V., Ed.; Westview Press: Boulder, CO, USA, 1995; pp. 79–98.
24. Clement, G. *Care, Autonomy, and Justice: Feminism and The Ethic of Care*; Routledge: New York, NY, USA, 1996.
25. FitzGerald, M. Reimagining Government with the Ethics of Care: A Department of Care. *Ethics Soc. Welf.* **2020**, *14*, 248–265. [CrossRef]
26. Tronto, J. *Caring Democracy: Markets, Equality, and Justice*; New York University Press: New York, NY, USA, 2013.
27. Ghandeharian, S.; FitzGerald, M. COVID-19, the Trauma of the Real, and the Political Import of Vulnerability. *Int. J. Care Caring* **2022**, *6*, 33–47. [CrossRef]
28. Harris, C.P. (Caring for) the World that Must be Undone. *Contemp. Political Theory* **2021**, *20*, 890–925.
29. Held, V. Can the Ethics of Care Handle Violence? *Ethics Soc. Welf.* **2010**, *4*, 115–129. [CrossRef]
30. Robinson, F. *The Ethics of Care: A Feminist Approach to Human Security*; Temple University Press: Philadelphia, PA, USA, 2011.
31. Gibson, N.; Beneduce, R. (Eds.) *Frantz Fanon, Psychiatry and Politics*; Rowman & Littlefield: Lanham, MD, USA, 2017.
32. Turner, L.; Neville, H. (Eds.) *Frantz Fanon's Psychotherapeutic Approaches to Clinical Work: Practicing Internationally with Marginalized Communities*; Routledge: New York, NY, USA, 2019.
33. Paré, É. A Theory of Actionality: Fanon, Racial Injuries and Existentialist Phenomenology. In Proceedings of the 2022 Canadian Political Science Association Annual Conference, Online, 30 May–3 June 2022.

 *philosophies*

Article

Ontology and Attention: Addressing the Challenge of the Amoralist through Merleau-Ponty's Phenomenology and Care Ethics

Anya Daly

Philosophy and Gender Studies, School of Humanities, University of Tasmania, Launceston, TAS 7250, Australia; anya.daly@utas.edu.au

Abstract: This paper addresses the persistent philosophical problem posed by the amoralist—one who eschews moral values—by drawing on complementary resources within phenomenology and care ethics. How is it that the amoralist can reject ethical injunctions that serve the general good and be unpersuaded by ethical intuitions that for most would require neither explanation nor justification? And more generally, what is the basis for ethical motivation? Why is it that we can care for others? What are the underpinning ontological structures that are able to support an ethics of care? To respond to these questions, I draw on the work of Merleau-Ponty, focusing specifically on his analyses of perceptual attention. What is the nature and quality of perceptual attention that underwrite our capacities or incapacities for care? I proceed in dialogue with a range of philosophers attuned to the compelling nature of care, some who have also drawn on Merleau-Ponty and others who have examined the roots of an ethics of care inspired or incited by other thinkers.

Keywords: ontology; attention; perception; ethics; ethics of care; phenomenology; Merleau-Ponty; ethical motivation; the amoralist

Citation: Daly, A. Ontology and Attention: Addressing the Challenge of the Amoralist through Merleau-Ponty's Phenomenology and Care Ethics. *Philosophies* **2022**, 7, 67. https://doi.org/10.3390/philosophies7030067

Academic Editors: Maurice Hamington and Maggie FitzGerald

Received: 15 May 2022
Accepted: 7 June 2022
Published: 16 June 2022

Publisher's Note: MDPI stays neutral with regard to jurisdictional claims in published maps and institutional affiliations.

1. Setting out the Problem

Bernard Williams, in his *Ethics and the Limits of Philosophy*, sets out the philosophical parameters of the challenge of the amoralist—why should the amoralist follow the requirements of morality? Williams begins with a discussion of the exchange in Plato's *Gorgias* between Socrates—who poses the foundational ethical question of "what should I do?"—and Callicles, the amoralist—who counters that there is nothing anyone *should* do, morality is a mere convention and cannot be justified. Williams writes:

> … even if there is something that the rest of us would count as a justification of morality or the ethical life, is it true that the amoralist, call him Callicles, ought to be convinced? Is it meant only that it would be a good thing if he were convinced? It would no doubt be a good thing for us, but that is hardly the point. Is it meant to be a good thing for him? Is he being imprudent, for instance, acting against his own best interests? Or is he irrational in a more abstract sense, contradicting himself or going against the rules of logic? And if he is, why must he worry about that? [1] (p. 26)

Williams points to two principal philosophical approaches that aim to address this challenge through establishing an Archimedean point; the first Archimedean point is to be found in Aristotle's elucidation of *Eudaimonia*, and the second in Kantian rational agency. Williams finds both of these approaches inadequate to the task of awakening the amoralist to the value of an ethical life. He suggests rather that what is needed is an extension of the imagination and an enhancement of the sympathies of the amoralist rather than an argument rationally demonstrating the folly of his ways. In concordance with the direction of Williams' thinking, I suggest that the seeming philosophical intractability of this challenge to the ethical life is overcome with the aid of Socrates himself; however, a Socrates in alliance with phenomenology and the ethics of care. Socrates famously declared "No one knowingly does wrong" [2], and this has been given the epithet 'Socratic Intellectualism', that wrongdoing is an error of judgment, not having sufficient knowledge of the full consequences of one's actions. While there is a case to be made for this interpretation, the failure in knowledge, I propose, is not so much in the details of decisions, motivations, actions, events, and consequences, as in the failure to recognize the ontological interdependence of individuals (or nations) existing in a world defined by radical contingency. The failure in knowledge of the amoralist (or the tyrant) is a failure of misrecognition, not an epistemological failure against the rules of logic, but a failure to recognize the relational ontological structures underpinning existence. The antidote to this misrecognition is not to be found in rational cognition and arguments, but rather in attentive percipience, a form of care. [1] This proposal is the underlying direction of the discussions and arguments I now present, engaging with a number of noteworthy thinkers along the way.

2. The Philosophical Roots of Care

The ethics of care initially emerged from the inadequacies of male-oriented approaches to moral theory to speak to women, to their personal and *particular concerns as women* with a different manner of being in the world. While the history of this style of thinking reaches back at least to sentimentalism, care ethics as a known philosophical school was launched with the work of Carol Gilligan and Nel Noddings. Within current feminist scholarship, there is some controversy surrounding the work of Gilligan. Some have accused her of essentializing women's experiences. It is beyond the scope of this paper to address this accusation fully here, however, suffice to say that I do not think this is the case. There are known facts, both biological and socio-political, that female bodies experience, undergo, and suffer situations that male bodies for the most part do not. Gilligan's work, in my view, is not only attuned to the feminist project, but is arguably crucial to it to the same extent as the work of other ground-breaking women thinkers—Mary Wollstonecraft, Jane Addams, Simone Weil, Simone de Beauvoir, Nel Noddings, Iris Marion Young, Helen Longino, Donna Haraway, Joan Tronto, to name but a few. Also, it would seem that this criticism springs from a concern to defend what is claimed to be a non-essentialist form of gender identity—that there can be entirely fluid identities, untied to the realities of the biological body—a kind of untethered consciousness that seems to be in danger of replicating all the problems of 'the ghost in the machine' as interrogated by Descartes.

What is without question is that Gilligan's ground-breaking critique of Kohlberg's 'stages of moral development', *In a Different Voice* [3], galvanized what is now recognized as one of the principal approaches to ethics across diverse domains. That *Care Ethics* has withstood the test of time is a testament to its pertinence, and it continues to evolve with the work of women and men philosophers who, as Annette Baier expressed it, "stand on the shoulders of us older ground-clearing women" [4] (p. xiii).

Baier explores the philosophical affinities of the ethics of care with the work of the Scottish Sentimentalist philosopher, David Hume, and she uses these commonalities to support and enrich the critiques of feminist care ethicists against the mainstream rationalist views. [2] [5]. Hume famously exposed the impotence of reason in confronting the amoralist standpoint, declaring: "it is not against reason that I prefer the destruction of half the world to the pricking of my little finger" [6]); the preference is entirely within the domain of sentiment, not reason. Due to Hume's foregrounding of sentiment and the particular, Baier describes him as the "women's moral theorist", proposing that "Hume turns out to be uncannily womanly in his moral wisdom" [7] (p. 62). Hume champions sentiments over reason and rationality as the basis for ethics; he is concerned with the cultivation of the capacities for sympathy, or fellow-feeling, rather than the reduction of morality to the dogmatic adherence to rationally determined rules or principles. Nonetheless, the sympathy he promotes is one susceptible to the constraints of reason when conflicts inevitably arise in our diverse sympathies; sympathy and sentiments, while paramount for ethical insight and motivation, are not sufficient for a full ethical theory. Anticipating phenomenologist Edith Stein's early 20th century work on the problems of empathy, Hume writes:

> When I am at a loss to know the effects of one body [material entity] upon another in any situation, I need only put them in that situation and observe what results from it. But should I endeavour to clear up after the same manner any doubt in moral philosophy, by placing myself in the same case with that which I consider, 'tis evident this reflection and premeditation would so disturb the operation of my natural principles, as must render it impossible to form any just conclusion from the phaenomenon. [6] (p. xix)

Standing in the shoes of another, while potentially useful in awakening sympathies to appreciate the other's experience, is not entirely reliable; it is still the subject standing in the other's shoes and with this come all the expectations, life history, personality, and unconscious biases of that subject. The *exact* experience of the other remains inaccessible—we might only approximate it or may even potentially distort it. Nonetheless, what still stands is *our susceptibility to sympathetic interconnectedness* and this is pervasive in Hume's account of morality, revealing both "mutual vulnerability and mutual enrichment" [7] (p. 63). Hume's writings are, moreover, replete with very particular illustrations from life, not with abstract, absolutist pronouncements. [3] Therefore, we can see there are commonalities that are key themes in both sentimentalism and the ethics of care—feeling, emotion, personal connections, situatedness, and vulnerability as against the rationalist ethics focused on autonomy, rules, and rights. Gilligan reflects on the difference between rationalist and sentimentalist approaches with the lived experience of women:

> Since the reality of connection is experienced by women as given rather than as freely contracted, they arrive at an understanding of life that reflects the limits of autonomy and control. As a result, women's development delineates the path not only to a less violent life but also to a maturity realized through interdependence and taking care. [3] (p. 172)

Simply, Gilligan here is giving formal recognition to the fact that we are born into sociality, into connection. [4] [8,9]. While for men the circumstances of birth are still determining, nonetheless, we cannot deny that the opportunities and freedoms afforded men have historically and culturally been more expanded than those of women. The circumstances of birth and of giving birth confront women directly with the realities of connection, interdependency, and contingency; men are able to avoid and ignore these realities, and so are more susceptible to building rationalist dreams which may be co-opted for violent ends. [5] And this is why Gilligan can say that women's moral development is likely to lead to a less violent life. The idea of a maturity grounded in recognition of the interdependencies mentioned above is developed in specific directions by a number of later thinkers, such as Tove Pettersen, who highlights the importance of reciprocity within the caring relation; both carer and the one cared for must be attended to in the care [10]. We can say then, the care must be recipient-appropriate, and that self-sacrificing care is not a viable option. Recipient-appropriate care recognizes that to wilfully impose care on those who reject such gestures or on those whose long-term autonomy and freedom would be better served by withholding care, would run contrary to the very telos of care—the other's wellbeing. At the other pole of the relation, to sacrifice one's own wellbeing in favour of another's wellbeing may appear morally worthy, but, in reality, can be a distortion of care; the care offered serves the ego project of the carer by inflating their sense of moral superiority. It is only when the deep interdependencies are integrated into the sense of self, that the other's welfare is equal to one's own or an essential component of one's own self-understanding and values, such as is often the case with parents and their children, that the gestures of what appears as self-sacrifice are appropriate and morally praiseworthy. In all expressions of skilful care, the wider intersubjective, social domains are also implicated, and the relation itself remains paramount.

While contemporary care ethics derive from feminist theorizing, and phenomeno-logical ethics from the aim to give philosophical significance to the interrelated concerns of the first-person perspective, the body, perception, and intersubjectivity, both can retro-spectively find points of concordance with the sentimentalist philosophers of the Scottish Enlightenment—not only Hume, but also his teacher Francis Hutchinson, Anthony Ashley-Cooper, Adam Smith, and some of the common-sense philosophers, notably, Thomas Reid and later Dougald Stewart. [6] (see also [11–17]).

Care ethics and phenomenological ethics are bottom-up ethics as opposed to top-down ethics dependent on universal principles of utility, duty, or virtue; they begin with the particular, the first-person perspective, the body, situatedness, affectivity, and relationality or intersubjectivity. [7] [18,19]. Care ethicist, Maurice Hamington, stresses that the ethical significance of the "body's perceptive, expressive, epistemic and responsive capacities" dispose the individual towards a caring orientation, which is more primary than any moral–theoretical deliberations [20] (pp. 46, 50, 92); [21]. While both care ethics and phenomenological ethics have found generally positive receptions within the feminist community of scholars; nonetheless, some criticize the focus on embodiment as missing crucial ethical considerations; that it cannot adequately take account of systemic power differences; that the body can be the basis for discrimination and that the focus on the body overlooks socio-political dimensions. However, others (and I find agreement with them) argue that the particular percipient body is the essential starting point and is able to build towards the wider concerns of the socio-political. [8] [11,17]. Increasingly, care ethicists are taking the themes of care into the wider socio-political sphere to address issues, such as justice, precarity, environmental issues, human rights, and democracy. [9] (see also [11,22–31]).

3. Merleau-Ponty's Analyses of Perception and Attention

The notion of attentiveness plays a key role in the ethics of care literature, with Noddings referring to "receptive attention as a fundamental characteristic of caring" [32] and Tronto proposing that "attentiveness is the first phase of care" [33] (p. 127). However, a phenomenological analysis of the role of perceptual attention in care and in morality is underexplored.

Perception inherently depends on the body and vice versa, and percipient bodies are situated in the phenomenal realm as well as the intersubjective realm. Merleau-Ponty em-phasized there are no isolated sense-data. These are unfindable in experience; objects always exist within a field, whether a visual field, an auditory field, a tactile field, etc., [34] (p. 4). So too there are no disembodied, solitary, world-less, subjects; subjects as percipient bod-ies exist within both the phenomenal field and the intersubjective field, including the shared world.

> Merleau-Ponty grounds morality unequivocally in perception. Perception opens up to the infinity of perceptual perspectives of all potential and historical others (and even future others), so that we inhabit a multiplicity of perspectives. [10] [35]. Nonetheless, *the view from everyone* does not elide our differences; while I am always on this side of my body both physically and culturally, I am no longer the impenetrable interiority as advanced in Cartesianism; there are exchanges and intertwinings between subjects and the world. This is how Merleau-Ponty is able to universalise his ethics and thereby avoid reduction to a relativist monocular perspective. Moral consideration is, thus, never a purely internal and private deliberation, but already implicates a multiplicity of perspectives. [11]

Importantly, Merleau-Ponty argues for a normativity within perception itself in con-trast to traditional ethical theories, which conceive normativity as supervening on the event, action, and person according to the particular moral principle invoked—virtue, utility, or duty. Rather, every percept has the structure of a gestalt; it is wholistic and functions according to the structures of figure–ground, the ground prescribing how the phenomena and other subjects are perceived. In the phenomenal realm, the ground/environment prescribes how the figure/object is perceived. For example, a misty atmosphere prescribes whether the landscape is perceived more or less determinatively. In the intersubjective realm, the context of others, the socio-political environment, and the culture prescribe more or less determinatively how the individual's self-perception, other-perception, and behaviour are perceived. [11] [36]. It is through the shifting attention between the figure and ground, between the object and environment, between the self and other, between the self and socio-political environment that the specific form and conditioning power of normativity can be recognized.

How can we better understand the role of attention in perception? Merleau-Ponty is able to offer crucial insights, and these insights give further support to the establishment of his non-dual, relational ontology. [12] (see [11,17,37–41]). He begins firstly with critical analyses of the two predominant accounts of attention in the history of philosophy, empiricism, and rationalism, exposing the limitations of these 'attention as searchlight' accounts before setting out his own.

The initial aims of Merleau-Ponty's analyses of attention are, thus, directed at refuting empiricism and intellectualism (rationalism) and their accounts of objectivity. Merleau-Ponty writes: " … empiricism deduces the concept of attention from the 'constancy hypothesis', that is, the priority of the objective world . . . . [and attention] is a spotlight illuminating pre-existing objects hidden in the shadows" [34] (p. 28). The attention described in this account is disinterested, neutral to its objects, and so it is difficult to account for consciousness, that there can be a connection between the object and the subject. And relatedly, how can a particular object be chosen among the multitudinous objects on offer? It would, as Merleau-Ponty describes, be necessary to show the power of "perception to awaken attention, and then how attention develops and enriches this perception" [34] (p. 29). On his account, empiricism, thus, has no resources to tackle these issues, because it relies solely on external connections. Merleau-Ponty then begins his consideration of the opposing account of intellectualism, drawing on Descartes' example of the piece of wax, which reveals that consciousness either grasps its object with clarity or with degrees of confusion. The form, for example, of either the candle shape of the wax, or of a geometric circle of a plate, exists only because consciousness already put it there. He sums up his rejection of both accounts thus:

> What was lacking for empiricism was an internal connection between the object and the act it triggers. What intellectualism lacks is the contingency of the opportunities for thought. Consciousness is too poor in the first case and too rich in the second for any phenomenon to be able to *solicit* it. Empiricism does not see that we need to know what we are looking for, otherwise we would not go looking for it; intellectualism does not see that we need to be ignorant of what we are looking for, or again we would not go looking for it. [34] (p. 30)

The upshot is that both empiricism and rationalism pre-suppose a pre-existing objective world for which attention provides neutral access, either directly through sensations or indirectly through representations. However, Merleau-Ponty proposes that *attention transforms the experience* and is a "new way for consciousness to be present toward its object … .", and it does this by creating a perceptual field according to the specificities of the exploratory perceptual organ [34] (p. 31). If the object includes features, such as colour, light, and form, then the field created depends on visual exploration; if it includes features of sound, tone, and rhythm, then the field created depends on auditory exploration. Until attention is directed towards the object within a sensory or thematic field, both the object and the field remain indeterminate. The subject is unable to identify, to understand, or to make sense of the perceived object until attention is solicited, and the field, even if not the focus of attention, is still an active presence within the perceptual encounter [34] (p. 33). The "perceptual syntax" includes attention, which brings the "constellation of givens" together, gives them sense, and, moreover, guarantees they can have sense [34] (p. 38).

Attention is, thus, neither neutral nor indifferent to its objects; attention transforms the experience of the object.

4. Presence with Objects and Presence with Other Subjects

In this section, I draw together a few ideas from diverse perspectives to make sense of the idea of attentive presence, and to demonstrate its importance in the ethical encounter. Within the literature on care ethics, there are various approaches to what is generally referred to as 'attentiveness' and the diversity of definitions, and the diversity of domains of applications of 'attentiveness' betray a certain level of conceptual imprecision. Sometimes, the term 'attentiveness' is used as an equivalent to 'attention'. An article titled "Attentiveness in care: Towards a theoretical framework" [42] seeks to offer clarification on 'attentiveness', particularly in the health care setting, and it delineates a number of useful distinctions. However, from a phenomenological perspective, a few imprecisions persist. The authors state in the abstract that they will argue that "attentiveness is constitutive for good care, as it can create a space in which a relationship may arise" [42] (p. 686). While it is indisputable that attentiveness is essential for good care, that it allows the space for a quality of care that is deemed 'good', nonetheless, the relationship is primary, it pre-exists the attentiveness. The relation is non-negotiable; however, attentiveness is not guaranteed. There is a similar confusion in the analysis of Noddings [32] when she writes: "we listen or observe receptively and then we feel empathy; that is attention precedes empathy". In my view, observing receptively is in fact a manifestation of empathy. Noddings refers to "a chain of events in caring"—a causal chain. This in my view, is not correct; the understanding of attention is imprecise due to the failure to take account of the ontological, and so I turn now to phenomenology and cognitive science, which offer insightful analyses addressing this issue.

Attention for Merleau-Ponty is transformative, in that it gives the subject *"a new way of being present to objects"*. However, in the intersubjective domain, I am proposing that attentive presence is triply transformative, giving *simultaneously a new way to be present to the subject's sense of self, to the sense of other subjects, and to the other subject's sense of self*. Attentiveness, or caring attention, therefore, transforms the self-experience of the carer, the experience of the other, and the self-experience of the recipient of the care simultaneously. Reciprocal attentiveness, thus, allows for mutuality in understanding each other's affectivity, situatedness, and historicity. This supports a calibrated responsivity that is suitably attuned to the experiential specificities of the other, and guards against dominating stances which would overtly and covertly bend the interactions to the agenda of the carer. Presence is not the outcome of attention, nor attention the outcome of presence—they are co-arising, and attentive presence is transformative. Concisely, we can say that intersubjective attentive presence is a reversible relation, and this is why it serves to both underwrite and illuminate the authenticity of the ethical encounter.

Interestingly, Baart [43,44] picks up on the idea of presence as being significant for care. He presents the idea that presence is something that is cultivated through attuning to another's "tempo, goals, work rhythm, language, work style, interest, perspective, etc . . . , the practitioner of presence offers, in addition to (professional) knowledge and experience, him- or herself" [42] (p. 687). And while this points to the potential of cultivation in presence, what it does not fully grasp is that antecedent to any of these cultivations, there is the need for the 'practitioner' to be present to themselves, present in their own skin, and present in their world. Without self-presence, any attempts to 'attend to' or to 'be present for' others remain artificial, fragile, and lacking authenticity. While this is hinted at, Baart and Klaver do not give a full analysis of the requirements of presence and the underlying supports for presence. And this is where a return to phenomenology would be useful.

The difference between *being present to* objects and *being present to* other subjects is also supported by recent studies in the psychology of perception. Shaun Gallagher, in his chapter 'Inside the Gaze' [45], draws out some interesting philosophical insights in considering the case of a young man whose perceptions were affected due to pressure impacts of deep-sea diving followed by a long flight [46]. In brief, in his perception of objects, he enacted a two-step process—firstly, the visual identification of the object, and then secondly, he needed to become aware that it was in fact he who was visually identifying the object. The sense of the self-ownership of the experience was not automatic. However, this young man was on psychometric testing entirely 'normal', and his social interactions and recognition of others were not impacted at all. Gallagher, while acknowledging further empirical research is needed, suggests that whereas the perceiving subject seems to have an inalienable sense of 'mineness' in the experience of another perceiving subject, with objects, this may dissolve [45] (p. 100). While objects are capable of defining a different vantage, it is only with other perceiving subjects gazing back that a subject has the irrefutable sense of 'me', that 'I am present', that 'I exist', and that 'this experience is mine'. [13] [47–49]. And Merleau-Ponty remarkably anticipated the significance of these recent scientific findings through his philosophical analyses:

> Just as perception of a thing opens me up to being, by realizing the paradoxical synthesis of an infinity of perceptual aspects, in the same way, the perception of the other founds morality by realizing the paradox of an alter ego of a common situation, by placing my perspectives and my incommunicable solitude in the visual field of another and all the others. [35] (pp. 26, 70). [14] [50]

Perception is thus fundamental to the establishment of a phenomenological ethics because perception itself has a normative dynamic due to the figure–ground structure inherent to any percept. Things are always situated in a field; so too subjects always belong to a broader sphere of sociality, 'whether at the level of family, community, nation, species, or animality'. Because of the shifting attention from self to other and back again, perception is fundamentally relational. 'This inherent relationality of perception is what underwrites the meaningfulness of physical events and social encounters' [36] (p. 147). Therefore, Merleau-Ponty demonstrates in two senses that even in mere object perception, others are implicit; firstly, while perception of an object includes various potential vantages for myself, these potential vantages are also available to other possible subjects; and secondly, because we are born into a world that is already inhabited, the meaning and use-value of objects is conveyed to us by others. The perceptible world already implicates others because we are *of* the world; the social and the perceptible worlds are interpenetrated through and through. 'Moral consideration is thus never a purely internal and private deliberation but already implicates a multiplicity of perspectives, 'all the others'; historical, present, and even future perspectives' [36] (p. 149).

While in the absence of empirical studies, the following thoughts remain purely philosophical even speculative, nonetheless, we can evaluate them on their coherence and explanatory value. So, we might ask, is this issue of attentive presence where the amoralist fails? He is unable to summon sufficient or *any* attentive presence. Perhaps he is absorbed or distracted elsewhere; perhaps his sense of embodiment is impaired, his sense of self is 'in his head' not grounded in body, place, and relation. Simply then, he is absent for himself, absent for others, and, arguably, absent for life. Merleau-Ponty describes the gaze of the amoralist as an inhuman gaze, through which he regards others as mere insects by withdrawing into the core of his thinking nature [34] (p. 378). And crucially, if the amoralist is incapable of an attentive reciprocal gaze, incapable of attentive presence with others, then we might well ask whether he has a sense of self-ownership within the experience of the encounter. Are encounters with other fellow creatures merely experientially equivalent to an encounter with an object that can potentially be dissolved? That the amoralist does not receive the affective payoffs of self-affirmation and self-transformation in his interpersonal interactions may be why he can remain impervious to the morality or immorality of his actions. Relations with others are perhaps, in a sense, experimental, and he is like an observer in his own life. Will there be a pay-off? Will my sense of self be affirmed? No? Well, let us try this then. [15]

Back to Bernard Williams— have his concerns as set out in the beginning of this paper been addressed? Williams considers whether Callicles, the amoralist, ought to be convinced that an ethical life is justified. "Is he being imprudent, for instance, acting against his own best interests? Or is he irrational in a more abstract sense, contradicting himself or going against the rules of logic? And if he is, why must he worry about that?" [1] (p. 26).

The first question requires us to answer with further questions—what in fact are Callicles' best interests? Who decides? If he prefers 'the destruction of half the world to the pricking of his little finger', then one could assume that putting existential survival under threat would be 'imprudent', but it may still not dissuade Callicles. Imprudence, why care? Why should he care? Why should anyone care about caring? And with that too, we could say he would not be persuaded by rational means. Ought Callicles, the amoralist, be convinced that an ethical life is justified? 'Ought' most certainly implies 'can'; perhaps Callicles is simply a defective, deficient, and possibly irredeemable human being. I have argued that the defects and deficiencies concern the capacities for attentive presence, for care, afforded to each subject in virtue of being embodied, percipient, and intersubjectively constituted beings. Until there is 'something in it for him' at a deeper level than mere rational and prudential considerations, there is no reason why Callicles ought to be convinced of the value of an ethical life. Until his likely fragile, small sense of self gains, a measure of robustness through attentive presence to his own embodied experience, supported by the gazes and gestures of care of others, living in his skin will remain an empty experience—the life of an amoralist (as too the tyrant) is devoid of authentic vital connection, and because of this, it is devoid of joy. Being attentively present to such a one demands the scope of care of a Bodhisattva.

Funding: This research received no external funding.

Conflicts of Interest: The author declares no conflict of interest.

Notes

1. In looking into the etymology of the term 'attention', I was struck by how many historical sources and definitions there were. Attention is variously described as: the active direction of the mind upon some object or topic; the power of mental concentration; the steady application of the mind; a giving heed; a stretching toward; consideration; observant care; the act of tending; being present at; care. We can see here that it is both a capacity and an act, it might be purely cognitive or affective, or both. Sometimes, the term 'attentiveness' is used in the care ethics literature, which, while highlighting a particular affective quality, does not capture all the senses of 'attention', which I aim to pursue in this paper.

2. See also Michael Slote's *The Ethics of Care and Empathy* [5], in which he traces key commitments in care ethics back to the sentimentalists.

3. Astonishingly, for an 18th century male thinker, from either his sympathetic imagination or from real-life experience, Hume sympathetically presents the case of a young Parisian woman of means who selects a sexual partner to father her child, and thereafter pays him an annuity to stay out of their lives, so she would not need to live under the tyranny of a husband.

4. This idea of an inherent sociality is one explored at length over the history of phenomenology. See Magri and Moran [8] and Jardine [9] for some recent contributions to this scholarship.

5. And the dreams of reason bring forth monsters as Francisco Goya depicted. Available online: https://www.19thcenturyart-facos.com/artwork/dreamsleep-reason-produces-monsters (accessed on 3 March 2022).

6. Additionally, both ethical traditions find significant points of agreement with American pragmatism (notably, Jane Addams [12] for Care Ethics, and William James, Charles Sander Pierce, and others for phenomenology). Both are also to a certain extent attuned to Asian philosophy; Confucianism for Care Ethics (see Li [13,14]), Buddhism and Taoism for phenomenology (see Loughnane [15] and May [16]). Feminist Care Ethics can also be regarded as a moral descendent of phenomenological ethics, in that feminist theorists have drawn on key ideas in phenomenology to build their own accounts. For a sustained discussion of these intellectual debts, see [11,17].

7. While these commonalities are significant, there are differences within each tradition (see Held [18] (pp. 9–15) and Engster and Hamington [19]) and between the traditions. Many key thinkers in the tradition of Care Ethics defend particularism and reject the attempts to reconcile 'care' with 'justice', which they propose requires a universalism. Nonetheless, there are equally defenders of the opposite view. As Engster and Hamington stress, "what binds care theories together, as with other schools of thought, is not a doctrinaire commitment to a singular understanding of the theory, but a general endorsement of a number of different theories"; they note in this regard the relational basis of morality, responsivity, context, and situatedness. They also cite Joan Tronto, who

declared that Care Ethics was committed to crossing boundaries; boundaries such as morality and politics, disinterested ethical theory versus approaches that demand attention to the particular and the boundaries between the private and the public.

8 The phenomenological position espoused in key statements from Merleau-Ponty allows for a multiperspectivalism that underwrites the universalising of an ethics based on the percipient body—and these are his non-absolutist universal ethics. See Daly [11,17].

9 See Young [22,23]; Tronto [24]; Hamington [25]; Robinson [26]; Engster [27]; Laugier [28]; Daly [11]; Fitzgerald [29,30]; Hamington and Flower [31].

10 Daly is citing the work of Merleau-Ponty [35] (pp. 26, 70) here.

11 See Daly [36] for an extended discussion of Merleau-Ponty's analyses of the normativity of perceptual gestalts and the implications for intersubjectivity, ethics, and ethical failure. Normativity has throughout history derived from religious traditions or universal principles (e.g., human flourishing, duty, human rights, etc.)—Merleau-Ponty's ground-breaking account of normativity brings it down to earth, to the situated, percipient, and embodied subject.

12 Many feminist philosophers have invoked relational ontologies to address their philosophical and ethical concerns (see Robinson [37] (p. 29); Butler [38]; Witt [39]; Brubaker [40]; Jenkins [41]). However, few have acknowledged how it is that Merleau-Ponty's ground-breaking work on the body, perception, and intersubjectivity that grounds such ontologies. Merleau-Ponty's work on the reversibility thesis and the later notion of 'flesh' articulates his relational ontology across these domains. See Daly [11,17].

13 Gallagher explores this case in more depth in his 2015 paper 'Seeing without an I: Another look at immunity to error through misidentification' [47]. Therefore, the impact of solitary confinement is known to have devastating impacts on the sense of self—see Gallagher [48] and Guenther [49].

14 For extended discussions of the significance of this short but powerful statement, see Daly [50] (pp. 17–19).

15 Moriarty in the latest film version of *Sherlock Holmes* fits this characterisation perfectly.

References

1. Williams, B. *Ethics and the Limits of Philosophy*; Routledge: London, UK; New York, NY, USA, 2011.
2. Plato. *Protagoras*; Vlastos, G., Ed.; Jowett, B.; Ostwald, M., Translators; Bobbs-Merril Company: New York, NY, USA, 1956.
3. Gilligan, C. *In a Difference Voice: Psychological Theory and Women's Development*; Harvard University Press: Cambridge, MA, USA, 1982.
4. Baier, A.C. *Moral Prejudices*; Harvard University Press: Cambridge, MA, USA, 1995.
5. Slote, M. *The Ethics of Care and Empathy*; Routledge: New York, NY, USA; London, UK, 2008.
6. Hume, D.; Selby-Bigge, L.A. *A Treatise of Human Nature*; Selby-Bigge, L.A., Nidditch, P.H., Eds.; Clarendon Press: Oxford, UK, 1978.
7. Baier, A. Hume, the Women's Moral Theorist? In *Moral Prejudices*; Harvard University Press: Cambridge, MA, USA, 1995.
8. Magri, E.; Moran, D. (Eds.) *Empathy, Sociality, and Personhood: Essays on Edith Stein's Phenomenological Investigations*; Series: Contributions to Phenomenology; Springer: Cham, Switzerland, 2017; Volume 94.
9. Jardine, J. *Empathy, Embodiment, and the Person: Husserlian Investigations of Social Experience and the Self*; Series: Phaenomenologica; Springer: Cham, Switzerland, 2022; Volume 233.
10. Pettersen, T. The Ethics of Care: Normative Structures and Empirical Implications. *Health Care Anal.* **2011**, *19*, 51–64. [CrossRef]
11. Daly, A. A Phenomenological Grounding of Feminist Ethics. *Soc. Br. J. Phenomenol.* **2019**, *50*, 1–18. [CrossRef]
12. Addams, J. *Women and Public Housekeeping*; National Women Suffrage Pub. Co., Inc.: New York, NY, USA, 1910.
13. Li, C. The Confucian Concept of Jen and the Feminist Ethics of Care: A Comparative Study. *Hypatia* **1994**, *9*, 70–89. [CrossRef]
14. Li, C. Does Confucian Ethics Integrate Care Ethics and Justice Ethics? The Case of Mencius. *Asian Philos.* **2008**, *18*, 69–82. [CrossRef]
15. Loughnane, A. Merleau-Ponty and Nishida: "Interexpression" as Motor-Perceptual Faith. *Philos. East West* **2017**, *67*, 710–737. [CrossRef]
16. May, R. *Heidegger's Hidden Sources: East-Asian Influences on His Work*; Routledge: Abingdon, UK; New York, NY, USA, 1996.
17. Daly, A. The Declaration of Interdependence: Feminism, Grounding, and Enactivism. *Hum. Stud.* **2021**, *44*, 43–62. [CrossRef] [PubMed]
18. Held, V. *The Ethics of Care: Personal, Political, and Global*; Oxford University Press: New York, NY, USA, 2006.
19. Engster, D.; Hamington, M. *Care Ethics and Political Theory*; Oxford University Press: Oxford, MS, USA, 2015.
20. Hamington, M. *Embodied Care: Jane Addams, Maurice Merleau-Ponty, and Feminist Ethics*; University of Illinois Press: Bloomington, IN, USA, 2004.
21. Hamington, M. Resources for Feminist Care Ethics in Merleau-Ponty's Phenomenology of the Body. In *Intertwinings: Interdisciplinary Encounters with Merleau-Ponty*; Weiss, G., Ed.; State University of New York Press: New York, NY, USA, 2008; pp. 203–220.
22. Young, I.M. Throwing Like a Girl: A Phenomenology of Feminine Body Comportment Motility and Spatiality. *Hum. Stud.* **1980**, *3*, 137–156. [CrossRef]
23. Young, I.M. *Responsibility for Justice*; Oxford University Press: Oxford, UK, 2011.
24. Tronto, J. Care Ethics: Moving Forward. *Hypatia* **1999**, *14*, 112–119. [CrossRef]

25. Hamington, M. Care Ethics and International Justice: The Composition of Jane Addams and Kwame Anthony Appiah. *Soc. Philos. Today* **2007**, *23*, 149–160. [CrossRef]
26. Robinson, F. Global Care Ethics: Beyond Distribution, Beyond Justice. *J. Glob. Ethics* **2013**, *9*, 51–64. [CrossRef]
27. Engster, D. *The Heart of Justice: Care Ethics and Political Theory*; Oxford University Press: Oxford, UK, 2007.
28. Laugier, S. The Ethics of Care as a Politics of the Ordinary. *New Lit. Hist.* **2015**, *46*, 217–240. [CrossRef]
29. FitzGerald, M. Reimagining Government with the Ethics of Care: A Department of Care. *Ethics Soc. Welf.* **2020**, *14*, 248–265. [CrossRef]
30. FitzGerald, M. Precarious Political Ontologies and the Ethics of Care. In *Care Ethics in the Age of Precarity*; Hamington, M., Flower, M., Eds.; University of Minnesota Press: Minneapolis, MN, USA, 2021; pp. 191–209.
31. Hamington, M.; Flower, M. *Care Ethics in the Age of Precarity*; University of Minnesota Press: Minneapolis, MN, USA, 2021.
32. Noddings, N. *A Relational Approach to Ethics and Moral Education*, 2nd ed.; University of California Press: Berkeley, CA, USA, 2013.
33. Tronto, J. *Moral Boundaries: A Political Argument for an Ethics of Care*; Routledge: New York, NY, USA, 1993.
34. Maurice, M.-P. *The Phenomenology of Perception*; Landes, D., Translator; Routledge: New York, NY, USA; London, UK, 2012.
35. Merleau-Ponty, M. *The Primacy of Perception and Other Essays*; Edie, J., Translator; Northwestern University Press: Evanston, IL, USA, 1964.
36. Daly, A. The Inhuman Gaze and Perceptual Gestalts: The Making and Unmaking of Others and Worlds. In *Perception and the Inhuman Gaze: Perspectives from Philosophy, Phenomenology, and the Sciences*; Daly, A., Moran, D., Cummins, F., Jardine, J., Eds.; Routledge: New York, NY, USA, 2020; pp. 143–157.
37. Robinson, F. *The Ethics of Care: A Feminist Approach to Human Security*; Temple University Press: Philadelphia, PA, USA, 2011; pp. 41–62.
38. Butler, J. Sexual Ideology and Phenomenological Description: A Feminist Critique of Merleau-Ponty's Phenomenology of Perception. In *The Thinking Muse: Feminist and Modern French Philosophy*; Allen, J., Young, I.M., Eds.; Indiana University Press: Bloomington, IN, USA, 1989.
39. Witt, C. *Feminist Metaphysics: Explorations in the Ontology of Sex, Gender and the Self*; Springer: New York, NY, USA, 2011.
40. Brubaker, D. Care for the Flesh: Gilligan, Merleau-Ponty and Corporeal Styles. In *Feminist Interpretations of Merleau-Ponty*; Olkowski, D., Weiss, G., Eds.; Pennsylvania University Press: Philadelphia, PA, USA, 2006.
41. Jenkins, K. What Women are For: Pornography and Social Ontology. In *Beyond Speech: Pornography and Analytic Feminist Philosophy*; Mikkola, M., Ed.; Oxford University Press: New York, NY, USA, 2017; pp. 91–112. [CrossRef]
42. Klaver, K.; Baart, A. Attentiveness in care: Towards a theoretical framework. *Nurs. Ethics* **2011**, *18*, 686–693. [CrossRef]
43. Baart, A. *Een Theorie van de Presentie (A Theory of Presence)*; Lemma: Den Haag, The Netherlands, 2001. (In Dutch)
44. Baart, A.; Grypdonck, M. *Verpleegkunde en Presentie. Een Zoektocht in Dialog naar de Betekenis van Presentie voor Verpleegkundige Zorg. (Nursing and Presence. A Search for the Meaning of Presence of Nursing Care)*; Lemma: Den Haag, The Netherlands, 2008. (In Dutch)
45. Gallagher, S. Inside the Gaze. In *Perception and the Inhuman Gaze: Perspectives from Philosophy, Phenomenology and the Sciences*; Daly, A., Moran, D., Cummins, F., Jardine, J., Eds.; Routledge: New York, NY, USA, 2020; pp. 99–108.
46. Zahn, R.; Talazko, J.; Ebert, D. Loss of the Sense of Self-Ownership for Perception of Objects in a Case of Right Inferior Temporal, Parieto-Occipital and Precentral Hypometabolism. *Psychopathology* **2008**, *41*, 397–402. [CrossRef]
47. Gallagher, S. Seeing without an I: Another Look at Immunity to Error through Misidentification. In *Mind, Language, and Action, Proceedings of the 36th International Wittgenstein Symposium, Kirchberg am Wechsel, Austria, 11–17 August 2013*; Moyal-Sharrock, D., Coliva, A., Munz, V., Eds.; Walter de Gruyter: Berlin, Germany, 2015.
48. Gallagher, S. The Cruel and Unusual Phenomenology of Solitary Confinement. *Front. Psychol.* **2014**, *5*, 585. [CrossRef] [PubMed]
49. Guenther, L. *Solitary Confinement: Social Death and its Afterlives*; University of Minnesota Press: Minneapolis, MN, USA, 2014.
50. Daly, A. *Merleau-Ponty and the Ethics of Intersubjectivity*; Palgrave Macmillan: London, UK, 2016.

philosophies

Article

An Ethics of Needs: Deconstructing Neoliberal Biopolitics and Care Ethics with Derrida and Spivak

Tiina Vaittinen 

Global Health and Social Policy, Health Sciences, Faculty of Social Sciences, Tampere University, 33100 Tampere, Finland; tiina.vaittinen@tuni.fi

Abstract: The body in need of care is the subaltern of the neoliberal epistemic order: it is that which cannot be heard, and that which is muted, partially so even in care ethics. In order to read the writing by which the needy body writes the world, a new ethics must be articulated. Building on Jacques Derrida's philosophy of deconstruction, Gayatri Chakravorty Spivak's notions of subalternity and epistemic violence, critical disability scholarship, and corporeal care theories, in this article I develop an ethics of needs. This is an ethical position that seeks to read the world that care needs write with the relations they enact. The ethics of needs deconstructs the world with a focus on those care needs that are presently responded to with neglect, indifference, or even violence: the absence of care. Specifically, the ethics of needs opens a space—a spacing, an *aporia*—for a more ethical politics of life than neoliberal biopolitics can ever provide, namely, the politics of *life of needs*.

Keywords: ethics of care; corporeality; the body; biopolitics; deconstruction; ethics of needs

Citation: Vaittinen, T. An Ethics of Needs: Deconstructing Neoliberal Biopolitics and Care Ethics with Derrida and Spivak. *Philosophies* **2022**, 7, 73. https://doi.org/10.3390/philosophies7040073

Academic Editors: Maurice Hamington and Maggie FitzGerald

Received: 26 February 2022
Accepted: 27 June 2022
Published: 30 June 2022

Publisher's Note: MDPI stays neutral with regard to jurisdictional claims in published maps and institutional affiliations.

1. Introduction

"The challenge is to think of 'need' in terms other than mere lack, as other than a barrier to well-being. Can we think of a need as something other than what must be overcome or satisfied prior to engaging in activities that provide real rewards?" [1] (p. 468)

"Care is omnipresent, even through the effects of its absence." [2] (p. 1)

At the time of writing this article, we have lived with the COVID-19 pandemic for over two years. In the privileged part of the world where I live, everyone over 12 years old who wants the vaccine has been vaccinated. Booster vaccinations have been delivered. With high vaccine rates turning the disease into something less incapacitating and deadly, the threat of health systems collapsing starts to pass. Societies are being re-opened, and traveling is again possible. But at the same time, these very same vaccines that protect our privileged bodies and societies remain inaccessible in large parts of the world. According to the World Health Organization, just under 71% of the population in high-income countries were fully vaccinated in mid-February 2022, while in lower-income countries, the same applied to only 6.6% of the population [3]. This is not because of any lack of capacity in the world to produce an adequate number of vaccines to get everybody protected. No—it is just world politics as usual, in an era of transnational capitalism.

Even though the only way out from the pandemic would be to vaccinate the entire global human population as a matter of urgency, many wealthy states have chosen a route of 'vaccine nationalism': i.e., hoarding vaccines for their own populations, opposing the waiving of patents from Pfizer, BioNTech and Moderna (the pharmaceutical companies that produce the vaccines) and refusing to push companies to share their technology and know-how in order to increase vaccine production. While these companies have benefited from public finances when developing their technologies, their monopoly and private profits are now protected by states, in the spirit of a (neo)liberal laissez-faire ideology.

Simultaneously, the three vaccine-producing companies have been seen to gain profits at the rate of 1000 US dollars per second, and in 2021, the companies made some $34 billion in profit between them [4].

For those familiar with the transnational distribution of care resources and the political economies involved, there is nothing new in this situation of global vaccination politics. In an era of transnational neoliberal capitalism, care provision and access to adequate care are increasingly dependent on the worth of the care-receiver to the accumulation of transnational capital. This means that those bodies and populations whose care needs are tied to financialized health care systems feeding into transnational capital accumulation are responded to with care. Simultaneously, the needs of those *whose care-receiving cannot be turned into profit* are increasingly neglected in the field of public care provision. In fact, neglect itself may be required for transnational capital accumulation to thrive. This is the case, for instance, when a pandemic is *not* stopped by providing vaccines to the entire global population, even though this would be possible. This protracts the pandemic, while at the same time creating a continuous need for new vaccines and boosters, which in turn increase the profits of the vaccine producing companies.

The same logics of care delivery apply across the world and in all societies, where the needs of those whose care adds to transnational capital accumulation are responded to with care, and the needs of the rest are likely to be neglected. In different public health systems, the accumulation of financial profit through care provision takes place through various mechanisms, for example through insurance, public-private partnerships, and private companies selling care services to public providers responsible for the care of the population [5–7]. Neoliberal states allow this to happen when governing care resources in ways that are in line with this neoliberal reasoning. Thus, even in the world's strongest welfare states today, tax-funded public care services are increasingly being commodified and marketized, which allows financialized care-selling corporations to make profits for transnational shareholders on the back of the citizens' care needs [5,6,8].

In terms of COVID vaccines, it would appear that the vaccines were developed quickly for this very same reason, and then distributed unevenly and selectively among the global population. The bodies that have now been protected with the vaccines are part of wealthy populations in states, whose public health financing benefits transnational capital in various ways, including globally privileged vaccine deals. In contrast, as Dr. Ayoade Alakija (Co-Chair of the African Union's Vaccine Delivery Alliance) put it in an interview with the BBC after the Omicron variant was first detected in South Africa: "Had the first SARS-CoV virus originated in Africa, the world would have locked us away and thrown away the key. There would have been no urgency to develop vaccines because we would have been expendable [9]". Her claim might not apply to all of the heterogeneous countries and populations in the African continent, yet it is likely to apply to those health systems that are unable to finance vaccine development and delivery in ways that benefit transnational capital.

As pointed out by several political theorists of care [5–7,10–12], such entanglements between care provision, care needs, and transnational capital accumulation are embedded in neoliberal discourses of governmentality that emphasize individual autonomy, independence, and rational economically driven decisions regarding the care of the self. In her recent work, Joan Tronto has criticized care theorists for treating this hegemony of neoliberal political ideology as invincible, and for not looking for alternatives. She proposes that instead of "trucking with *Homo oeconomicus*"—the economic man—as the main form of political subjectivity, we should envision "a different route to greater political concerns, one that begins with acknowledging that humans are essentially, in the plural, *Homines curans*, 'caring people'" [13] (p. 28).

I side with Tronto's call for alternatives. However, while I support and agree with her visions of *Homines curans* as providing a basis for the kind of political order that she calls a "caring democracy" [13,14], in this article, I seek to provide a theory that erodes the neoliberal ideological hegemony from a slightly different, although complementary angle.

Building on my previous deconstructive readings of feminist care ethics [6,15], I argue that care ethicists' traditional focus on practices of car*ing* as the origin of moral and political relations is inadequate in resisting neoliberal forms of governmentality.

Inspired by the work of moral psychologist Carol Gilligan [16], classic care ethics argues that caring practices provide a unique source of moral and political thinking, that recognizes human vulnerability and interdependency as existential and shared conditions of humanity. While challenging the (neo)liberal political ideal of individuated autonomous subjectivity, the emphasis in care ethics on practices of car*ing* as originary to political and moral relatedness makes it incapable of fully addressing the political power of bodies that depend on care. I have demonstrated this previously through a close reading and deconstruction of Joan Tronto's groundbreaking work in the field of political theory [17,18]. However, the same argument would apply to all care ethics that put car*ing*—rather than e.g., embodied practices and habits of needs—at the center of the theory. This could include, for example, the important work of scholars such as Selma Sevenhuijsen [19], Victoria Held [20], and Fiona Robinson [21,22] to name but a few.

In this article, I continue to argue that an ethic of care that is framed through car*ing* cannot fully address the power that bodies have—as bare, needy bodies—in shaping the world. This is because the kind of care ethics that emphasizes car*ing* ties political agency (even of *Homines curans* in caring democracies) to subjects that have the capacity to care. Thereby, the political power of "bare" bodies in need of care is written out and left unexamined, inarticulate, and mute. To tackle this problem, in this article I develop an *ethics of needs* that is complementary (i.e., not to be seen as a replacement) to the traditional accounts of care ethics. The concept of "bareness" in my ethics of needs comes from Giorgio Agamben's view of the modern politics of life (i.e., modern biopolitics), where life is divided between *bios,* the politically valuable life, and *zoē,* the "bare" life that lacks political value and can be killed without sacrifice [23]. In the neoliberal epistemic order, those lives in need of care, whose care cannot contribute to transnational capital accumulation seem to emerge as bare life that can be left to die without wider loss. To resist this particular violence imbued in the neoliberal ideology, I believe we need an ethics of needs, where the field of *zoē* emerges not as *bios,* but as something politically valuable in its own terms.

In my article *The Power of the Vulnerable Body: A New Political Understanding of Care,* I deconstructed Tronto's political care ethics through Giorgio Agamben's biopolitics in the context of the Finnish welfare state [15,23]. There, I analyzed the Finnish aging population as "bare life" that is capable of exerting "pressure on the sovereign power" in its bare neediness, while "forcing the state to adjust its political economy to the citizen's bodily needs of care". In this account, the "frail" and seemingly apolitical care-dependent bodies of persons with dementia emerged as politically powerful—even though they no longer have the capacity to practice car*ing*. Their "bare" yet globally *powerful needs* were capable of putting in motion entire political processes that could, by means of transnational biopolitics, reach across the globe [15] (p. 100). This happens, for instance, through the practices of global nurse recruitment, the political economies of which I analyzed in detail in *The Global Biopolitical Economy of Needs* [6].

These deconstructive readings of classic care ethics have given rise to my definition of *care understood as a corporeal-relation-enacted-by-needs* (see also [11]). This conceptualization of care decenters the focus of care ethics from practices of car*ing* and care labor while bringing the power of the needy body to the center of political life. Effectively, in deconstructing care ethics at its own limits, the concept of care as a corporeal-relation-enacted-by-needs opens a space for an ethics and politics where the body in need of care gains a central position, while at the same time decentering the subjective, caring "I".

In this article, I continue this previous work and start to develop a new care-ethical thinking, namely, an ethics of needs. It is, however, important to note I am deeply embedded in the classic tradition of care ethics, too. My aim is not to replace it with a better theory, but to extend and complement it through deconstructive exercises. The aim of deconstructive reading is never to destroy or reject the tradition one writes about. Rather, it is about undertaking an intimate reading of the tradition from within, with the aim of working with its margins, exclusions, and intimate others, and with the ethical motive of rewriting the tradition into something more inclusive of its previous Others [24,25]. As such, deconstructive reading is never seen as complete, but as an opening that calls for further deconstructions. Thus, while in this article I seek to rewrite the ethics of care into an ethics of needs, the new ethics of care needs that I propose is but an *aporia* or a call for further deconstructions of care from the perspective of bodies in need of care.

The ethics of needs that I propose emerges along with my theorization and writing. Yet, some (arguably limited and deficient) elaborations and delineations are required vis-à-vis previous literature. To begin with, the ethics of needs that I propose in this article is not the first theory for an ethics of needs in the field of care theory. Sarah Clark Miller's book *The Ethics of Need: Agency, Dignity, and Obligation* (2012) is an extensive philosophical exploration of the nature of needs, about "which needs are morally salient", and "what moral agents ought to do when they encounter others with such needs" [26] (p. 1). To determine our moral obligations to meet the needs of others, she combines Kantian ethics with feminist care ethics. In so doing, her account, too, thus eventually focuses on the moral-political power of car*ing*, whereas I seek to focus on the power of embodied needs. Furthermore, I draw on theories that could be seen as being critical of Kantian positions, and our philosophical premises are therefore somewhat different and possibly in tension with one another. Due to these differences, and since my account of the ethics of needs has emerged independently from Miller's work, in this article I do not engage with this previous ethics of needs, as much as I would love to. I do believe such engagements would be useful and necessary. However, I must reserve the space of this article to develop and articulate my own theory for an ethics of needs, however lacking it may be in its engagement with previous discussions on the politics and ethics of care needs. Thus, the time and space needed to engage my ethics of needs with that of Sarah Clark Miller, or to discuss the relationships of these with, say, Nancy Fraser's "politics of needs interpretation" [27] will hopefully come later, in some other space of writing and thinking.

The ethics of needs that I propose in this article is radically corporeal. As such, it complements the work of those care theorists and care ethicists who explicitly focus on the body. Maurice Hamington, for instance, has argued that "care cannot be fully understood without attending to its embodied dimension", and that "care can be a viable component of social ethics only if its corporeal aspect is addressed" [28] (p. 4). For Hamington, care is "corporeal potential realized through habits" [28] (p. 5), and it "flows from the knowledge manifested in the body" [28] (p. 39). I believe that such embodied understandings of care are entirely compatible with the ethics of needs that I propose in this article.

Other care theorists who focus on the body have emphasized the material dimensions of care as body-work that necessitates physical proximity, co-presence, and corporeal encounters [29,30]. My ethics of needs is deeply influenced by these accounts, yet emphasizes embodied needs over embodied work and care encounters. Furthermore, building on biopolitical theories of governmentality, my ethics of needs draws corporeal connections between micro-level care encounters and macro-structures of transnational capitalism. Similar to Fiona Robinson's critical global ethics of care [21], the ethics of needs thus emphasizes that the corporeal relation of care is never just a linear, singular trajectory that connects the caregiver with the care-receiver, but rather a vehicle through which needs shape political economies, both locally and across societies [6,11,15,31].

Importantly, however, my ethics of needs also emphasizes that care, when seen as a corporeal-relation-enacted-by-needs, always entails the potential of neglect or other forms of violence. Siding with recent accounts in care theory and critical disability studies, the ethics of needs recognizes that care can be non-innocent [32] and destructive [33], may include violence [34,35], or at least is "not necessarily diametrically opposed" to it [36]. In this regard, the ethics of needs goes beyond care and does not necessarily reach the event where car*ing* takes place. Thus, in contrast to the ethics of care that tends to begin its analysis of political relatedness at the point where caring happens, the ethics of needs begins where bare needs exist—also when they fail to enact relations of care. Following Erin Manning's theory of relatedness, the ethics of needs thus understands the relations that our needy bodies enact as "reaching-toward" intervals that are "active with tendencies of interaction", while never limited to interaction. They are "quantum leaps occur[ing] in a fractal mode of relation where events build on events" influencing simultaneously both the body in need and its environment" [37] (p. 34).

Philosophically, the ethics of needs that I propose builds on Jacques Derrida's philosophy of deconstruction, as well as Gayatri Chakravorty Spivak's notions of subalternity and epistemic violence. Much like critical disability scholars such as Margrit Shildrick [38,39] and Robert McRuer [40], I argue that in the present era of transnational capitalism, the body in need of care is the subaltern of neoliberal biopolitics and remains partially so even in the traditional accounts of the ethics of care. To provide an alternative account that complements this lack in care ethics, I embark on the task of writing the "subject under erasure" in relations of care. This, I argue, is necessary for us to envision how our embodied care needs and their power differentials silently write the world while subverting both the rational *Homo oeconomicus* as well as the subjective, caring "I".

The ethics of needs is therefore an ethical urge to try and read the needy bodies' subaltern writing of the world, even when its messages remain inarticulate and incomprehensible in the prevailing epistemic orders and the care relations inscribed in (and by) these orders. In developing the ethics of needs, my ambition is to provide an additional alternative to the hegemonic ideology of neoliberalism: a new grammar by which the neoliberal readings of the world could be eroded—also at times and places where care relations do not emerge when they should or could emerge. The erosive work of *Homo egēnus*, the needy human being, therefore complements the work of *Homines curans*, in striving for a political order where caring *and* care needs are central elements of emergent political orders.

My theorization in this article proceeds as follows. In the section following this lengthy introduction, I discuss Derrida's philosophy of deconstruction and *différance*, in order to provide the conceptual tools needed for reading my theory of an ethics of needs (which I admittedly still struggle to articulate comprehensively and in an elaborate fashion). Here, I also sum up Spivak's theory of subalternity and epistemic violence, drawing on Spivak's engagements with Derridean philosophy. In the third section, I return briefly to the Foucauldian vision of neoliberal biopolitics, so as to describe how life (including care and care needs) is governed in the present neoliberal episteme of transnational capitalism. Siding with critical care ethicists, I argue that a most pressing moral problem with neoliberalism is that it cannot caringly accommodate death in its politics, nor other forms of life that are dying and fully dependent on care for survival. This renders bodies-in-need as an excess to politics and political life in neoliberal capitalism, or as "bare life" in terms of Agambenian biopolitics [15,23].

The body in need of care thus emerges as the subaltern of our present epistemic order: it is that which cannot be heard and is muted, partially so even in care ethics. In order to read the writing through which needy bodies write the world, a new ethics of needs must be articulated. To do so, and following the Derridean philosophy of deconstruction, I write the subject under erasure in the fourth section of the article. Here, I articulate my theory of the lowest common denominator of embodiment, which demarcates the level of embodiment where the power differentials between bodies' needs can be traced and analyzed. As elaborated in the fifth and concluding section of the article, the body-organism-in-need-of-care now emerges as *différance*, and as a bundle of relatedness that constantly (re)writes the world with the corporeal relations of care *and* neglect it enacts. To conclude, I articulate the ethics of needs as an ethical position that seeks to read the world that embodied care *needs* write, with a particular focus on those needs that are presently responded to with neglect, indifference, or even violence, rather than care. Specifically, the ethics of needs opens a space or *aporia* for a more ethical politics of life than neoliberal biopolitics can ever provide, namely, the politics of *life of needs*.

2. Conceptual Tools

2.1. Deconstruction, General Writing and Différance

To understand how the ethics of needs demands an ongoing deconstruction of the world through silenced care needs, we need to return to the philosophy of deconstruction, as devised by Jacques Derrida. In short, deconstruction denotes the practice of a systematic tracing of silences and suppressed meanings in texts. It involves a double move, where one first works with the dichotomies on which meaningfulness relies to undo and displace their hierarchical opposition so that in the second move of deconstruction, the terms of these hierarchical binaries can eventually be situated anew [24,25,41,42].

In the philosophy of deconstruction, the term "text" does not refer to writing in the traditional sense of the term. For Derrida, "there is nothing outside the text" [41], and following Saussure [43], Derrida understands language as a system of differences. Words have a meaning only in relation to other words in a system that is defined in relation to other signs. Unlike Saussure who considered the spoken word as the pure form of language, for Derrida, this applies not only to spoken language. In fact, Derrida criticizes Saussure for prioritizing speech over writing. He shows how by suppressing the written language, Saussure prioritizes the Western metaphysics of presence over absence: disregarding that which is written out from language for the language to make sense in the first place. As a system of differentiation, language can only make sense in relation to the absented [41] (p. 46). In deconstructing Saussure's work (and thereby, the entire Western metaphysics of presence), Derrida produces an understanding of generalized or *arche*-writing. Here, generalized writing does not just take over the primary position that was previously held by spoken language, but rather spoken language already belongs to what Derrida calls generalized writing [41] (p. 55).

Due to its emphasis on "texts" and "writing", Derrida's deconstructive philosophy is all too often (mis) understood as dealing merely with language and textuality on an "immaterial" level of discourse. However, to understand how the care needs of bodies write the world, it is pivotal to understand that the generalized writing does not refer to writing in its literary, generic sense [44–46]. As Vicki Kirby elaborates, with reference to Derrida [41] (p. 9), when writing is understood in its most general sense, even "'the most elementary processes of the living cell' are a 'writing' and one whose 'system' is never closed" [45] (p. 61). Thus, Derrida's "notion of writing cannot be understood literally", and this is because "it makes the world, objects and relations possible; it structures and gives the world and its contents meaning and value" [44] (p. 84). Such writing of the world (s) "is itself a silent play" [47] (p. 5).

The ethics of needs is therefore an ethics attuned to reading the writing or 'play' of the most silent and silenced care needs, with an effort to decipher *how these needs and their power differentials write* the world we live in. To comprehend this playfulness of generalized writing, we need to revisit the meanings of *différance* in Derrida's philosophy. *Différance* is a neologism coined by Derrida, which denotes the vehicle of temporized movement of differentiation in texts. It is "literally neither a word nor a concept" [47] p. 3, and the precise definition of *différance* is therefore impossible. "[T]here is nowhere to begin to trace the sheaf or the graphics of *différance*", Derrida writes, "[f]or what is put into question is precisely the quest for a rightful beginning, an absolute point of departure, a principal responsibility" [47] (p. 6). Nevertheless, to elaborate on the play of *différance*, one must begin somewhere.

A semantic play on the French verb *différer* forms the (impossible) origin for the (non)concept of *différance*. While in English, the Latin verb *differer* translates into two separate words—to defer and to differ—in Derrida's mother tongue, French, the single verb *différer* holds both these meanings. Here, the first refers to deferring, i.e., "the action of putting off until later, of taking into account, an operation that implies an economical calculation, a detour, a delay, a relay, a reserve: temporization" [47] (p. 8). The second meaning of *différer* is then the "more common and identifiable one: to be not identical, to be other, discernible etc."—i.e., to differ. Thus, the word *différence* with an e cannot simultaneously refer to both deferral and difference. As an economic compensation for "this loss of meaning", Derrida invented the neologism *différance* (with an a), which, he maintains, "can refer simultaneously to the entire configuration of its meanings" [47] (p. 8). *Différance* can do so precisely because it is not a concept or a word, but a "sheaf": a bundle, which has a "complex structure of a weaving, an interlacing which permits the different threads and different lines of meaning—or of force—to go off again in different directions, just as it is always ready to tie itself up with others" [47] (p. 3).

Influenced by the writing of feminist disability scholar Margrit Shildrick [38], I suggest in this article that the body in need of care is *différance*, operating in exactly these ways when enacting corporeal relations (of care and neglect) in the world. Through the relations it enacts with others, the needy body is always a "sheaf" and a "complex structure of weaving" that is "ready to tie itself up" with other corporeal beings [47] (p. 3). As I shall explain, this is so also when the response-able respond with neglect, indifference, or even violence, and when the epistemic order makes it impossible to recognize the needy as needy, or worthy of caring responses. To understand these relations of neglect—care in its absence—and the silencing of needy bodies in neoliberal biopolitics as well as in the ethics of care, it is necessary to return to Spivak's notions of the subaltern and epistemic violence.

2.2. The Subaltern and Epistemic Violence

As emphasized by Derrida in the entirety of his corpus, and Spivak in her introductory essay to *Of Grammatology*, the sheaf of *différance* takes as many forms as there are texts [48] (p. xv). This means that in all texts there is a silenced, heterogeneous Other, whom deconstruction aims to reveal as an erased trace in the text, with attempts to rewrite the text into a form that makes space for the Other in its complete Otherness. This haunting, speechless Other is *the subaltern*. It is marginalized and written out, and yet crucially central to the text since, without its very erasure, the epistemic order of the text would fall apart, and the text would no longer make sense. Thus, the erasure of the subaltern gives epistemic orders their sense and meaning.

In her widely cited essay "Can the subaltern speak?", Spivak argues that the subaltern subjectivity is utterly incomprehensible, irrational, and mute [49]. As a constitutive outside of the episteme, the subaltern cannot speak without simultaneously being assimilated into the order and thereby losing their own history and being. This is because to be comprehensible, the subaltern ought to speak the language of the hegemonic episteme—a language that remains intact only insofar as the world of the subaltern Other is muted. In this order, the voice of the subaltern does not make sense: it is irrational. The epistemic order is, by definition, an order that is founded on constant *epistemic violence* against the subaltern Other, who must forcefully and repeatedly be written out from the order, so as for the order to make sense in its constitutive rationality.

The trace of the absented subaltern is, therefore, *différance* written throughout the sign system of the episteme, as that which temporally defers and spatially differs from what makes sense. From the margins of the order, it continues to haunt the order's completeness. This play of the subaltern as *différance* has compelling consequences for the ethical reading of the world-as-text(s) [24,50,51]. As a trace of the haunting absence of the subaltern that differs in meaning and defers presence, for the meaningful presence of something/someone/somebody else to be possible, it keeps the material-discursive world open for subversion. *Différance*, in other words, renders all writings of the world incomplete and open, making it possible and obligatory to read the world otherwise or in a different way. Especially, it underlines the necessity to deconstruct all texts and all epistemic orders, including the ethics of care.

In Spivak's reading of Derrida, it is revealed how the philosophy of deconstruction is an ethical response to epistemic violence. Here, deconstruction begins with the recognition that all texts are based on an epistemic violence that erases the Other, while rendering the subaltern mute. Margins are never simply margins, but simultaneously "the silent, silenced center" [49] to which deconstruction recurrently returns. Deconstruction is about the constant imperative to recognize the muteness of the subaltern as "that interior voice that is the voice of the other in us" [49] (p. 294). Deconstruction means constantly working with epistemic violence, while simultaneously "presupposing [a] text-inscribed blankness" [49] (pp. 293–294) that is the silenced subaltern.

However, this does not mean that deconstruction can remove all epistemic violence to make space for the subaltern to speak. The deconstructed new epistemic orders will also only make sense by writing something (and consequently some bodies) out, by erasing the constitutive outside that is its subaltern Other. Here again, the subaltern remains mute and continues to haunt the deconstructed realities. In an interview with Jean-Luc Nancy, Derrida himself underlined that the "deconstructive gesture" is always summoned by a "surplus of responsibility", which "is excessive or it is not a responsibility" [52] (p. 286). Here, justice is never completely gained, which means that the responsibility to try and hear the Other remains, even when they cannot speak. In the following section, I turn to elaborating how the body in need of care in neoliberal biopolitics is the subaltern that cannot speak—and whose writing of the world we are nevertheless excessively responsible to try and read.

3. The Body in Need as the Subaltern of the Neoliberal Episteme
3.1. Neoliberal Biopolitics and Care

In his 1978–1979 lectures on the birth of biopolitics, Michel Foucault critically describes the operation of neoliberal governmentality as a means of governing populations through the embodied lives of individuals, and their rational choices [53]. Governmentality for Foucault is the "art of government", which directs the subjects to lead their lives in certain ways rather than others. The technologies of governmentality involve not only legislative and disciplinary measures but also discursive tools that govern "the conduct of conduct" among the population, in the most mundane, embodied level of their everyday practices of living.

The Foucauldian account of neoliberal biopolitics provides a useful prism through which to criticize the politics of care in neoliberal societies in an era of transnational capitalism [10,11]. Emphasizing the governance of populations through the practices of self-care of autonomous individuals, Foucauldian biopolitics reveals how in its recognized spheres of governmentality, neoliberalism mutes the recurrent moments of life where subjects are care-dependent and needy of others. This makes the neoliberal worldview illusionary and unrealistic [10–13]. Care ethicists, for instance, have emphasized that human beings are vulnerable and factually dependent on receiving care from each other [1,17,18,21]. Embodied care is a species activity without which humankind would perish [18]. While this should be self-evident, unfortunately the neoliberal politics of social and health care are increasingly being framed as if this were not the case, and as if the subjects and objects of care were disembodied, independent, self-caring actors making rational decisions throughout their autonomous lives, irrespective of their decaying bodies that start to age from the moment they are born [10,11].

The critical disability theorist Robert McRuer has argued that global neoliberal capitalism relies not only on an illusion of disembodied political subjectivity but simultaneously on the "compulsory nature of ablebodiedness". He explains that our freedom to sell our labor and flourish through economic productivity effectively means that we are "free to have an able body but not particularly free to have anything else". However illusory it may be in our factually needy material existence, this compulsory ablebodiedness is deeply interwoven with neoliberal conceptions of how to be human, and consequently with the political orders that emerge with those conceptions. "[W]ith the appearance of choice", writes McRuer, compulsory ablebodiedness covers over "a system in which there actually is no choice" [40] (p. 8).

However, as Tronto's notion of *Homines curans* emphasizes, in political life there always is a choice and an alternative [13]. Together with Hanna-Kaisa Hoppania we have argued that the very paradox within neoliberal governmentality—between the real-life embodied demands of care and the illusionary neoliberal rationality by which care is governed—reveals the fragility of neoliberalism. Producing endless frictions in the neoliberal political economy, the embodied demands of care and the relations they enact keep the order open for the political, open for change, and open for alternative orders [11]. In this article, I wish to emphasize that if we take seriously both the ethics of care and ethics of needs and understand that *Homines curans* are also *Homines egeni* (needy human beings)—the change can be something radically different from the neoliberal episteme. However, to deconstruct neoliberal biopolitics, we need to first start re-reading it from within its own limits.

3.2. Dying Life as the Limit of Neoliberal Biopolitical Governmentality

In his last lecture of the 1975–76 lecture series *Society Must be Defended*, Foucault defines biopower as "the right to make live and let die" [54] (p. 241). In the same context, he indirectly comes to elaborate on why neoliberal biopolitics cannot accommodate life that is dependent on care from others for its own survival. He does this by addressing the question of aging and death as limits to biopower and neoliberal governmentality. These limits illuminate how the body-in-need is the subaltern of the neoliberal epistemic order.

Foucault argues that when calculating and preparing for moments where an individual becomes "incapacitated", put "out of circuit" and "neutralized", the neoliberal governance of life necessarily intervenes in certain universal and irreducible conditions of our embodied vulnerability. These risks, he argues, are particularly "the problem of old age, of individuals who, because of their age, fall out of the field of capacity, of activity" [54] (p. 244). In the field of care, we know that such "incapacitation" can happen at any time for various other reasons such as disablement, illness, or injury, and not just due to old age. Some individuals may never in their lifetime fall *in* to "the field of capacity", as defined by neoliberal reason.

According to Foucault, the neoliberal solution to this universally ever-present possibility of the body's "incapacitation" is the securitization of life through insurance and collective savings [54] (p. 244), which ensures the means to access care when the person is no longer capable of autonomous self-care. This is the logic by which present-day neoliberal societies operate in their governance of care. In her diagnostics of the neoliberal ideology in the politics of care, Joan Tronto names the techniques by which care needs are met in neoliberal societies: namely through responsibilizing individuals to take care of their own needs, by turning to markets for care provisioning, or by relying on family members to provide for their needs [13] (pp. 29–30). However, none of these technologies of governing life can fully tame the biopolitical problems of aging, dying, and "incapacitation". Ultimately, even if a body is insured for the costs of care, capable of purchasing care from the market, or able to rely on familial relations of care, eventually all life must give way to death. And death and the processes of dying—the moments where self-caring capacities and the productive potential of life escape governmentality—are something that neoliberal biopolitics cannot fully capture. Death becomes "the most private and shameful thing of all", the ultimate taboo "to be hidden away". In neoliberal biopolitics, death emerges as "the moment when the individual escapes all power, falls back on himself and retreats, so to speak, into his own privacy. Power no longer recognizes death. Power literally ignores death" [54] (pp. 247–248).

In erasing death and dying life from within its governmental reason, the neoliberal politics of life (i.e., biopolitics) is incapable of addressing those bodies that would die without care, and that may no longer be made to live the kind of life that is productive, and therefore, a life that is worthy of living according to the neoliberal rationality. Neoliberal biopolitics, I argue, relegates these lives to the apolitical field of *zoē*, as bare life, whose life and death are a matter of indifference: they can be killed—or, as Foucault might put it, be let to die—without sacrifice to the capitalist economy, which is the *sine qua non* of neoliberal governmental reason [6,15,23]. Consequently, in the epistemic orders of neoliberalism, the bodies in need of care—i.e., those who can no longer be turned into rational, choice-making, self-caring subjects—cannot speak.

Thereby, the dependent body-in-need emerges as the subaltern of the neoliberal episteme: Whenever a person is fully dependent on care for survival, without the promise to re-establish themselves as a self-caring subject, they become potentially meaningless for today's hegemonic epistemic order. This is particularly so at times when the body's care needs are no longer capitalizable and economically productive [55], for instance, when care-provisioning does not add to transnational capital accumulation through insurance or commodified care markets. Thus, in the present neoliberal epistemic orders, embodied care needs often operate as an erased yet central trace, or as Spivak might put it, as "an always already absent present, of the lack at the origin that is the condition of thought and experience" [49].

That care-dependent bodies are muted by default enables injustices in the field of care, which cannot be tackled because they are non-representable. Therefore, the subalternity of the body in need of care is a central field of epistemic violence in our neoliberal times. This concerns us all. Namely, when it is recognized that all human beings live in body organisms that cannot survive without care from others, the lives of us all are partially written out from the neoliberal political order, even if differentially so. In the following section, I elaborate on how care ethics also partakes in this epistemic violence by partially muting the subaltern body-in-need.

3.3. The Muted Body in Care Ethics

Rosi Braidotti has described the Foucauldian biopolitics' incapacity to address death and dying life as a "residual type of Kantianism" that emphasizes the individual's responsibility for the self-management of one's health, care, and life in general. Braidotti goes on to argue that the downside of this position is that such a politics of life "perverts the notion of responsibility towards individualism, in a political context of neo-liberal dismantling of the welfare state and increasing privatization" [56]. Care ethics provides a realistic challenge to this individualism, and thereby to the neoliberal rationalities by which societies (and the care relations therein) are presently being governed. Showing how morality and politics do not ensue from rational decisions over self-care made by illusionary, disembodied, autonomous individuals, but from situated and embodied practices of care in response to the needs of others, care ethics uncovers the paradoxes and limits of neoliberalism. Thereby, care ethicists continue to challenge the very foundations of (neo)liberal thinking.

There is, however, still a particular liberal "residue" in care ethics. Namely, as discussed in the introduction, by emphasizing moral and political relations as emanating from the practices of car*ing*, care ethics foregrounds the political value of those embodied actors in care relations who have the individualized capacity to care and respond caringly to the needs of others. Chris Beasley and Carol Bacchi have argued that this "paternalist protectionism that care ethics invokes represents only a partial challenge to neoliberal thinking" [57] (p. 285). To me, this is symptomatic of the ways in which many classic accounts of care ethics also mute the needy body. To further examine this argument, let us briefly return to Spivak's theorization of subalternity.

In elaborating on the muteness of the subaltern, Spivak has argued that the hegemonic episteme operates "its silent programming function" through all segments of the population [49] (pp. 282–283). This means that all subjects of the epistemic order partake in muting the subaltern, and insofar as one seeks to make sense within the grammars of the order, it is practically impossible not to. In her essay "Can the subaltern speak?", Spivak showed how even the allegedly most radical critics of the prevailing episteme can end up silencing the subaltern. Spivak's criticism in the essay was addressed to two celebrated leftist intellectuals, namely Foucault and Deleuze, and through their example, more generally to the Western post-structuralist theory. In her essay, Spivak deconstructs the seeming transparency of these intellectuals' positions in their critiques of the capitalist world order. She argues that even as Foucault and Deleuze in their critical analyses of the modern society seek to appreciate "the discourse of the society's Other", they still tend to present the Other as a monolithic mass—for instance, as "the Maoists" and "the workers".

Thereby, while the intellectuals were capable of naming and differentiating between other (Western) intellectuals, they simultaneously failed to recognize the heterogeneity within the otherness they claimed to know [49] (p. 272).

The danger (and epistemic violence) here is "the first-world intellectual masquerading as the absent nonrepresenter who lets the oppressed speak for themselves" [49] (p. 292), while the subaltern Other is simultaneously excluded from the epistemic order within which the critique is being articulated. Spivak emphasizes that whereas the transparent intellectual may well be in a position to make visible the mechanisms of oppression, rendering the oppressed 'speakable' within the prevailing discourse is a whole other question, and impossible in the case of the subaltern.

It is important to bear in mind the decolonizing aims of Spivak's argument, which I do not wish to co-opt in this article when utilizing her argument to underline the muting of the needy body in the neoliberal episteme. Spivak's discussion of the subaltern unmasked the role that the transparent Western intellectual has had (and often continues to have) in upholding the postcolonial economic order and its international division of labor. Spivak drew particular attention to the constitution of the female colonial subject as the Other in the specific context of India, asking whether and how she could speak. These aims in Spivak's theory of the subaltern are also crucial to the ethics of needs, even if the scope of this article is inadequate to fully account for the postcolonial and economic dimensions of this ethic. Consequently, in several ways (and not least due to the citational politics of this article), the ethics of needs as articulated in this article remains open for decolonial critique.

For the moment, however, I wish to utilize Spivak's critique of the Western intellectual to elaborate on the relationship between care ethicists as intellectuals, and the position of bodies-in-need within the ethics of care. As a care ethicist, I am also the object of my own critique. Indeed, my very need to try and articulate an ethics of needs as a supplement to the ethics of care is triggered by the necessity to address the needs of those bodies that my care ethics will inescapably always exclude, that is, the bodies that will forever haunt the ethics of care. The question here is: When the ethics of care sees moral and political relations as enacted by the practices of car*ing* (i.e., by the practices of subjects with an adequate psycho-social capacity to care), *can the bare-bodies-in-need speak in this ethics?*

Care ethicists recognize care recipients and their responses as central to the "circles of care" [18], and caring practices in care ethics are always responses to the needs of particular others. This gives care ethics its distinctive situated and political character, different from moral theories of abstracted justice. It would therefore be misleading to argue that care-recipients are entirely absent in the ethics of care. They are there, possibly in the margins as far as political agency is concerned, yet present as objects of caring practices as well as through their subjective demands and responses to care. However, when caring is a response to the needs of particular others, only those bodies-in-need whose *subjectivity as a care recipient* is recognized become included in the ethics of care. Those bodies that need care, yet lack recognized subjectivity altogether, or subjectivity *as needy* in established epistemic orders of care provision (often including the embodied needs of the carers), become written out as a mass of Others. The needs of such bodies cannot quite speak, and in the ethics of care, *needs* as *bare needs* cannot speak—only the *subjects* of caring and care-receiving can.

Thus, as care ethicists and intellectuals, we may recognize and see bodies-in-need as a corporeal, fleshy mass, muted in the circles of care. Yet, we still fail to decipher the heterogeneity of this mass and the power relations and differentiations therein. Thereby, in its quintessentially ethical demand to respond to the needs of particular, individuated, personified others, the ethics of care remains tied with the liberal emphasis on *individuated subjectivity* when the question of needs is considered. To the extent that the body-in-need (as a "bare" body stripped from its subjectivity) is the subaltern of neoliberalism that cannot speak, the ethics of care operates within the same epistemic order. Consequently (like Foucault and Deleuze in Spivak's critique of the Western intellectual), as care ethicists and intellectuals, we are unable to fully escape the rationality of the epistemic order we seek to criticize. Instead, we remain partially and unintentionally "complicit" in upholding it, even in the most radical criticisms that emphasize caring as a challenge to *Homo oeconomicus* [13]. Therefore, as care ethicists, we need a complementary ethics: an ethics of needs that begins its analysis of relatedness from bare bodies deprived of their subjectivity.

Spivak writes: "In the face of the possibility that the intellectual is complicit in the persistent constitution of Other as the Self's shadow, a possibility of political practice for the intellectual would be to put the economic 'under erasure,' to see the economic factor as irreducible as it reinscribes the social text, even as it is erased, however imperfectly, when it claims to be the final determinant or the transcendental signified." [49] (p. 280) As care ethicists, we are inescapably complicit in the persistent constitution of the needy body as the Other in the shadow of our caring Selves. This inescapable othering is written in the grammar of the ethics of care, where it cannot be avoided. The subaltern Other here is not the *subject*-in-need, but the (bare) body-in-need, deprived of its individually articulated subjectivity. Therefore, the subject (and not simply the self-caring individual as in care ethicists' criticisms of neoliberalism) is the economic reason for our epistemic order that must be written under erasure. This I start to do in the next section.

4. Analyzing Relatedness on the Level of "Bare" Needs

4.1. The Lowest Common Denominator of Embodiment

In the ethics of needs, each embodied trajectory of human life is recognized as unique, defined by particular and innumerable corporeal relations of care and neglect, which are enacted by needs and that extend through both space and time. From one embodied life to another, the quality and quantity of care, as well as neglect, differs. This is because the care needs of our bodies are always differentially powerful when compared to all other bodies at any moment of time. The need is furthermore never entirely powerless. It always enacts a response, although at times the response is not a caring one. The care ethicist Fiona Robinson has argued, that "relations of care are constructed by relations of power determined by gender, class, and race. These are, in turn, structured by the discourses and materiality of neoliberal globalization and historical and contemporary relations of colonialism and neo-colonialism. In this view, thinking about care in the context of global politics and security cannot posit a universal need for care as unproblematic and undifferentiated; *needs are themselves constructed and produced by a wide range of relationships and structures.*" [21] (p. 5, emphasis added).

Here, the ethics of needs adds that care needs are not only *produced by* various relationships and structures. Embodied care needs and the power differentials between them are also *productive of* the governmental relationships and structures, within which they demand a response. These power differentials are not only inscribed in our intersectionally gendered bodies, but our needy bodies also inscribe the world. They do so by the relations that their differentially powerful needs enact with other bodies. Yet, it is only after recognizing all bodies as existentially needy of care from one another, and all needs of care as capable of enacting political relatedness, that analyzing the differential power of needs becomes possible at all. When it comes to the body's ontological need for care, there is no exception, only difference.

But how does one read difference and the power relations between "bare" bodies-in-need, that lack the subjectivity of a speaking, caring "I"? The ethics of needs focuses on needy bodies at the level of what I call *the lowest common denominator of embodiment*. Hamington has pointed to the body as a common denominator of humanity [28] (p. 39), but in the ethics of needs, I wish to be somewhat more specific. With the lowest common denominator of embodiment, I refer to the fact that as living and dying organisms, all human bodies have certain mundane material needs that make humans existentially dependent on one another. The lowest common denominator of embodiment is the physiological fact that, in addition to breathing, all human bodies need to be fed and watered, and need to digest, as well as discharge urine and excrement in an adequately hygienic way. These are daily needs, and when they are not met, the human being in question will eventually die.

While there may also be other fundamental needs, I refer to these needs because they apply to every single living/dying human body, every day, from the moment of birth to death, regardless of age, gender, sexuality, race, ethnicity, class, status, or any other attributes of identity, or social position. Indeed, if we just focus on the fact that every human body must eat, drink, urinate, defecate, stay clean, and require care from others when not capable of doing so independently, the lowest common denominator of embodiment marks a space where differentiating the feminine from the masculine is not self-evident. Hence, as living organisms on the level of the lowest common denominator, our bodies are needy as well as fundamentally queer.

From this perspective, no human being can exist without care provided by other bodies: and in order to survive, we all must eat, drink, poo and wee, and stay clean, and over a lifetime this requires care from other bodies if we are to stay alive. Just as in the ethics of care, in the ethics of needs, care is still a species activity [18], and we are never truly individual as our inescapable neediness makes us always already related to and with other bodies. With this argument, the ethics of needs seems no different from the ethics of care. And yet, it is. Namely, in the ethics of needs, also other responses to needs than the caring ones count: Neglect is a species activity, too.

The lowest common denominator of embodiment refers to the analytical level of embodied human life where we are stripped of articulate subjectivity—i.e., our capacity to speak, either literally, or in ways that make sense in the prevailing epistemic order. It is a level of analysis where we are but animals, with the socio-political relationality of humankind, and its structures of governance emerging with the relationships that our barest needs enact with other bodies in the world. Such relations emerge because bodies can be (and very often are) both needy and caring. That our lives are defined by such a dimension of bare-neediness on the level of the lowest common denominator does not mean that other levels and dimensions of life would simply go away. As Hamington puts it: "there is no essence of embodiment that can be separated entirely from the body's transactions with society" [28] (p. 41). Hence, on the level of the lowest common denominator, the subject positions and identities that are inscribed on bodies by the governmentalities of prevailing epistemic orders still matter. The power differentials between bodies' capacities to enact relations of care or neglect are shaped by such inscriptions.

For instance, if inscribed with the citizenship and rights of a welfare state, a person living with a severe form of dementia, having lost their capacity to articulate themselves as a speaking, thinking, or caring "I" in the prevailing episteme, may still rather powerfully enact care relations. They may put in motion political processes and legislative changes by demanding the publicly funded services that bodies in that society are eligible to access. Their bodies (even if in many ways muted) demand vaccines and boosters in the time of a pandemic, when entire populations elsewhere are left without. Such bodies can make care resources and caring bodies move, even across the globe, when migrant care workers are utilized to fill care deficits in aging welfare states [6,15]. These powerfully needy bodies are not only inscribed with rights and wealth, however. Their bodies, as bare, needy bodies, are also inscribed with *value* that can be extracted for transnational capital accumulation through care provision. This is because the neoliberal welfare state's financialized structures of care provision tie public spending on care needs with the interests of transnational shareholders of care-selling businesses. This may apply to situations where financialized care providers owned by private equity firms sell services to the public sector [5,8], as well as to global vaccine politics where vaccine nationalism feeds into the profits of pharmaceutical companies.

In short, bodies' care needs in this world are attended to when the needs are both economically and politically valuable. And yet, even here amongst an allegedly privileged mass of aging citizenry in wealthy welfare states, needy bodies are never just about a bare mass. There are power differentials between needy bodies, where inscriptions of racial, class, gender, and other differentiations come to matter, in just *how* powerfully the bare-body-in-need (on the level of the lowest common denominator) can make the different networks, circles, and processes of care move, both globally and locally.

Simultaneously, a body-in-need with dementia also demands care in some less care-privileged part of the world (e.g., a homeless person in the US, or some other care-poor society where bodies are *not* inscribed with rights to care or publicly funded social security). This body, too, enacts relatedness by its very needs. The body may not have the power to instantiate relations of car*ing*, or when they do, the relations enacted are understood to be about charity rather than political rights. Yet, the body is still not powerless in its needs. Its needs, too, write the world with political relatedness. For instance, the body's needs feed into the structures of the transnational political economy. While responding to the body's needs *caringly* may not add to transnational capital accumulation, neglecting the needs may well do just that, when enforcing the idea that one should be financially insured against the need for care—or is not eligible to care at all. However violent and unjust it may sound, these relations matter, both politically and economically. Each relation enacted by needs is a unique opening and reaching out to the world that inscribes the world and its order at that particular moment of time. If we only focus on caring relations as is the case in traditional accounts of care ethics, we miss the opportunity to even try and read what other types of relations of needs do to and with the world.

I argue that as care ethicists, it is our moral responsibility to develop ways to read the relatedness that all kinds of needy bodies write, or else we remain partially complicit in reproducing the position of these bodies in neoliberal epistemes as a bare abandoned mass outside the established governmentalities of care. When focusing on human relatedness on the level of the lowest common denominator (bare speechless bodies related to one another through needs), the network of relatedness that emerges is quite different from the one that emerges with relations of car*ing*. To envision this network of needs, we need to move away from caring and write the subject temporarily under erasure.

4.2. Writing the Subject under Erasure

The ethics of needs focuses on bare-bodies-in-need-of-care. Such a focus necessarily obscures the psyche and lived experiences of those subjective persons who need care. In various seminars and conferences, I have been interrogated about this ethical dilemma: Is it not *unethical* to talk about the bodies in need without addressing their subjective agency? Does that not lead to a position where the needy are treated as a faceless mass lacking agency altogether? These concerns are undoubtedly valid in an epistemic order where the body-organism "in itself" (without a subjective speaking, caring, thinking "I", or as bare life of *zoē* outside the politically valuable life of *bios*) is understood as lacking political agency. However, the ethics of needs attempts to go beyond such an epistemic order. It seeks to comprehend the power relations and politics of human life in the field of *zoē*, where we always already exist as meaningful, unique, and valuable body-organisms, both before and after articulating ourselves as a subjective, caring "I".

The trouble is that once we let the subject speak, addressing the moral and political relations between bare-bodies-in-need in the field of *zoē* becomes impossible. After all, once addressing the subject, we are back in the field of *bios*—the political life constituted by muting the subaltern life of bare-bodies-in-need. Thus, for an ethics of needs to emerge, the subject must be decentered. The point here is not to deny subjective agency from the care-dependent, or to say that their lived experiences do not count—they do! In the ethics of needs, disregarding subjectivity is merely an analytical tool that might (just might) allow us to read power differences between the relations that differential care needs enact in the world. This, in turn, might (just might) also allow us to analyze those moral and political relations that bare bodies enact when caring fails to take place where it should or could but does not. In other words, it might allow us to examine the shadows of care ethics, where the moral power of care is present only in its own absence.

However, writing the subject under erasure is difficult. The Cartesian mind/body split remains one of the most powerful binaries defining our political life. As a remnant of modernity in supposedly postmodern times, our political thinking continues to be organized as if our subjective minds owned our bodies: the (feminized) body remains devalorized and muted so the (masculine) rationality of the subjective mind can be cherished and heard. To an extent, this remnant is visible in the ethics of care, too. Even if it does not emphasize the rational mind of the *thinking* "I", the focus on a car*ing* "I" still remains.

In reality, of course, as feminist theorists of the body have shown for decades, the mind—be it caring, embodied, thinking, or speaking—does not own the body. We may well speak of the body "to others as of a thing that belongs to us; but for us it is not entirely a thing; and it belongs to us a little less than we belong to it" [58] (pp. 398–399, cited in [45] (p. 65)). However much we may valorize the rational mind as the basis of political subjectivity, our bodies continue to both pre-live and outlive the speaking "I" that is commonly understood as subjectivity: first in infancy and later for example through dementia or other cognitive conditions, illnesses, or injuries, or simply through the decay of the aging and dying brain. When subjectivity is considered from the perspective of embodied care, as pointed out by Hamington, "subjectivity is physical" [28]. The human body organism is an agent of its own, and also at times when it lacks an articulate subjectivity: an "I" that can speak, think, articulate, take responsibility, and care.

The ethics of needs focuses on political relations that bodies-in-need-of-care enact in the world. This is an ethics that seeks to read the ways in which care needs, as "bare" needs, write the world, and how they do it even when they lack articulate subjectivity. Because the writing of needy bodies takes place through the corporeal relations that *differentially powerful* needs enact in the world, the ethics of needs requires tools to read the power differentials between bodies as "bare" bodies that lack the subjectivity of the speaking "I". For this purpose, the ethics of needs focuses on bodies-as-organisms-in-need-of-care at the level of the lowest common denominator. This demands me *not* to address the subject.

Let us now briefly return to Derrida's philosophy of deconstruction, which involves the technique of writing "under erasure"—i.e., a writing method where words are struck out. As elaborated by Spivak, ~~the trace of the other~~ in the text for Derrida is not simply the inarticulable. It is *différance*, "the mark of the absence of a presence, an always already absent present, of the lack of the origin that is the condition of thought and experience" [49] (pp. xvii–xviii). In the predominant epistemic orders, ~~the-body-in-need~~ is ~~the-trace-of-the-other~~ that cannot speak and cannot be heard. The ethics of needs is therefore a deconstructive move, where the binary between the Subject and its ~~erased Other in the body-in-need~~ is subverted by putting the subject temporarily under erasure.

To make space for writing about and with the body as a body organism with nothing but an ~~erased trace of subjectivity, the subject~~ must be written under erasure in the ethics of needs. This does not make questions about ~~subjectivity~~ and ~~subjective agency~~ simply go away, and indeed, it is imperative to address the subjective agencies of care-dependent persons in the wider politics and ethics of care. As elaborated in the previous section, from within the margins of the ethics of needs, various inscriptions of a ~~subject position and identity~~ on the body-organism still shape the power that bodies-in-need have in enacting relatedness with the world. The erasure of ~~the subject~~ is therefore but a deconstructive move that the ethics of needs makes for analytical purposes. The ethics of needs fails to make it a double move, in that the binary hierarchy between ~~the subject~~ and the body-in-need would be "'reinstated' with a complete "reversal that gives it a different status and impact" [25] (p. 150). Nevertheless, to examine how the body as "bare", needy human life writes the world with the relations its *needs* enact—and to read the power differentials between *needs* rather than ~~subjects~~—in the ethics of needs we must refuse to talk about ~~the subject~~ or *its* embodiments, even if that means leaving the deconstruction of embodied subjectivity in suspension. Consequently, the new episteme that emerges when ~~the subject~~ is written under erasure is an order written by the relations that needy bodies enact in the world in their differential power to enact relations of care. I will elaborate on this *différantial* play in the following concluding section, where I try and articulate how the ethics of needs challenges the prevailing neoliberal episteme, while still complementing the moral positions of the ethics of care.

5. Conclusions: Envisaging a World Written by Care Needs

Vicki Kirby has argued that as scenes of writing, bodies always operate as their "own historical and cultural context[s]" [45] (p. 62). They carry within themselves complex trajectories of the past, while simultaneously being capable of enacting new histories and alternative politics when relating to other bodies. Here, some bodies are marked with subjective affects and pasts that help the bodies to enact relations of care, rather than neglect. However, due to a range of historical reasons and governmental technologies, revolting affects [59] such as disgust, abjection, and indifference easily stick to the bodies of others, which then tend to enact an aversive corporeal relatedness rather than relations of care. In such an affective "cultural politics of emotion" [60], historicized postcolonial scripts of racial difference materialize on bodies and their interrelations, as do scripts of dis/ability, gender, sex, age, wealth, and various other entanglements of differentiation.

Thus, each human body in this world produces and carries with it an ever-changing bundle of corporeal relatedness, enacted by differentially powerful care needs, while mobilizing care resources, caring bodies, and political economies of care in some ways rather than others. Be it the body of a globally mobile care worker, that of an older citizen with rights to public care in a welfare state, or a person dying of COVID-19 because of wealthy states' refusals of patent waivers; each human body lives and moves as an intersection of a unique set of corporeal relatedness. This weave of relatedness that the body both carries and produces simply through the processes of living is defined by the needs of the body itself, by the needs of the multiple others who demand care elsewhere, by the governmentalities that organize care relations and resources, as well as by the global political economies where needs are defined as differentially valuable for capital accumulation.

As further elaborated below, the ethics of needs is a theory of care that can be extended to more-than-human worlds [2]. Yet, if we stay with the human population for now, the ethics of needs imagines it as being comprised of almost eight billion bundles of relatedness, also known as human bodies. Of these billions of bodies, each needs care, and the needs of every one enact political relatedness in the world. As the innumerable corporeal relations elicited by these billions of needy bodies intertwine, each relation in tension with one another, what emerges is an entirely new articulation of the world and its changing relations of power. Here, the "bare" human body that is existentially in need of care from other bodies is no longer about a faceless, apolitical mass residing in the field of *zoē*. Instead, the body-in-need emerges as *différance*, that is, a sheaf of relations and potential politics that "permits the different threads of force to go off again in different directions, just as it is always ready to tie itself up with others" [47] (p. 3).

When understood as *différance*, the needy body is not just "a sign, a function of discourse", as bodies are sometimes understood in the feminist theorization of body writing [61] (p. 63). As *différance*, the needy body is a vehicle of temporized movement of differentiation. This means that the differentiation between bodies worthy of care and those that deserve mere neglect is never inscribed on the body from outside, in an over-arching manner. Rather, as a vehicle of temporized movement of differentiation, the body in need of corporeal others constantly calls for the recognition of its needs. Importantly: *the differentiation takes place not on the body, or in the body, but in the emergent corporeal relations that the body in need enacts.*

Derrida argued that "*différance* is also the element of the same (to be distinguished from the identical) in which oppositions are announced" [62] (p. 9). This supports my understanding that power relations between bare-bodies-in-need should be examined at the level of the lowest common denominator, where all bodies need to stay adequately clean, eat, drink, urinate and defecate, and need care to do all that, albeit differentially so. In their power to enact relatedness, all needy bodies are the same, yet never identical.

Ultimately, when deprived of articulate subjectivity, it is on this level of political life where oppositions and differences become inscribed and reiterated. When all needy bodies as *différance* enact their relatedness in and with other bodies of *différance*, what emerges is a *différantial* play or struggle over which needs come to matter as relations of care, and which materialize as relations of neglect. In this play of *différance*, new epistemic orders are always in the making, as bodies expose themselves to each other as vulnerable and needy. Here, the body in need as *différance* emerges as "the most general structure of the economy" without which "there is no economy" [62] (p. 9).

Différance is always about a constant struggle over the (re)appropriation of value that draws the boundaries of epistemic orders. In the ethics of needs, the needy body that enacts relatedness emerges as such a shifting point of origin for economic and political orders: As *différance*, the bare-body-in-need demanding relatedness with other bodies is the most general structure of the economy or of emergent economies in the making. Without it, there would not be the differentiation camouflaged as sameness that all economies depend upon.

The economies that emerge with relations enacted by needy bodies are always open for the political. This is because the needy body as *différance* obligates a constant rewriting of human relatedness. Or to put it otherwise, the needy body as *différance* writes the world with the relations its needs enact. Writing here, of course, is not to be confused with its literal meaning, but is to be understood in the general sense, where bodies are never simply passive surfaces of inscription, or texts to be written from "outside" by historicized discourses of differentiation. In the general sense of writing, "nature scribbles" and "flesh reads" [45] (p. 127), and in the ethics of needs, it does so because the human body is existentially dependent on care provided by other carnal human beings. While some bodies' needy scripts are readable in the predominant epistemes as demands of care, the scripts for care written by needy others are ~~erased~~ and rendered incomprehensible and ~~unspeakable~~.

It thus becomes possible to see that, when ~~the subject of the speaking, thinking, and caring "I"~~ is written under erasure, the differentially powerful embodied needs make a language of (and for) the world that is not written in words but in corporeal relations of care, as well as neglect, the absence of ~~caring~~. Here, all human bodies are the same at the level of the lowest common denominator, and yet always *différantial* in power. The ethics of needs is a call to deconstruct this kind of body writing, with respect to the subaltern bodies that cannot write the kinds of texts that those response-able can or are "programmed" to read in the prevailing epistemic order.

Epistemic orders and existing governmentalities still matter. Bodies' capacities to write the world with the relations their needs enact does not mean that corporeal relations of care and neglect would just freely float in the air or cut through political voids. Relations and trajectories of care both disrupt and are limited by the changing structures of governance and governmentality. The corporeal relations of care and neglect are also never simply relations between two individuals. They always cut through, and emerge with, a range of biopolitical governmental orders, which aim to manage the very relatedness of life, across populations. These governmentalities operate as grammatical rules by which bodies' writing of relatedness must abide if their writing is to make sense. The various grammars of neoliberal governmentality still issue their subtle instructions on how to relate with the needy, and which needs are to be read as response-able. However, due to the body's capacity to operate as *différance*, these grammatical rules of governmentality are recurrently rewritten and subverted, as needy bodies enact relatedness in unexpected ways. To cite Vicki Kirby, in such a "rhythm of *différance* the body is never not musical. The body is the spacing, the ma(r)king of an uncanny interlude" [45] (p. 63).

In Berenice Fisher and Joan Tronto's well-known definition, care is understood as "*a species activity that includes everything we do to maintain, continue, and repair our 'world' so that we can live in it as well as possible.* That world includes our bodies, our selves, and our environment, all of which we seek to interweave in a complex, life-sustaining web" [18] (p. 40, emphasis in original). The ethics of needs partially describes, and seeks to map, this very same web of relatedness, albeit from another perspective. Instead of focusing on caring as a "species activity", the focus is on the needs that enact or do not enact care. Then, it is not simply caring that does the weaving of the complex web of inter-connectedness, but also the embodied needs that ask for and demand a response—any response. However, the web that the needs weave is only partially the same as the life-sustaining web of care. Namely, the web that the needs interlace in the world also includes those relations where care needs are responded to with ignorance, neglect, or even violence. This web—or text of the world—exposes transnational (bio)political economies of care, making visible not only those relations where life is sustained but also those where it is not. At the level of the lowest common denominator, it reveals which bodies enact caring relations, and which enact only care that could have been. Furthermore, it exposes which "worlds are being maintained, and at the expenses of which others" [2] (p. 44).

Furthermore, since the ethics of needs writes ~~(human) subjectivity~~ under erasure, there is no need to limit the theory to apply only to the human species. Nonhuman bodies also enact relations of care, as well as relations of neglect and violence in this world, and today it is perhaps more urgent than ever to try and read those relations, their entanglements, and the hierarchies therein between relations of care and neglect. The ethics of needs is therefore about a politics of care beyond species activity, which could perhaps be utilized in developing a more planetary ethics of needs.

But how does the ethics of needs challenge the neoliberal politics of life? I have discussed how the incapacity of neoliberal biopolitics to engage with dying, decaying, and "incapacitated" life makes it a politics of life, where life is given care and made to live only insofar as the living body is either economically productive through its labor, or profitable through the sales or financialization of its needs. Beyond this, life may as well be let to die or killed without sacrifice, as far as neoliberal governmental reason is concerned. Such violence against economically non-productive bodies is immoral. Yet, in the prevailing epistemic order, there are far too many examples where we can see this rationality at play, ranging from the politics of eldercare to global vaccination politics, to the global distribution of human resources for health to the instrumentalization of non-human life in capitalist modes of production. For some reason, such a neoliberal politics of life seems very difficult to resist and subvert, regardless of its implicit (and often explicit) violence towards the presumably speechless bodies of the needy.

Here, the question remains as to whether a kind of transnational politics of life is possible, where care needs would be seen as valuable and capable of enacting caring responses also when the care involved would not feed into transnational capital accumulation. In lieu of a conclusion, I want to suggest that the ethics of needs provides an opening for a more ethical biopolitics. To the extent that neoliberal biopolitics is about a politics of making live and letting die, its moral problem derives from the corporeal ontology of care, and the material resources it requires: Making live and letting die involves decisions over which lives are worthy of life-sustaining care, and which can be left to die. After all, the ontological condition of care is that it is always a concrete relation with a particular body-in-need. Therefore, when taking care of somebody—or some population, as in nationalized vaccine politics for instance—it is an ontological necessity for the caring body or institution to turn its back on the needs of some others. This means that one may *care about* a whole variety of corporeal care needs in the world, but for no embodied subject, population, or institutional construct is it ontologically possible to concretely *take care of* and *provide care* for everybody at the same time [18]. In all care relations, some needs and some bodies remain neglected.

In the ethics of needs, corporeal relations of care and neglect are understood as being deeply entangled, and these entanglements emerge continuously and everywhere, through the choices and decisions as to whose needs count for the response-able. While the "bare" body-in-need, living in the field of *zoē*, is generally silenced in the neoliberal episteme, the ethics of needs allows us to see how within this allegedly non-political sphere of life, there are still relations of power. Some needy bodies, even at the level of the lowest common denominator, dominate others, and this is because the *needs* of some bodies are economically and politically more powerful, and hence more worthy, and speakable compared to the utterly muted bodies of Others. When care is defined as a corporeal-relation-enacted-by-needs, these material-discursive *hierarchies of needs become a property of care*. This means that the ~~care that is absented~~ to make care present elsewhere is part and parcel of the corporeal relations of care. Neglect is thus ~~care~~ that could have been. When analyzing the political relatedness that caring enacts in the world, it is perhaps a moral necessity of the care ethicist to also try and address the political relations of ~~caring~~ that are absented when care is provided. Such a perspective makes care (including the care of populations) always a potential question of neglect and/or violence.

When striving for a more caring politics of life than neoliberal biopolitics, we must remember that regardless of all the possible care in the world, eventually, all living bodies become incapacitated and die. Indeed, if its sole goal was to sustain life, care would be forever doomed to fail. But then, care is not only about the sustenance of *life* or economically productive life, even if neoliberal biopolitics may be directed to this end. Rather, care is about responding to the needs of others, and about corporeal relations where *needs rather than life itself* count as originary starting points for politics. Consequently, *the biopolitics of needs* that emerges with the ethics of needs is not just about the politics of *life*. The biopolitics of needs is, literally, about the *politics of life of needs*. In such biopolitics, it is recognized that needs have a life of their own, even when the life of a speaking, caring, and economically productive subject seizes to exist. In the politics of life of needs, it becomes possible to address the care needs of the dying and the incapacitated, also when the care provided fails to accumulate capital, or be otherwise economically productive in the capitalist sense of productivity.

In other words, for care to be an ethical relation, we must let the needs of Others haunt the epistemic orders and governmental relations that are understood as presumably caring. The potentially more caring biopolitics that emerges with the ethics of needs requires a recognition of the difficult choices that care, as a response to the *living needs* of others, always demands. It is an approach that necessitates a constant deconstruction of care relations. The ethics of needs always returns to those to whom our backs were turned when care was provided, and to the excess of life that the neoliberal biopolitics fails to account for. It returns to the impossible limits of ~~care~~—recurrently, persistently, chronically.

Funding: This research was funded by The Academy of Finland, grant number 321972, and The Kone Foundation, grant number 201802636.

Acknowledgments: I want to thank the editors of this special issue, Maggie FitzGerald and Maurice Hamington, for inviting me to participate in the issue, and thereby providing the opportunity to finally try and put this argument together in an article format. Thank you for your patience with my delays and requests for deadline extensions. Heartfelt thanks also go to the two anonymous reviewers for their encouraging and thought-provoking comments on the first draft of the manuscript. I think I have failed to fully respond to all your critiques, but I promise to go back to them when developing my ethics of needs further in the future. It is certainly still a theory that is underdeveloped and full of gaps, but thanks to your caring labor of reading and commenting, this first version of the argument is now "out there". The article is based on my PhD Thesis *The Global Biopolitical Economy of Needs: Transnational Entanglements between Ageing Finland and the Global Nurse Reserve of the Philippines* (University of Tampere, 2017). While this article is a theoretical piece of writing, I remain indebted for my research participants for helping me to think through the ethics of needs empirically, as transnational phenomena. Thank you also for the examiners of the thesis, Eeva Jokinen, Fiona Robinson, and Anna Agathangelou for all your feedback and encouragement at the time when the theory was first developed. Anna "Imppu" Rajala, dear friend and colleague: Thanks for reading this thing through and just being the wonderful colleague and person you are. Sanna Melonen and Sini Rönnqvist: Thank you for all the cooking and feeding and care you provided me with during our week in Levi, Lapland, during the writing-retreat-with-some-skiing, when I finished the first and the most difficult draft of this manuscript. I needed the get-away, the food, and your company, and it was all just lovely.

Conflicts of Interest: The authors declare no conflict of interest.

References

1. Kittay, E.F.; Jennings, B.; Wasunna, A.A. Dependency, Difference and the Global Ethic of Longterm Care. *J. Polit. Philos.* **2005**, *13*, 443–469. [CrossRef]
2. Puig de la Bellacasa, M. *Matters of Care: Speculative Ethics in More than Human Worlds*; University of Minnesota Press: Minneapolis, MN, USA, 2017.
3. WHO Coronavirus (COVID-19) Dashboard with Vaccination Data. Available online: https://covid19.who.int/table (accessed on 18 February 2022).
4. Pfizer, BioNTech and Moderna Making $1000 Profit Every Second While World's Poorest Countries Remain Largely Unvaccinated—World. Available online: https://reliefweb.int/report/world/pfizer-biontech-and-moderna-making-1000-profit-every-second-while-world-s-poorest (accessed on 18 February 2022).
5. Hoppania, H.-K.; Karsio, O.; Näre, L.; Vaittinen, T.; Zechner, M. Financialization of Eldercare in a Nordic Welfare State. *J. Soc. Policy* **2022**, 1–19. [CrossRef]
6. Vaittinen, T. *The Global Biopolitical Economy of Needs: Transnational Entanglements between Ageing Finland and the Global Nurse Reserve of the Philippines*; Tampere University Press and Tampere Peace Research Institute: Tampere, Finland, 2017.
7. Neilson, B. Ageing, Experience, Biopolitics: Life's Unfolding. *Body Soc.* **2012**, *18*, 44–71. [CrossRef]
8. Horton, A. Financialization and Non-Disposable Women: Real Estate, Debt and Labour in UK Care Homes. *Environ. Plan. A Econ. Space* **2022**, *54*, 144–159. [CrossRef]
9. Global Justice Now Omicron Covid Variant—BBC Interview with Dr Ayoade Alakija. Available online: https://www.youtube.com/watch?v=e97XFM2WmQ0 (accessed on 18 February 2022).
10. Dahl, H.M. Neo-Liberalism Meets the Nordic Welfare State—Gaps and Silences. *NORA—Nord. J. Fem. Gend. Res.* **2012**, *20*, 283–288. [CrossRef]
11. Hoppania, H.-K.; Vaittinen, T. A Household Full of Bodies: Neoliberalism, Care and "the Political". *Glob. Soc.* **2015**, *29*, 70–88. [CrossRef]
12. Hoppania, H.-K. Politicisation, Engagement, Depoliticisation—The Neoliberal Politics of Care. *Crit. Soc. Policy* **2019**, *39*, 229–247. [CrossRef]
13. Tronto, J. There Is an Alternative: Homines Curans and the Limits of Neoliberalism. *Int. J. Care Caring* **2017**, *1*, 27–43. [CrossRef]
14. Tronto, J.C. *Caring Democracy: Markets, Equality, and Justice*; NYU Press: New York, NY, USA, 2013.
15. Vaittinen, T. The Power of the Vulnerable Body. *Int. Fem. J. Politics* **2015**, *17*, 100–118. [CrossRef]
16. Gilligan, C. *In a Different Voice: Psychological Theory and Women's Development*; Harvard University Press: Cambridge, MA, USA; London, UK, 1982.
17. Tronto, J.C. *Moral Boundaries: A Political Argument for an Ethic of Care*; Routledge: New York, NY, USA; London, UK, 1993.
18. Fisher, B.; Tronto, J.C. Toward a Feminist Theory of Caring. In *Circles of Care: Work and Identity in Women's Lives*; Abel, E.K., Nelson, M.K., Eds.; State University of New York Press: Albany, NY, USA, 1990; pp. 35–61.
19. Sevenhuijsen, S. *Citizenship and the Ethics of Care: Feminist Considerations of Justice, Morality and Politics*; Routledge: London, UK, 1998.
20. Held, V. *The Ethics of Care: Personal, Political and the Global*; Oxford University Press: Oxford, UK; New York, NY, USA, 2006.
21. Robinson, F. *The Ethics of Care: A Feminist Approach to Human Security*; Temple University Press: Philadelphia, PA, USA, 2011.
22. Robinson, F. *Globalizing Care: Ethics, Feminist Theory, and International Relations*; Westview Press: Boulder, CO, USA, 1999.
23. Agamben, G. *Homo Sacer: Sovereign Power and Bare Life*; Stanford University Press: Stanford, CA, USA, 1998.
24. Critchley, S. *The Ethics of Deconstruction: Derrida and Levinas*, 1st ed.; Blackwell: Oxford, UK; Cambridge, MA, USA, 1992.
25. Culler, J.D. *On Deconstruction: Theory and Criticism after Structuralism*; Routledge: London, UK, 1998.
26. Miller, S.C. *The Ethics of Need: Agency, Dignity, and Obligation*; Routledge: London, UK, 2012.
27. Fraser, N. Talking about Needs: Interpretive Contests as Political Conflicts in Welfare-State Societies. *Ethics* **1989**, *99*, 291–313. [CrossRef]
28. Hamington, M. *Embodied Care: Jane Addams, Maurice Merleau-Ponty, and Feminist Ethics*; University of Illinois Press: Champaign, IL, USA, 2004.
29. Twigg, J. *Bathing—the Body and Community Care*; Routledge: London, UK, 2000.
30. Isaksen, L.W. Toward a Sociology of (Gendered) Disgust: Images of Bodily Decay and the Social Organization of Care Work. *J. Fam. Issues* **2002**, *23*, 791–811. [CrossRef]
31. Zechner, M.; Näre, L.; Karsio, O.; Olakivi, A.; Sointu, L.; Hoppania, H.-K.; Vaittinen, T. *The Politics of Ailment—A New Approach to Care*; Policy Press: Bristol, UK, 2022.
32. Slater, L. Learning to Stand with Gyack: A Practice of Thinking with Non-Innocent Care. *Aust. Fem. Stud.* **2021**, *36*, 200–211. [CrossRef]
33. Varfolomeeva, A. Destructive Care: Emotional Engagements in Mining Narratives. *Nord. J. Sci. Technol. Stud.* **2021**, *9*, 13–25. [CrossRef]
34. Kelly, C. Care and Violence through the Lens of Personal Support Workers. *Int. J. Care Caring* **2017**, *1*, 97–113. [CrossRef]
35. Clifford Simplican, S. Care, Disability, and Violence: Theorizing Complex Dependency in Eva Kittay and Judith Butler. *Hypatia* **2015**, *30*, 217–233. [CrossRef]
36. FitzGerald, M. Violence and Care: Fanon and the Ethics of Care on Harm, Trauma, and Repair. *Philosophies* **2022**, *7*, 64. [CrossRef]

37. Manning, E. What If It Didn't All Begin and End with Containment? Toward a Leaky Sense of Self. *Body Soc.* **2009**, *15*, 33–45. [CrossRef]
38. Shildrick, M. *Embodying the Monster: Encounters with the Vulnerable Self*; SAGE: London, UK, 2002.
39. Shildrick, M. *Dangerous Discourses of Disability, Subjectivity and Sexuality*; Palgrave Macmillan: London, UK, 2012.
40. McRuer, R. *Crip Theory: Cultural Signs of Queerness and Disability*; New York University Press: New York, NY, USA, 2006.
41. Derrida, J. *Of Grammatology*, 1st ed.; Johns Hopkins University Press: Baltimore, MD, USA; London, UK, 1977.
42. Norris, C. *Deconstruction: Theory and Practice*; New Accents; Taylor and Francis: Abingdon, UK, 2003.
43. De Saussure, F. *Course in General Linguistics*; Bally, C., Sechehaye, A., Riedlinger, A., Eds.; McGraw Hill: New York, NY, USA, 1966.
44. Grosz, E. *Volatile Bodies: Toward a Corporeal Feminism*; Theories of representation and difference; Indiana University Press: Bloomington, IN, USA, 1994.
45. Kirby, V. *Telling Flesh: The Substance of the Corporeal*; Routledge: New York, NY, USA; London, UK, 1997.
46. Irwin, J. *Derrida and the Writing of the Body*; Ashgate: Farnham, UK; Burlington, VT, USA, 2010.
47. Derrida, J. Différance. In *Margins of Philosophy*; University of Chicago Press: Chicago, IL, USA, 1982.
48. Spivak, G.C. Translator's Preface. In *Of Grammatology*; Johns Hopkins University Press: Baltimore, MD, USA; London, UK, 1977; pp. ix–xc.
49. Spivak, G.C. Can the Subaltern Speak? In *Marxism and the Interpretation of Culture*; Nelson, C., Grossberg, L., Eds.; University of Illinois Press: Urbana, IL, USA; Chicago, IL, USA, 1988; pp. 271–313.
50. Cornell, D. *Philosophy of the Limit*; Routledge: New York, NY, USA; London, UK, 1992.
51. Derrida, J. Force of Law: The Mystical Foundation of Authority. In *Deconstruction and the Possibility of Justice*; Cornell, D., Rosenberg, M., Carlson, D.G., Eds.; Routledge: New York, NY, USA; London, UK, 1992; pp. 3–67.
52. Derrida, J. 'Eating Well', or the Calculation of the Subject. In *Points . . . Interviews 1974–1994*; Weber, E., Ed.; Stanford University Press: Stanford, CA, USA, 1995.
53. Foucault, M. *The Birth of Biopolitics: Lectures at the Collège de France, 1978–1979*; Senellart, M., Ed.; Palgrave Macmillan: New York, NY, USA, 2008.
54. Foucault, M. *Society Must Be Defended. Lectures at the Collège de France, 1975–1976*; Bertoni, M., Fontana, A., Eds.; Penguin Books: London, UK; New York, NY, USA, 2004.
55. Green, M.; Lawson, V. Recentring Care: Interrogating the Commodification of Care. *Soc. Cult. Geogr.* **2011**, *12*, 639–654. [CrossRef]
56. Braidotti, R. *The Posthuman*; Polity: Cambridge, MA, USA, 2013.
57. Beasley, C.; Bacchi, C. Envisaging a New Politics for an Ethical Future: Beyond Trust, Care and Generosity—Towards an Ethic of 'social Flesh'. *Fem. Theory* **2007**, *8*, 279–298. [CrossRef]
58. Valéry, P. Some Simple Reflections on the Body. In *Fragments of a History of the Human Body Part Two*; Urzone: New York, NY, USA, 1989.
59. Tyler, I. *Revolting Subjects: Social Abjection and Resistance in Neoliberal Britain*; Zed Books: London, UK, 2013.
60. Ahmed, S. *The Cultural Politics of Emotion*, 2nd ed.; Routledge: New York, NY, USA, 2015.
61. Dallery, A.B. The Politics of Writing the Body: Écriture Feminine. In *Gender/Body/Knowledge. Feminist Reconstructions of Being and Knowing*; Rutgers University Press: New Brunswick, NJ, USA; London, UK, 1992; pp. 52–67.
62. Derrida, J. *Positions*; The University of Chicago Press: Chicago, IL, USA, 1972.

philosophies

Article

Caring for Whom? Racial Practices of Care and Liberal Constructivism

Asha Leena Bhandary

Department of Philosophy, College of Liberal Arts and Sciences, University of Iowa, Iowa City, IA 52242, USA; asha-bhandary@uiowa.edu

Abstract: Inequalities in expectations to receive care permeate social structures, reinforcing racialized and gendered hierarchies. Harming the people who are overburdened and disadvantaged as caregivers, these inequalities also shape the subjectivities and corporeal habits of the class of people who expect to receive care from others. With three examples, I illustrate a series of justificatory asymmetries across gender and racial lines that illustrate (a) asymmetries in deference and attendance to the needs of others as well as (b) assertions of the rightful occupation of space. These justificatory asymmetries are cogent reasons to evaluate the justice of caregiving arrangements in a way that tracks data about who cares for whom, which can be understood by the concept of the arrow of care map. I suggest, therefore, that the arrow of care map is a necessary component of any critical care theory. In addition, employing a method called living counterfactually, I show that when women of color assert full claimant status, we are reversing arrows of care, which then elicits resistance and violence from varied actors in the real world. These considerations together contribute to further defense of *the theory of liberal dependency care*'s constructivism, which combines hypothetical acceptability with autonomy skills in the real world. Each level, in turn, relies on the transparency of care practices in the real world as enabled by the arrow of care map.

Keywords: interpersonal justification; contract theory; care ethics; care theory; liberalism; constructivism; intersectionality; women of color feminism; critical care theory; the arrow of care map; John Rawls

Citation: Bhandary, A.L. Caring for Whom? Racial Practices of Care and Liberal Constructivism. *Philosophies* **2022**, *7*, 78. https://doi.org/10.3390/philosophies7040078

Academic Editors: Maurice Hamington and Maggie FitzGerald

Received: 25 March 2022
Accepted: 25 June 2022
Published: 5 July 2022

Publisher's Note: MDPI stays neutral with regard to jurisdictional claims in published maps and institutional affiliations.

1. Introduction

Prior to the advent of care ethics [1–4], mainstream western political philosophy had been written by people from populations who were socially licensed to receive care without providing it. For these people, usually western white males, the fact that they received both material caregiving and epistemic leaning[1] [5] was narrativized in ways that made it invisible or justified. These authors' conceptions of themselves were no doubt shaped by the ease with which they accessed care, the lack of a felt debt to their caregivers, and the deftness with which they grafted the substance of others onto themselves.

Care ethicists destabilized those dominant understandings by articulating the moral significance of the human need for care, thereby asserting the robust and nonreductive value of caring relationships. In addition to articulating care as a moral value and as a competing moral theory, care theorists have also articulated its political value [6–8]. I shall employ the term "care ethics" to refer to the moral theory, and "care theory" to characterize a broader swath of theories that address the value of care while combining them with other values in ways that do not always prioritize care as value or practice. Thus, "care theories", as I use this term, are indebted to care ethics, but they can also conflict with some of the core tenets of care ethics, such as the relational self or the rejection of the justice paradigm [9].[2] For instance, an account may endorse care ethics and show how it relates to other doctrines [10]. Alternatively, a theory may evaluate the subject of care from within an alternative framework, as for instance is the case with liberals theorizing about care and feminist liberals[3] [11–14].

This article advances a form of care theory that is decidedly liberal. It is informed by feminism, but it departs from canonical feminist liberalisms and maintains distance from a set of care ethical tenets, such as positing the self-in-connection as the base unit for analysis. "Liberalism" characterizes a cluster of political theories, and the form of liberalism I advance [12] is a political theory that prioritizes the autonomy and freedom of the individual within the constraints required for fair social cooperation [15,16], (p. 65, [17]), [18,19]. In addition, I insist on conceptual separation between the *philosophical* doctrine of liberalism and the *actual* western societies where the doctrine has been developed, but which have never yet conformed to its ideals due to histories of racism, colonization, and gender oppression [19,20]. Thus, I develop a new variety of the philosophical doctrine of liberalism, one that enables critique of the real societies where the doctrine has been dominant. In addition, the critical theoretical lens I devise for these societies sheds light on the distributive inequalities that inform care ethical claims about entitlements to receive care. I thereby expand the toolkit of care theory to include liberal constructivism and a methodology I call "living counterfactually". In doing so, I further advance the theory of liberal dependency care (LDC), which is an anti-oppression liberalism that evaluates care through the lens of a minority woman, thereby bringing together care theory, liberalism, and women of color feminism(s).[4]

The *theory of liberal dependency care* evaluates the justice of a society by including caregiving arrangements in its system of practices [16], and in doing so, it evaluates these arrangements in a way that is open-ended about the practices and caregiving relationships compatible with a just society [12,28–30].[5] Crucially, the arrow of care map [28] supplies a conceptual scaffold for evaluating group-based patterns of caregiving. The arrow of care map demands that we track who receives care, and from whom, in ways that evaluate system-wide patterns and distributive inequalities. Thus, it requires an assessment of racialized distributions of caregiving, requiring that we ask the question: do people of color provide most of the care, and whose care needs are met at a high level? Due to the abstract nature of the mapping function, which takes human beings as the first basis for analysis and then asks social scientists to track data about care received and provided according to, first, salient social groups, and then, to continue looking for underacknowledged categories, the map is a vital tool for care theory to identify unequal distributions of care given and received, indirectly revealing who is considered worthy of care. The arrow of care map's reliance on individual human beings as the basic units of analysis leaves open that people may have conceptions of the self and lives of value that are deeply intertwined with others. When we track care in a way that does not already assume the legitimacy of the relations of care, local patterns of caregiving and globalized care become evident. Tracking quantities of hands-on caregiving in intensity and duration, the map yields information about how care practices are arranged within and across societies. It brackets people's relationships, and in doing so it provides an independent source of data, one that does not embed background inequalities into its justification.[6] Understanding our care relations through the data as it is organized by this concept is necessary for a comprehensive evaluation of the justice of the system of practices. It brings into view the way that, for instance, in the U.S.—where emotional care, social cohesion, and material care are invisible and undervalued—these forms of labor are often assigned to the intersectional groups with the least social power, namely, women of color and individuals who do not have the security of citizenship.

If liberal theorists are to arrive at considered judgements about what a fair set of social practices will be, we must include our practices of caregiving.[7] But to include caregiving practices is to unearth a form of labor that generally coincides with subordinate status in social hierarchies. Because the subjectivities of people in hierarchical social forms are relationally linked, asserting full claimant status for women of color, who are socially assigned the role of serving as the repository of society's needs, will threaten the subjectivities of many men of all races and white women. To evaluate the obstacles to this change, and the nuances required for it, I employ a methodology for a care theory of justice defined by women of color called "living counterfactually".

2. Living Counterfactually

Imagine a world in which people of color refuse to care for white people.[8] Call this world W'. In this (logically) possible world, the normative intuition that we are entitled to the care we need would no longer be a care ethical foundation. Instead, the referent for "we" would become more specific. With the global white population unable to access caregiving by people of color, they would have to meet their own care needs. Here, world W' is one in which, at first, whiteness is accompanied by the habits, entitlements, and subjectivities present in the current world. In this world, I expect that a social norm would gradually emerge requiring an explicit justification on occasions when one is called upon to care for others. This justification would likely be grounded in a variety of reciprocity (see [29] for an account of interpersonal reciprocity).

By entertaining this possible world, I am not necessarily recommending the corresponding action-guiding measure of not caring for others across asymmetric racial lines. There are contexts in which such care is legitimate. In addition, in the real world, people of color often experience negative repercussions when we withhold socially expected care and deference. Of particular interest to the present inquiry is that these negative consequences are not limited to persons who are formally employed in the caregiving sector. For instance, the mammification of women of color in white collar workplaces creates conditions under which minority women who are not caring and deferential are quickly labeled "uncollegial" [33–35].[9] A final reason to refrain from recommending the global withdrawal of racialized care is that a global withdrawal of racialized care might create economic hardships for people of color who are caregivers for white families in the short term [36].[10] Therefore, although I do assert the normative aim of upending racialized care as a system, and consider a commitment to achieving it essential to egalitarianism, determining how to achieve that aim will require the expertise of interdisciplinary scholars, including historians, anthropologists, policy makers, activists, and care workers.[11]

In world W', where racialized material caregiving has been upended, we would see modifications to the selves of people who have long received care without reciprocating it. Those changes, in turn, would liberate women of color because an expansive freedom for women of color will become possible only when we are no longer perceived as people who exist for use by others. I investigate obstacles to *this ideal state of affairs*—ideal in the sense that it is an ideal to be aimed for—through a method I call "living counterfactually". *Living counterfactually*, as a minority woman, means asserting full claimant status in micro-interactions in the workplace, in one's community, and generally, in both private and impersonal spaces. Because Latinas, Black, Indigenous, Arab, Asian, and/or Brown Women are expected to attend and defer to others, asserting full claimant status from this social position reveals the challenges and the requirements of the radical social change required to achieve justice.[12] To move through the world as a woman of color asserting full claimant status is like sending iodine contrast into an artery prior to a CT scan. It supplies contrast through which to diagnose the system. And, like this injection, it can burn.[13]

When we live in this way, we live *as if* the world were otherwise. We thus act as if affordances are present when they do not exist. As a result, the subjectivities that supervene on the unjust caregiving arrangement become evident through their holders' expressions of rogue emotions, violence in its many forms, microaggressions, and confusion.

Mainstream egalitarian political philosophy does not *assert* the non-personhood of women of color, but actual social practices in the U.S. do not fully grant equal personhood status to women of color [38].[14] We see this phenomenon clearly in the case of caregiving injustice in the U.S., which, together with racial and gender hierarchies, is a vital part of the social forms where philosophical liberalism is endemic.

For women of color, then, to withdraw care and deference from impersonal others is to reveal how the social form depends on these hierarchies. A social form's underlying social inequalities and sedimented hierarchies are revealed through overt violence as well as microaggressions which together betray "liberal" societies' caregiving substructures.

Thus, when women of color live as full persons, which is to act on reasonable entitlements and to expect the corresponding affordances, allowances, and responses, we are living counterfactually, in ways that are not permitted, and also in ways that do not exist in a robust ontological sense based on the social meanings of our actions. We are living as if we are in a possible world, one that does not currently exist. Therefore, the completion of a woman of color's expression of agency does not occur; our expressions of full claimant status do not receive the proper uptake, and we do not have the affordances that should accompany it, such as walking on a trail unmolested or having our workplace contributions recognized.[15] Although we put forth an action, the meaning we enact is not the meaning it gains from dominantly positioned interlocutors, and therefore the action is not completed.[16] Moreover, these actions' lack of intelligibility to socially dominant subjects can elicit negative social and legal consequences.

Because caregiving demands that the carer direct their attention and actions toward others, there is both conceptual and practical overlap between hierarchical deference and caregiving. The umbrella concept for these dispositions and actions is an "other-directed orientation" (p. 79, [12]), because they each demand a psychological stance in which a person is oriented toward the other, permeating both what they pay attention to (other-directed attention) and the target of their action (other-direct action) (pp. 79–89, [12]).[17] "Whiteness" depends on caregiving in many forms by non-whites. In the absence of these relations of care, deference, attendance, and epistemic prioritization, whiteness would not exist in its hierarchical form, which is as a world-structuring form of illegitimate privilege.[18] I define "privilege" here as society-relative disproportionate ease of access to fundamental primary goods, to one's relationships, and to a cultural context in which one feels at home, relative to others in the society where you live.

Correspondingly, the non-proximate cause of resistance to the freedom of women of color is that caregiving arrangements continue to be largely invisible, and that, in these invisible arrangements, which are in place to secure vital needs, women of color are the default repository for the needs of others. Some of these "needs" are deeply felt by individuals as emanating from their sense of themselves. They shape their understanding of what is just and right. For example, a man living in, and shaped by, a patriarchal society feels he should have his preferences met at a high level. In the space of social affordances, these preferences transmute into needs. Less well understood in the context of theories of justice is how white women's subjectivities are also predicated on caregiving injustice in social practices that subordinate women of color. For this reason, I dissect this relational micro-construction below.

When violence against women of color is perpetrated by white women, the language with which to articulate this violence, and the means for receiving acknowledgement, justice, and repair, are often absent. The lack of uptake often occurs when the authority to which the complaint is addressed is a white man, because the dyadic relationship between white women and white men includes an affordance of femininity and vulnerability for white women not extended to women of color.

Let us now consider three cases of violence, where the violence is physical or epistemic, and it becomes obvious when viewed through an analytic lens that prioritizes the autonomy of women of color.[19] My argument throughout relies on the undefended premise that women of color have legitimate needs to receive sensitivity, consideration, and epistemic leaning from others.[20]

In the first two examples, which I include in the mode of personal narrative, or what Bat-Ami Bar On calls public autobiography, [46–48][21] I describe a minority woman who is ambiguously raced enacting an entitlement to run on a public trail, and that act being received with anger and violence when it conflicts with the claims of a white woman and her dog. The third example supplies an additional data point to fill in the larger social context in which microaggressions and racism take on varied forms. In that example, I rely on testimony to describe an Asian family living in a large home in a predominantly white historic district enacting a legal entitlement to build a fence.

Although these cases can be profitably analyzed by the sophisticated discourse on epistemic injustice,[22] I include them here in defense of LDC's anti-oppression liberalism. I leave open, at this stage, whether the phenomena of epistemic injustice more generally can serve as decisive considerations in favor of the two-level contract theory of the theory of liberal dependency care [12]. To ascertain whether that is the case, a comparative analysis is required to evaluate the theory of liberal dependency care against other leading theories, such as deliberative democracy, the capabilities approach, Rawls's own view, and other varieties of constructivism (e.g., [49]). Here, though, I employ these examples in a more limited way to illustrate how particular cultural constructions inform needs-interpretations, entitlements, and accountability—all of which are rooted in the injustice of caregiving arrangements.[23]

3. The Trail

3.1. Episode 1. Epistemic Leaning and Interspecies Status Hierarchies

On a beautiful spring day in April 2020, during the beginning of the COVID-19 pandemic, in Iowa, I left my house for my routine run in a local park one mile away. I wore a neck gaiter, sunglasses, and a baseball cap. As I ran up a hill on a trail, a route I took that day to avoid the more heavily traveled main trail, I moved to its right edge to attempt a 6-foot separation from a couple and their two dogs, simultaneously pulling up my gaiter as a mask to avoid breathing on them.

The smaller of the two dogs, held on a retractable leash by a tall and slender white woman likely in her 40 s, silently darted horizontally across the trail and bit my leg, decisively. Evidently, the leash had not been retracted to control the dog's range. Shocked by the attack, I stopped in my tracks, and said, "Your dog just bit me!" The woman quickly picked up the dog and continued walking, her male companion and his dog also proceeding down the hill on the path. I remained in place, examining my leg, and told them that I was looking at whether he had broken the skin with his bite. The man then paused, as the woman continued forward. I continued with, "It really hurts", and then, "Maybe I should get your contact information, just in case". The man, registering the words, reached into his pocket to procure his business card, which he handed to me. This was the first moment at which it seemed like they (or rather, he) registered that I was a person. He said, now displaying either distress or concern, "Oh yes, we would be happy to send you all of her vaccination records, whatever you need". Then they continued their walk, proceeding down the hill.

Gathering myself, I continued my run, only to pause at the top of the hill, in pain from the injury, my calf throbbing and pulsing, realizing that I could not continue as planned because I was, in fact, injured. Thus, I deviated from my route, taking a shortcut through the trail system back to the cemetery adjacent to it, which is my standard point of entry and exit for this trail run. During this injured interlude, I had been thinking about the need for moral repair, and that I was not interested in litigation as its instantiation. Consequently, when I saw the couple in front of me, I resolved to approach them, and in so doing, to initiate a process of moral repair.

Upon seeing me approach, when I notified them of my presence by saying, "Coming up behind you", the woman accused me of chasing her. What I did next was an attempt to hold them accountable[24] for their actions and in doing so, to extract from them recognition of my status as a person. Maddeningly, the woman (henceforth Woman A) persisted in refusing to acknowledge that I had suffered, and her level of concern toward me seemed no different from what I imagine it would have been if her dog had chased, and bitten, a squirrel. Woman A exhibited resistance. The form her resistance took included repeatedly stating that her dog had never bitten anyone *before*. When stating this claim, she appeared to believe it erased what had just occurred.

Although I had been cooperative and neighborly at first, and perhaps even conciliatory, offering the possible exculpation that perhaps the dog had bitten me because my face was covered by my mask, sunglasses, and hat, woman A's demeanor and persistent denial led me to begin to suspect I was not being engaged with as a full person with the attendant claimant status and rights. The physical space my body occupied was not being granted to me.[25]

I began to suspect that the most likely explanation for their attempt to convert my status was racism,[26] and that our interpersonal conflict was resulting from my refusal to accept this ascription. Some readers will not need further justification of this claim, but others may find it helpful. As an ambiguously brown, biracial, Asian-American woman, living in the Midwest during the years of Donald Trump's presidency and subsequent popularity has given me an education in recognizing the signs of racism. Nonrecognition of my bodily integrity and deafness to my claim-making are typical indicators of racism. So, after Woman A repeated, yet again that he has never bitten anyone before, I lobbed an exploratory rhetorical charge by saying, "Then why did he bite me? Has he never seen a brown person before?"

In response to my exploratory antagonism, her thin veneer of reasonableness dissolved. At this moment, a second conversion occurred. Woman A was now the aggrieved party. To attempt to substantiate her victimhood, she shouted: "I'm from South Africa. I'm sick of you Americans talking about race". And then, "If we are going to talk about race, we have to talk about gender . . . . And, I am a *woman*".

She seemed to think the fact of her womanhood made *her* a victim in relation to *me*, both of us appearing as cis women. Her complete lack of grasping the irony of this claim betrayed her lack of knowledge of Sojourner Truth's 1851 "Ain't I a Woman" speech. Unfortunately, her proclamation was not without cultural meaning in 2020, for Woman A's claim was presumably intelligible against the backdrop of racialized tropes of hardiness of black and brown women, which are employed as contrastive to fragile white femininity. The phenomenon that white women understand themselves through their victimhood in relation to white men certainly extends beyond Woman A, and can occupy a salient locus in their cognitive schemas, thereby facilitating evasiveness about the evils they perpetrate [55].[27]

However, this woman's attempt at feminine victimhood was so inane, given the facts of the situation, that it merely increased her male companion's discomfort with the situation, who spoke alternating to each of us, saying to her, "Please, [her name], please, stop", and saying to me, "Please, please, what can we do, we want to make it right". After he convinced her to walk away with their offending dog, we continued to speak (he was holding their other dog), and he ultimately acknowledged their wrongdoing, saying that how they responded was not right, and they had not displayed the proper concern for me.[28]

Over the following several days, the bite turned red, and a bruise appeared in the shape of half a dog's jaw (half of the jaw because, when it occurred, I was wearing a capri length legging, so that the dog's lower jaw made contact with my bare skin, while the upper half of the jaw made contact with the legging). Ultimately, it took about three weeks to heal. I was reluctant to report the incident, but my friend, a responsible dog owner, and (whom I will note is also a white woman) repeatedly urged me to report it, so I called the city, who verified that the dog's vaccinations were up to date, and they then contacted the owners to require the dog to stay indoors for a monitoring period at their home.

Although not initially concerned about me, the man was permeable to my demand for interpersonal recognition and justification. When prompted, he emerged from his state of inattention. The woman, in contrast, was entirely impermeable to the fact of my physical suffering. Instead, she appeared concerned about her dog, whom she held in a protective way. I think it would be accurate to say that she cared quite a lot about her dog, and not at all about me. What this example demonstrates is the importance of asking the questions, "Care *for whom*?" about the broader social system, in terms of distributive arrangements, and then, as a matter of evaluating individuals, "Toward *whom* does a person activate their caring disposition?" This second question, an ethical one as it occurs in an individual's action-guided life, supervenes on the background distributive arrangement. The example also illustrates how the moral tenor of social relations in social forms such as ours are structured by caregiving and deference by people of color, where this deference and attendance is a matter of race and gender. The inequalities of concern and access to physical space described above become more pronounced when considered in light of a second case [56,57].[29]

3.2. Episode 2. Equations of Space

Let us return to the same park nearly a year later, where one can run about seven miles without repetition, and where, throughout the pandemic, a proliferation of (in my experience, only white) dog owners have flouted the laws requiring dog leashing in this park. In February (2021), I encountered a different woman (Woman B) who refused to move her dog from the trail so that I could pass. This event took place when we had about 1.5 feet of snow on the ground, with a narrow trail of packed snow. She had moved aside, and two of her dogs had accompanied her, but a third dog remained squarely blocking the only passable trail. I asked her whether she was going to move this dog from the trail or leash him so that I could pass, to which she responded by looking at me, and saying that, no, she would not leash her dogs, and also, "I've been walking like this with my dogs for 20 years". With a pause and then a hardened stare, she stated the challenge, "Have my dogs been aggressive toward you?"

When this woman failed to comply with the law, she demanded that I give a reason for expecting lawful access to natural space while she continued to act illegally. Moreover, rather than eventually conveying any regard for my mental and affective states, she demanded that I *epistemically lean* toward her dog. In fact, after I had stopped running, and was engaged in this interaction, and it seemed that the dog was quite elderly and somewhat confused and not that alert, whereas the woman had a hard stare and an aggressive stance. My conclusions resulted from attuning myself to the situation, which took effort, and was not the type of cardiovascular effort I planned to expend during my trail run.

Woman B, like Woman A, attempted to invert who was aggrieved, and her attempt to assert victimhood involved demanding a justification from me for my reasonable claims.[30] And, although in fact I did offer a reasonable justification, it was not one that she accepted.[31] Tracing the gap between actual *acceptance* and what is *acceptable* often coincides with sedimented social hierarchies, and in the two cases above, the relevant social hierarchy was our visually coded races.[32]

3.3. Episode 3. We Don't Like Your Fence!

Finally, let us consider a third example, one experienced by my family in 2020. I am not home when it happens, and so this is reconstructed through testimony of my family and our friend. In this example, an elderly white woman walks past our stately home in a left-leaning historic neighborhood in which property ownership is buttressed by generational wealth. Although we know many of our neighbors, we are unacquainted with this woman.

My partner is outside in the yard, which is situated on the top of a hill. This woman, let's call her Woman C, blurts from the sidewalk, launching her words up over our fence, says: "The neighbors have been talking about it. And you know, we don't like your fence". My partner responds with something like, "Well, it's nice to own a house and make those decisions for oneself". Because my partner is a physician who specializes in cognitive decline, he tempers his internal response with compassion, in case she suffers from some sort of decline of her cognitive capacities. His response is restrained, but she perseveres, now saying, "Where are you from?", to which he responds, "We're American". And she says, "No, you're Indian".

Our daughters observe, eyes wide, watching the strange lady. It is only when she has seemingly sated herself that she promptly turns toward our friend, a neighbor walking her dog, to chat with her in a friendly and fully lucid manner. Registering the episode with amazement, this friend tells us it gave her a new appreciation of the existence of racism. Upon discussing the example at the dinner table two years later in order to write this article, my teenage daughter reported frankly that this experience is one of her core childhood memories.

These three episodes sample a larger experiential trove. And although the discerning reader may have already noted that all three conflicts occur as competitions over physical space, I set aside those questions about claims to territory to assess the significance of racialized asymmetries in justification for a theory of justice.

As one can see in each of the preceding cases, a normative presumption in favor of interpersonal justification will disadvantage the person occupying the position I occupied nearly every time. Asymmetric demands for justification, paired with the impermeability of the interlocutors to my reasonable demands, serve as considerations against an account of justice that is limited to what actual people agree to. Therefore, the discursive inequalities I have described are among the inequalities that make it essential to a theory of justice to augment actual discursive exchanges with an abstract account of entitlements and distributions. The value of interpersonal justification must be tempered by attention to the social hierarchies that shape the justificatory exchange. Repeated demands for justification can create an atmosphere of terror, as they do when domestic abusers gaslight their victims, slowly dwindling the energetic reserves of the person from whom justifications are demanded. Even when these asymmetric demands for justification do not rise to the level of psychological abuse, the phenomenology of being asked for justifications consumes the time and attention of the person who is interrogated for living as a full claimant.

4. Entitlements Are Relational

Moreover, if we are to think about the distribution of care, we must also think about inequalities in entitlements and the political and deliberative procedures by which we arrive at these judgements. We must ask, "When a person is entitled to care, from whom are they entitled to receive it?" There is always a person who will provide that care. "Entitlements", are needs that are socially sanctioned, with corresponding affordances. They are construed as legitimate needs in dominant discourses.[33] Legitimate needs are standardly indexed to an individual. However, because social inequalities enter into functionings at an early stage, reifying these inequalities must be avoided. Moreover, what others will grant as a legitimate need for a particular person will be infused with the dominant social narratives about the groups of which they are a part.

For a need to be understood as a need, and not a mere preference, it requires the uptake of others. The others, though, are not necessarily those who understand the form of life from which the need emanates.[34] So, for instance, a non-white family living in a racist state might need a yard fence to create greater visual and physical separation from strangers who might pass by wearing MAGA hats for young brown children playing in a sprinkler. Their white neighbors will not construe visual and physical separation as a *need*. Instead, they might interpret it as a mere preference for privacy, and even as a deficit in neighborly virtue. In contrast, other minority families who experience the social form in ways that are relevantly similar will recognize it as a need, and this uptake, by others who experience the social form in ways that are relevantly similar, is essential to well-being and to mitigate stress. In addition, though, the need must be approved by an authority if it is to be effectively construed as a legitimate need. Correspondingly, those who must endorse it as a need rather than as a preference are *powerful others* who occupy a dominant position (dominant in some sense—as the makers of theory, as government officials or legislators, as community members with the power to shape community understandings).

A need for safety is understood in its general form as a legitimate need. But when a person has an *effective entitlement*, particular instances of their needs are comprehended by others as violations in the social system. For instance, when someone has an effective entitlement, legal remedies and/or interpersonal and community rectification are available to that person when her effective entitlement is violated. So, for instance, laws forbidding sexual assault and harassment now make it the case that women are protected qua women in the workplace. But as Kimberle Crenshaw has shown, women of color are disadvantaged in the law because they have to articulate claims in terms of either gender or race [39].

Furthermore, if we are entitled to something, someone else is accountable to us if they violate our ability to satisfy that entitlement. Thus, entitlements are linked to accountability. Accountability for harm and entitlements are relationally defined in a distributive arrangement. Who meets the need that we assert is ours? Against whom do we assert our entitlements? Complicating the relational nature of entitlements are the ways the founding inequalities of a society render privileged members unaccountable in practice when their actions violate the objective needs of others.

Consequently, when entitlements are asserted by people of color, and when we enact them without even a superficial nod of deference to the global racial order, resistance arises, deriving from an implicit social norm that people of color should tend to whites. These demands are relationally inextricable from the content white care recipients insert into the notion of "reasonable life expectations". When whiteness is informed by a colonial subjectivity, it includes the ideas of dominating others and appropriating their labor (p. 47, [19]).[35] In western liberal societies informed by white supremacy, whites' responsiveness to the need of nonwhites carries the social meaning of charity or altruism rather than obligation or a kind of reciprocity demanded by relations of equality.

Thus, the injustice of existing caregiving arrangements informs the intuitions philosophers employ when theorizing about care. The caregiving structure shapes human entitlements, affordances, and accountability. The injustice of the caregiving structure has also informed intuitions in non-feminist moral and political theories, but the extent to which racialized care has informed theorizing about care by white feminists has also been underexplored.

Moreover, background maldistributions of care inform intuitions at *more than one* stage of theory development. They inform the modeling of persons and the assessment of distributive justice. Caregiving arrangements include social understandings about which (social) groups of people should act in materially caregiving ways, and for whom. Consequently, a theory of distributive justice is needed as a framework with which to evaluate competing claims, where the people charged with theorizing about justice should gather information about existing patterns using the arrow of care map methodology, which requires tracking the care people give and received to assess the overarching fairness of the distribution without assuming the legitimacy of the relationships where care occurs. The task for theorists of justice, therefore, properly includes when assessing social norms and expectations for social groups conflict with a distributively just society, defined as a basic structure that is acceptable when viewed from any social position.

5. Resetting Intuitions in Liberal Theories of Distributive Justice

Because care is the spine of culture [12,20], and how we are habituated in relation to care deeply informs our understandings of the world as well as our physical embodiment in it, many of the people who assert a commitment to equality in a cool and calm moment will not sustain that commitment in their bodily habit and comportment, when relations of care are stake [58].[36]

Therefore, to arrive at a *sense of justice* in a world with distributive injustice, one that avoids the reification of those very hierarchies, a method for thinking abstractly about distributions of care is needed.

My neo-Rawlsian theory, the "theory of liberal dependency care" (LDC), defends a variety of constructivism, one that makes explicit the background distributive patterns of caregiving. Rendering these practices transparent will also affect what counts as a "reasonable" claim and way of life. LDC's structure of justification consists of (1) an abstract form of evaluation to assess what we would agree to if we did not know our social position in society, and (2) a requirement for autonomy skills in the real world. The first aspect, this abstract module, is a version of the Rawlsian idea of hypothetical acceptability that embraces his earlier formulation of its subject of justice as the "system of practices" [16].[37]

LDC, like John Rawls's account of liberalism, is a form of constructivism. Constructivism embraces an iterative process of justification, whereby the intuitions employed are reflexively assessed in light of the account of justice that results. However, although the model of the person in Rawls's account of the original position is purportedly neutral, the way persons are understood is in fact informed by assumptions about what a reasonable life includes in social forms where caring labor has been invisible [20] (Chapter 4).

When we assess the social practices allocating space on a trail in the first two cases above, the unequal standing of persons along racial and species lines becomes clear. If this practice is to be justified, it will have to make explicit the claim that the white person's dog gets more consideration than a woman of color.[38] This aspect of social conventions about walking freely in a park then requires a justification in response to the question, "Why do these dogs get more freedom on the trail than this Asian woman?" As one can see in both cases above, actual interpersonal justification failed, but a removed form of impersonal justification will quickly falter when attempting to justify these encounters.

In addition, the constructivism I advance requires autonomy skills for people in the real world. Employing these autonomy skills, individuals who are treated as less than equal can articulate the harms they experience. The consequent augmentation of autonomy skills to hypothetical acceptability creates space for lived disagreements about values. It does so while retaining the need for an abstract form of evaluation to evaluate quantities of care received. Knowledge of the society's distribution of care given and received must be used to further evaluate judgements and intuitions.[39]

Consider how individuals' command of autonomy skills in the real world impacted the resolution of the first two cases above. The law was clear for both cases because this park is one where dogs must be leashed. In addition, after the second incident, city officials had recently benefitted from anti-racist trainings as the result of the Black Lives Matter movement. And, informed by that knowledge and the salience it conferred on possible instances of racism, they responded to my report of these incidents (and most likely, reports by others) by posting large signs throughout the park reminding people of the existing policies, and the fine of $195 for noncompliance. Notably, they did not trivialize the occurrences. I take the normatively satisfactory institutional response to be the indirect result of the combined effects of the agentic skills of the people who galvanized protests against racial injustice. Consequently, Woman A's inversion did not gain institutional traction. Instead of institutional complicity, city officials acknowledged the wrong.

The theory of liberal dependency care is a theoretical backdrop with which to interpret these cases, which illustrate how racist interpersonal hierarchies persist in ways that limit the freedom and autonomy of women of color.[40] The racist interpersonal hierarchies are thereby linked to global caregiving arrangements that designate women of color's role as one of attending to the needs of others.

Without experience in a fully just world to serve as the fount of intuitions, thought experiments, hypotheticals, and counterfactuals are needed to refine the intuitions that are to serve as inputs into this device. In addition, *living counterfactually* serves as a lived method of diagnosing the basic structure's incompatibility with the autonomy and freedom of members of subordinated groups. Living counterfactually supplies necessary insights to guide the intuitions needed to understand how racialized patterns of caregiving pervade the bodies and expectations of white people. The autonomy of women of color thus exerts pressure against the social forms that are premised on our appropriation.

The lived experiences described above also provide support for the need for critical evaluation of our social practices. An excavation of specific culture tropes reveals the ways that images and narratives inform assessments of reasonable expectations. Reasonable expectations are premised on a given identity group's assignment of a domain of self-sovereignty and autonomy. Here, I have shown how the relational expectations of members of other groups collectively crowd out the domain for self-sovereignty of women of color.

To discern the content of individuals' entitlements and affordances requires a cultural excavation. Abstract claims for entitlements and needs are always enacted in a particular context, and it is only through cultural critique that we can refine intuitions. This cultural critique is one that must be engaged in by people with autonomy skills and understanding of dominant cultural norms.[41]

The examples I have discussed offer additional reasons in favor of liberalism as the structuring distributive framework within which to evaluate a society's caregiving arrangement. Theorizing from the perspective of a woman of color, I insist on the enduring need to prioritize autonomy in a society with just care practices, and to maintain a conceptual distinction between autonomy and care (p. 10, [20]). For example, embedding a requirement of responsiveness to others, or of caring, into a conception of autonomy would inevitably transmute into more epistemic leaning and deference from people of color. Therefore, if we are to understand the nature of a liberal society with just caregiving practices, we must begin with the autonomy of women of color, for doing so reveals the hidden nature of western "liberal" social forms and the determinate content embedded in abstract accounts of accountability, entitlements, needs, care, and equality. To diagnose caregiving injustice through the lens of a woman of color is to present new reasons in defense of a liberal constructivism. This constructivism must interrogate (1) the relational nature of entitlements, (2) the distributive backdrop that shapes whether a claim is a need or a preference, and (3) how accountability can be evaded by perpetrators of violence, when the victim acts against dominant cultural constructions. I have taken a few steps toward explicating each of these three components, thereby demarcating the next stage of a critical liberalism to serve as a variety of care theory that begins from social understandings as a woman of color.[42]

Funding: This research was supported by a Career Development Award from the University of Iowa support from the Obermann Centers as a Fellow-in-Residence in Fall 2021. I thank the staff for their support, and the faculty for their comments. For administrative and technical support, I thank Erin Hackathorn.

Acknowledgments: This paper benefitted from discussion of a related paper at the Pacific APA 2021 in the White Care/Brown Care session, and at SAAP 2021 in the FEAST panel, where V. Denise James lucidly reflected back to me my claim about the connection between attending, caregiving, and deference. For comments on the written version of this paper, I thank Anthony Laden, Maurice Hamington, and Maggie Fitzgerald, the participants in the online lecture series in Social Rights and Relationship Goods funded by Rice University Creative Ventures fund, including Elizabeth Brake, Kimberly Brownlee, and Justin Clardy, and the Fall 2021 Obermann Faculty Fellows in Residence directed by Teresa Mangum. I also wish to thank Lina-Maria Murillo, Elizabeth Rodriguez Fielder, Deborah Whaley, Gigi Durham, Jessica Moorman, Jacki Rand, Meena Khandelwal, Ashley Howard, Richard Fumerton, Lori Gruen, and Diana Tietjens Meyers for stimulating comments/conversations on this subject. Although they do not necessarily share my views, they informed them directly or indirectly.

Conflicts of Interest: The author declares no conflict of interest.

Notes

1. "Epistemic leaning" is Sandra Bartky's term [5].
2. The way I am defining "care theory" is also implicit in Daniel Engster's work (see [9] and private correspondence).
3. For liberals theorizing about care, see all contributors) [11]. For a liberal theory of justice that incorporates care, see [12]. Because questions about care have overlapped with debates about women's labor and socialization, the liberal tradition of evaluating care has been shaped by Susan Okin's feminist liberalism [13]. For an overview of feminist liberalism, see also [14].
4. See e.g., Lorde [21], hooks [22], Lugones [23], Bhandary [12,20]. For decolonial feminisms, see Spivak [24], Narayan [25] and Khader [26]. See Welch [27] for a feminist discussion of autonomy in conditions of oppression.
5. See also [31,32].
6. See [6], (p. 11, Figure A) for the initial depiction of the arrow of care map.
7. For a range of liberal approaches to care, see [11].

8.   There are forms of racialized care that are not defined by whiteness in Asian and Arab countries. In Europe, as well, less advantaged white groups care for the white groups in power. In Asian and Arab countries, for instance, social hierarchies locate care in servant classes. Therefore, a global form of addressing whiteness would not remedy all caregiving invisibility.

9.   See Collins [33] for articulation and explication of the mammy trope. On controlling stereotypes for women of color in academia, see Muhs et al. [34]. On stereotypes for Asian and Muslin women, see Sheth [35].

10.   Eva Yguico [36] argues that eliminating the gendered division of labor may have harmful effects on, for example, relatives who submit remittances to family members in the Philippines.

11.   Of course, there will be many exceptions, based on part on the fact that many families are interracial. My recommendations apply to the overall global distribution of care and its systems, and not to legitimate relations of care. Families are the paradigm case for legitimate care relations.

12.   The experience for racially ambiguous mixed race cis women may be most relevant here. On Muslim women, see Sheth [37].

13.   I use "burn" here in a phenomenological sense to describe the sensation as reported anecdotally. The sensation is described medically as "warmth/feeling hot" [38] (Palkowitsch et al., 2014, Table 6).

14.   See Crenshaw (1991) [39] for the way these inequalities manifest in the law, and also for the concept of intersectionality.

15.   An "affordance", in Gibson's sense (1979) [40], is what the environment makes available to person X. See also Schapiro 2003 [41] (cited in [42] (p. 106, n 2). The minority person's agency is thwarted when an "act" is not completed, because uptake does not occur, and the intended meaning does not ensue.

16.   Compare [43] on the ways background social norms impede particular expressions of agency for disadvantaged persons. In this case, I prefer the language of "uptake" to Kukla's contextual analysis of background social norms. This is my account here centers questions about who it is who endorses and hears claims. This approach deliberately makes individuals the salient objects of analysis, and it is complementary to the analysis of caregiving injustice I make possible with the "arrow of care map" [28].

17.   See Tronto and Fisher for the original account of care as attentiveness and responsiveness (p. 127, [6]).

18.   Here I follow Charles Mills to define whiteness relationally in a status hierarchy that demands deference from non-whites (p. 71, [44]).

19.   See [45] for the argument that microaggressions cause stress, and the consequence of this stress is harm akin to violence against the body.

20.   By epistemic leaning, here, I mean that the person who epistemically leans considers the world from the perspective and priorities of the other.

21.   Susan Brison has argued forcefully for the relevance of personal narrative to the validity of the philosopher's claims [46]. (Bat-Ami Bar On calls the use of personal narrative a kind of "public autobiography" (p. 3, [47]). Anthropologists call a similar approach "auto-ethnography" (For a good example, see Chin [48]).

22.   See e.g., [49–53].

23.   These encounters have the significance I attribute to them because they are located among many other similar interactions, constituting a pattern. They can be understood, therefore, only through a form of pattern-recognition, one more readily accessible to others who also experience racism.

24.   On accountability, see hooks (p. 124, [22]), and (pp. 16–17, [54]).

25.   Perhaps this was a case of being rendered a "subperson", the category Charles Mills defines as characteristic of blackness (p. 6, [44]). Or perhaps I was being placed in a nonperson category. I will not take a position here on this question, because what is important for the present analysis is that these people were far more concerned with their dog than they were with me, and my claim to freedom had less uptake than their dog's claims to freedom of movement.

26.   As is the case with pattern recognition and instances of microaggressions, whether any one event is an instance of racism will be indeterminate. However, by evaluating the effects of multiple events on the target of the actions, we can see the accumulation of racist and intersectional microaggressions. It thereby paints a picture of the terrain an individual traverses in a particular social form.

27.   See Claudia Card's [55] groundbreaking work on evil for the argument that people who have suffered as victims can also perpetrate evil.

28.   I do not want this episode to be read as if the man was the reasonable adjudicator in a fight among women. Indeed, triangulation to men when they have de facto or de jure power is an anti-feminist dynamic in a patriarchal society. The details about the man are relevant to this example because he was the other witness to the exchanges.

29.   See (p. 63, [56]) on white privilege as habit. Literary affect studies critiques relations of sympathy and affection. For a recent work, see [57].

30.   There are ample examples of white ontological expansiveness resulting in an unjustified demand for a justification. To list just one, consider neighbors who used our driveway for several years for their bed and breakfast clients, both they and their customers looked surprised and alarmed when we asked them to stop, or to move their car so that we could rush to work. They simply replied that the previous owner of our house (who had been largely bedridden), had allowed them use of the driveway.

31   In fact, initially hoping for reasonableness, I explained to this woman that, although she knows her dogs, and may have a high degree of confidence that they will not bite other people, other people who do not know those dogs do not know that they won't bite them. However, her hostile gaze quickly made evident that the attempt was futile.

32   In the interactions described above, where it seemed that these women simply *could not* consider me, it seems that what is acceptable has to be indexed to someone with very different affordances than these women. This line of inquiry then points to the need to index affordances to the reasonable individual, where reasonableness is not only a set of beliefs, but it is also corporeal. I take up this question in [58]. On psychocorporeal agency and affordances, see [59]. See [60] on the Habermas-Rawls dispute about acceptance versus acceptability. On my view, if the clause is "acceptable *to her*", then the woman's "strong horizons" [61] are invoked.

33   See my (p. 3, [20]) for my discussion of legitimate needs. Cf. (p. 312, [62]) for the view that determining which needs are legitimate is a matter of politics.

34   My view here is in the same family as Charles Taylor's account of horizons of significance. On my view, though, these horizons occur within a particular social form, and salient social stratification creates regularities in our experiences of moral salience and significance.

35   Cf. Mills, who claims that the liberal idea of reasonableness is based on the perspective of the western white person, who have "traditionally thought of nonwhite assets as a common white resource to be legitimately exploited" (p. 47, [19]).

36   For further elaboration on this concept, see my monograph in progress, *Being at Home: Liberal Autonomy in an Unjust World*.

37   See [29]. For a feminist defense of hypothetical acceptability, see also [63].

38   Here I use "persons" in a way that is inclusive of the view that dogs are persons or members of families (for a species-inclusive view of who counts as a moral patient, see (p. 75, [64]).

39   See [12] for the comprehensive account.

40   For an alternative account, see [65].

41   For the original account of autonomy skills, see Meyers [66] and many other works. For a role for autonomy skills in a liberal care theory of justice, see [12]. The critical awareness of norms engendered by these skills overlaps with Medina's epistemic virtue of metalucidity [52]. For discussion, see [20].

42   For the first stage, see [12].

References

1. Gilligan, C. *In a Different Voice*; Harvard University Press: Cambridge, UK, 1982.
2. Noddings, N. *Caring: A Feminine Approach to Ethics and Moral Education*; University of California Press: Berkeley, CA, USA; Los Angeles, CA, USA, 1984.
3. Ruddick, S. Maternal Thinking. *Fem. Stud.* **1980**, *6*, 342–367. [CrossRef]
4. Ruddick, S. *Maternal Thinking*; Beacon Press: Boston, MA, USA, 1989.
5. Bartky, S. *Femininity and Domination: Studies in the Phenomenology of Oppression*; Routledge: New York, NY, USA, 1990.
6. Tronto, J. *Moral Boundaries*; Routledge: New York, NY, USA, 1993.
7. Kittay, E. *Love's Labor*; Routledge: New York, MY, USA, 1999.
8. Held, V. *The Ethics of Care*; Oxford University Press: New York, NY, USA, 2006.
9. Engster, D. *The Heart of Justice*; Oxford University Press: New York, USA, 2007.
10. Gouws, A. Toward a Feminist Ethics of Ubuntu: Bridging Rights and *Ubuntu*. In *Care Ethics and Political Theory*; Engster, D., Hamington, M., Eds.; Oxford University Press: New York, NY, USA, 2015.
11. Bhandary, A.; Baehr, A.R. Introduction. In *Caring for Liberalism*; Bhandary, A., Baehr, A., Eds.; Routledge: New York, NY, USA, 2021; pp. 1–21.
12. Bhandary, A. *Freedom to Care*; Routledge: New York, NY, USA, 2020.
13. Okin, S. *Justice, Gender, and the Family*; Basic Books: New York, NY, USA, 1989.
14. Baehr, A. (Ed.) *Varieties of Feminist Liberalism*; Rowman and Littlefield: Lanham, MD, USA, 2004.
15. Mill, J.S. On Liberty. In *On Liberty and Other Writings*; Stefan, C., Ed.; Cambridge University Press: Cambridge, UK, 1989; pp. 1–116.
16. Rawls, J. Justice as Fairness. *Philos. Rev.* **1958**, *67*, 164–194. [CrossRef]
17. Lloyd, S.A. *Varieties of Feminist Liberalism*; Baehr, A.R., Ed.; Rowman and Littlefield Publishers: Lanham, MD, USA, 2014; pp. 63–84.
18. Bhandary, A. Liberal Dependency Care. *J. Philos. Res.* **2016**, *41*, 43–68. [CrossRef]
19. Mills, C.W. *Black Rights/White Wrongs*; Oxford University Press: New York, NY, USA, 2017.
20. Bhandary, A. The theory of liberal dependency care: A reply to my critics. In *Critical Review of Social and Political Philosophy*; Mullin, A., Ed.; Routledge: London, UK, 2021. [CrossRef]
21. Lorde, A. *Sister Outsider*; Ten Speed Press: Berkeley, CA, USA, 1984; Random House: New York, NY, USA, 2021.
22. Hooks, B. *Racism and Feminism: The issue of accountability*, In *Ain't I a Woman: Black Women and Feminism*, 2nd ed.; Routledge: New York, NY, USA, 2015; pp. 119–158.

23. Lugones, M. *Pilgrimages/Peregrinajes*; Rowman and Littlefield: Lanham, MD, USA, 2003.
24. Spivak, G.C. 'Can the Subaltern Speak?: Revised Edition, from the 'History' Chapter of Critique of Postcolonial Reason. In *Can the Subaltern Speak?: Reflections on the History of an Idea*; Rosalind, C.M., Ed.; Columbia University Press: New York, NY, USA, 2010; pp. 21–78.
25. Narayan, U. *Dislocating Cultures*; Routledge: New York, NY, USA, 1997.
26. Khader, S. *Decolonizing Feminism*; Oxford University Press: New York, NY, USA, 2019.
27. Welch, S. *Existential Eroticism*; Lexington Books: Washington, DC, USA, 2015.
28. Bhandary, A. The Arrow of Care Map: Abstract Care in Ideal Theory. *Fem. Philos. Q.* **2017**, *4*. [CrossRef]
29. Bhandary, A. Interpersonal reciprocity. In *Caring for Liberalism*; Bhandary, A., Baehr, A., Eds.; Routledge: New York, NY, USA, 2021; pp. 145–167.
30. Bhandary, A. Arranged Marriage: Could it Contribute to Justice? *J. Political Philos.* **2018**, *26*, 193–215. [CrossRef]
31. Brake, E. *Minimizing Marriage*; Oxford University Press: New York, NY, USA, 2012.
32. Metz, T. *Untying the Knot: Marriage, the State, and the Case for their Divorce*; Princeton University Press: Princeton, NJ, USA, 2010.
33. Collins, P.H. *Black Feminist Thought: Knowledge, Consciousness and the Politics of Empowerment*, 2nd ed.; Routledge: New York, NY, USA, 2000.
34. Muhs, G.G.; Niemann, Y.F.; González, C.G.; Harris, A.P. (Eds.) *Presumed Incompetent: The Intersections of Race and Class for Women in Academia*; University Press of Colorado: Boulder, CO, USA, 2012.
35. Sheth, F. *Toward a Political Philosophy of Race*; SUNY Press: Albany, NY, USA, 2009.
36. Yguico, E. Unpublished Conference Paper. In Proceedings of the Pacific APA Meeting, Symposium on Gina Schouten's "Liberalism, Neutrality, and the Gendered Division of Labor", Online, 5–10 April 2021.
37. Sheth, F.A. *Unruly Women: Race, Neocolonialism, and the Hijab*; Oxford University Press: Oxford, UK, 2022.
38. Palkowitsch, P.; Bostelmann, S.; Lengsfeld, P. Safety and tolerability of iopromide intravascular use: A pooled analysis of three non-interventional studies in 132,012 patients. *Acta Radiol.* **2014**, *55*, 707–714. [CrossRef] [PubMed]
39. Crenshaw, K. Mapping the Margins: Intersectionality, Identity Politics, and Violence Against Women of Color. *Stanf. Law Rev.* **1991**, *43*, 1241–1299. [CrossRef]
40. Gibson, J.J. The Theory of Affordances. In *An Ecological Approach to Visual Perception*; Houghton-Mifflin: Boston, MA, USA, 1979.
41. Schapiro, T. Compliance, Complicity, and the Nature of Nonideal Conditions. *J. Philos.* **2003**, *100*, 329–335. [CrossRef]
42. Laden, A. *Reasoning: A Social Picture*; Oxford University Press: New York, NY, USA, 2012.
43. Kukla, R. Performative Force, Convention, and Discursive Injustice. *Hypatia* **2014**, *29*, 440–457. [CrossRef]
44. Mills, C.W. *Blackness Visible*; Cornell University Press: Ithaca, NY, USA, 1998.
45. Bhandary, A. "Being at Home", White Racism, and Minority Health. In *Applying Nonideal Theory to Bioethics*; Victor, E., Palgrave, Guidry-Grimes, L., Eds.; Springer Nature: Cham, Switzerland, 2021; pp. 217–234.
46. Brison, S.J. 'We Must Find Words or Burn': Speaking Out Against Disciplinary Silencing. *Fem. Philos. Q.* **2017**, *3*, 1–10. [CrossRef]
47. On, B.-A.B. *The Subject of Violence: Arendtean Exercises in Understanding*; Rowman and Littlefield Publishers: Oxford, UK, 2002.
48. Chin, E. *My Life with Things*; Duke University Press: Durham, NC, USA, 2016.
49. O'Neill, O. *Constructions of Reason: Explorations of Kant's Practical Philosophy*; Cambridge University Press: Cambridge, UK, 1990.
50. Fricker, M. *Epistemic Injustice: Power and the Ethics of Knowing*; Oxford University Press: New York, NY, USA, 2007.
51. Dotson, K. Tracking Epistemic Violence, Tracking Practices of Silencing. *Hypatia* **2011**, *26*, 236–257. [CrossRef]
52. Medina, J. *The Epistemology of Resistance*; Oxford University Press: New York, NY, USA, 2013.
53. Pohlhaus, G., Jr. Varieties of Epistemic Injustice. In *The Routledge Handbook of Epistemic Injustice*; Routledge: New York, NY, USA, 2017; pp. 13–26.
54. Allen, A.L. *Why Privacy Isn't Everything*; Rowman and Littlefield: Lanham, MD, USA, 2003.
55. Card, C. *The Atrocity Paradigm: A Theory of Evil*; Oxford University Press: New York, NY, USA, 2002.
56. Sullivan, S. *Revealing Whiteness: The Unconscious Habits of Racial Privilege*; Indiana University Press: Bloomington, India, 2006.
57. Greyser, N. *On Sympathetic Grounds: Race, Gender, and Affective Geographies in Nineteenth-Century North America*; Oxford University Press: New York, NY, USA, 2017.
58. Bhandary, A. *Being at Home: Liberal Autonomy in an Unjust World*; Department of Philosophy, College of Liberal Arts and Sciences, University of Iowa: Iowa City, IA, USA, 2022; in progress.
59. Meyers, D.T. Psychocorporeal Selfhood, Practical Intelligence, and Adaptive Autonomy. In *Autonomy and the Self*; Michael, K., Nadja, J., Eds.; Springer: Dordrecht, The Netherlands, 2013.
60. Laden, A. The Justice of Justification. In *Habermas and Rawls*; Finlayson, J.G., Freyenhagen, F., Eds.; Routledge: New York, NY, USA, 2011.
61. Taylor, C. *Sources of the Self*; Harvard University Press: Cambridge, MA, USA, 1989.
62. Fraser, N. Talking about needs: Interpretive contests as political conflicts in welfare-state societies. *Ethics* **1989**, *99*, 291–313. [CrossRef]
63. Stark, C.A. Hypothetical Consent and Justification. *J. Philos.* **2000**, *97*, 313–334. [CrossRef]
64. Gruen, L. *Ethics and Animals: An Introduction*; Cambridge, University Press: Cambridge, UK, 2011.
65. Jaggar, A. Thinking about Justice in the Unjust Meantime. *Fem. Philos. Q.* **2019**, *5*, 9. [CrossRef]
66. Meyers, D.T. *Self, Society, and Personal Choice. Part I. Personal Autonomy and Related Concepts*; Columbia University Press: New York, NY, USA, 1989.

Article

Wittgenstein and Care Ethics as a Plea for Realism

Sandra Laugier

Institut des Sciences Juridique et Philosophique de la Sorbonne (ISJPS, CNRS-Paris 1),
Université Paris 1 Panthéon Sorbonne, 75005 Paris, France; sandra.laugier@univ-paris1.fr

Abstract: This paper aims to bring together the appeal to the ordinary in the ethics of care and the 'destruction' or philosophical subversion which Wittgenstein references in his *Philosophical Investigations*: Where does our investigation get its importance from, since it seems to destroy everything interesting, all that is great and important? What we are destroying is nothing but houses of cards. The paper pursues a connection between the ethics of care and ordinary language philosophy as represented by Wittgenstein, Austin and Cavell, in particular in a feminist perspective. The central point of Carol Gilligan's *In a Different Voice* may not be the idea of a 'feminine morality' but a claim for an alternative form of morality. Gilligan's essay seeks to capture a different, hitherto neglected yet universally present alternative ethical perspective, one easy to ignore because it relates to women and women's activities. The ethics of care recalls a plea for 'realism'; in the sense given to it in Cora Diamond's *The Realistic Spirit* to mean the necessity of seeing (or attending to) what lies close at hand. Reflection on care brings ethics back to *everyday practice* much as Wittgenstein sought to bring language back from the metaphysical level to its everyday use.

Keywords: ethics of care; moral philosophy; Wittgenstein L.; Diamond C.; Gilligan C.; ordinary language philosophy

Citation: Laugier, S. Wittgenstein and Care Ethics as a Plea for Realism. *Philosophies* **2022**, *7*, 86. https://doi.org/10.3390/philosophies7040086

Academic Editors: Maurice Hamington and Maggie FitzGerald

Received: 4 May 2022
Accepted: 26 July 2022
Published: 4 August 2022

Publisher's Note: MDPI stays neutral with regard to jurisdictional claims in published maps and institutional affiliations.

1. Introduction

In recent moral philosophy, the importance of care has come into its own, both in relation to the moral attitudes upon which it draws, such as solicitude, attentiveness, and responsiveness, and to the practice of care in attending to the needs of others in everyday contexts. Yet the central point of Carol Gilligan's work, first established in her seminal study, *In a Different Voice*, remains to be fully appreciated. The idea of a 'feminine morality' is at once so provocative and so reasonable that we forget it is first and foremost feminist, and that it stakes a claim for an alternative form of morality. In searching for the latter, Gilligan's treatise seeks to capture a different, hitherto neglected yet universally present alternative ethical perspective, one easy to ignore because it relates to women and women's activities. In this respect, Gilligan's approach is not essentialist, but plainly ordinary, imbued with the perspective of the everyday.

The ethics of care recalls a plea for 'realism', in the sense given to it in Cora Diamond's *The Realistic Spirit* to mean the necessity of seeing (or attending to) what lies close at hand [1]. Reflections on care brings ethics back to its proper domain in *everyday practice* —much as Wittgenstein sought to bring language back from the metaphysical level to its everyday use, where words have ordinary meanings grasped by the speakers of the language.

According to this perspective, ethics is not grasped by reference to an enumerable set of preexistent rules—nor by attending to a metaphysically independent moral realm. Rather, it is embedded in human situations, affects and practices. The ethics of care affirms the importance of care and of paying attention to others; in particular, it concerns those whose life and well-being depend on particularized and sustained everyday patterns of care. Moreover, the ethics of care is based on an analysis of the historical conditions that have promoted a division of moral labor in which the activities of care are socially and morally devalorized. The relegation of women to the domestic sphere translates into the demotion of legitimate acts and concerns to the status of mere 'private' sentiments lacking moral and political weight. It is because the work and activities of care have traditionally fallen to women that care is first and foremost a women's issue.

By proposing to valorize moral values such as caring, attention to others, and solicitude, the ethics of care has contributed to modifying the dominant conception of ethics and how we understand morality. Moreover, the ethics of care introduces ethical stakes into politics by effecting a critique of prominent theories of justice [1]. However, most significantly, it has given due representation to *the ethically ordinary*. The ethics of care draws our attention to what we are unable to see, not because it is hidden or secret but sometimes because it is right before our eyes.

To locate this discussion, observe that the ethics of care as rooted in the ordinary adumbrates Wittgenstein's definition of *the ordinary*: "What we are supplying are really remarks on the natural history of human beings . . . observations which no one has doubted, but which have escaped remark only because they are always before our eyes" [2] (§415). Wittgenstein addresses the question of what is 'important' and the fact—concretely substantiated by the ethics of care—that what seems the most important often is not so in the least:

> Where does our investigation get its importance from, since it seems to destroy everything interesting, all that is great and important? What we are destroying is nothing but houses of cards [2] (§118).

This paper aims to bring together the 'destruction' or 'philosophical subversion' to which Wittgenstein refers and the appeal to the ordinary in the ethics of care. The paper pursues a connection between the ethics of care and my longstanding interest in ordinary language philosophy as represented by Wittgenstein, Austin, and Cavell, in particular in drawing upon their ideas to illustrate a feminist perspective.

2. Sensitive Subject

The subject of care is sensitive—it requires delicacy in its handling. In a first sense, broadly presented, reflection on care tends to give rise to objections or even to outright rejection. It is thus a 'sensitive' subject matter. At first encounter, it seems to oppose a 'feminine' and a 'masculine' conception of ethics. As discussed by Gilligan and Noddings, feminine ethics is defined by the concepts of attentiveness, concern for others, and a sense of responsibility reflecting close personal ties, while masculine ethics is defined by the concepts of rights, justice, autonomy. Much has been said about the difficulty of opposing feminine and masculine ethics as the ethics of care and the ethics of justice, respectively, which incurs the risk of reproducing the very ingrained prejudices that the ethics of care are meant to counter. As proposed by Gilligan (1987), one can argue that far from being oppositive concepts, care and justice are, in fact, complementary perspectives, the choice of which depends on the context of the application (in this respect, recalling the famous Wittgensteinian duck–rabbit model involving the voluntary switch of perspective from one possible representation to another). In another sense, then, the subject of care is a sensitive subject. Thus, care is fundamentally embedded in the form of life unfolding in a context of relationships—in the latter sense, the subject is defined as attentive and *caring*.

In addition to forming a new point of departure, reflection on care affects a transformation of the very status of ethics. The question of *sensitivity* is indeed at the core of care. It is therefore urgently necessary to understand what is meant by this important concept and what kind of sensitivity is involved in the ethics of care. Indeed, it is not so much *feeling*—in the sense of Hume, for example, to be contrasted with rationality—as *perception*. Yet it is but an *ordinary* perception. It is here that we must look for the starting point for a modification of the ethical framework.

Gilligan herself, in returning to the antithesis of care and justice perspectives in 1987, starts from the paradigmatic duck–rabbit illustration of perspectival switching ([3], p. 31). Like Wittgenstein, by drawing attention to it, she does not aim to introduce a relativism of moral perspectives but rather to indicate several important observations. First, it indicates the possibility of changing viewpoints, even if one viewpoint necessarily dominates the choice. Moreover, it indicates the necessity of attending to only one perspective at any given time (in other words, the impossibility of mastering both at any given instance). In the third place, it indicates the importance of context to the representation of choice. By analogy, ethics is a matter of highlighting, for each moral situation we examine, not only different visual 'orientations', but a *framework* of perception ([3], p. 32). Gilligan suggests a *Gestalt* approach to ethics, insisting, as did Wittgenstein, on the necessity of choosing aspects of salience against a given background.

Cora Diamond defines the *specificity* of this approach as follows:

> Our *particular* moral conceptions emerge against a more general background of thought and sensibility. We differ in the way we allow (or do not allow) moral concepts to shape our lives and our relationships with others, in the way these concepts structure our accounts of what we have done or experienced [4].

Such a perceptual approach will be not only situational and dynamic but particularistic. It is only through attending to the particular, as opposed to the general, aspects of a situation that we will find the right perspective in ethics—as in aesthetics, for that matter. Here there is a further reference to Wittgenstein, beyond the seminal duck–rabbit *Gestalt*, but to his particularism, the 'attention to the particular'.

He illustrated this with the example of the word 'game', in relation to which he observed (1) that even if there is something common to all games, it does not follow that this is what we have in mind when we call a particular game a game, and (2) that the reason why we call so many different activities games does not require that there be anything in common between them, but only, from one use to another, a 'gradual transition'. With regard to the word 'good', he observed that the different way in which one person, A, manages to convince another, B, that a certain thing is good, fixes, each time, what the meaning according to which 'good' is used in this discussion ([5], p. 104).

This observation is surprising if one considers that Wittgenstein's published writings contain relatively little that could pass for moral philosophy. In the *Tractatus logico-philosophicus* (1922), Wittgenstein took a firm stand against the very existence of moral philosophy, given that the purpose of philosophy is said to be the logical clarification of propositions. Philosophy itself is not a body of doctrine, but an activity, which consists in making our thoughts clear (*Tractatus*, [6] 4.112). From this description of the central task of philosophy, it follows that there can only be such a thing as 'moral philosophy' if there is a body of propositions such that it is the task of moral philosophy to clarify them. However, Wittgenstein also maintained, for reasons beyond the scope of this paper, that there can be no ethical propositions [6] (6.42). Yet taking a Wittgensteinian approach to morality does not entail subscribing to a relativistic or skeptical program. Wittgenstein described the *Tractatus*, which denied the existence of moral philosophy and ethical propositions, as having an ethical aim. In thus maintaining an ethical aim, he did not intend to represent the *Tractatus* as entirely silent on moral philosophy. Rather, Diamond writes, the point is that ethics does not arise from theorizing.

> His position was (then and later) that work, for example, a novel or a short story, could have a moral purpose even in the absence of any moral teaching or theorizing. Such work could help us to tackle the tasks of life in the required spirit. This was to be the effect of the *Tractatus* [7].

The aim of such thinking is not to reject the idea of morality, nor of moral philosophy understood in a specific sense, but of a systematic moral theory. It should be noted that a certain number of anti-theoretical thinkers will nevertheless find in these remarks a specific form of realism. This is evinced, for example, by John McDowell in his essays on Wittgenstein [8] and by Diamond herself in *The Realist Spirit* [1]. Even so, the proposed realism is to be discovered not in a metaphysical reality or a realm of moral objectivity but is produced by attention to details, to what is before your eyes Diamond and McDowell thus criticize the view *from sideways on* (as they term it). For example, we seek to determine the nature of the obligation inherent in a rule by relating it to something in reality rather than by looking at the ordinary way of saying what a rule requires. As Diamond puts it:

> We have, for example, the idea that we look at the human activity of *following a rule* 'from the side', and ask from this point of view whether or not there is something *objectively* determined that the rule requires to be done in the next application ([9], p. 30).

Thus, an examination of our particular moral practices is more *realistic* (in the sense proposed by Diamond) than the theoretical search for a moral reality. We are dealing here with a moral philosophy inscribed in our ordinary practices and emerging from particular questions. By proposing to valorize moral values such as caring and attention to others, the ethics of care has contributed to modifying a dominant conception of ethics and has profoundly changed the way we look at it. It has given voice to the ordinary. The ethics of care draws our attention to the ordinary, to what we are unable to see, to what is right before our eyes and is, for this very reason, invisible to us (see [10]. It is an ethics that gives voice and attention to humans who are undervalued precisely because they perform unnoticed, invisible tasks.

3. Ordinary Realism

Diamond's aim, drawing on Wittgenstein, is to define an ethics of (attention to) the particular, and this is a perspective shared by the ethics of care: attention to ordinary life. Realism in ethics, on this view, consists in *returning to* ordinary language, examining our words and paying attention to them, and taking care of them: taking care of our words and expressions, as well as of ordinary others.

Ordinary Language Philosophy teaches us indeed that our ethical lives cannot be captured with the traditional concepts of moral philosophy. In his later works, which converge with Diamond's ideas, Hilary Putnam proposed to abandon a certain form of realism in ethics and the possibility of a common ground for ethical discussions:

> Our ethical life cannot be captured by half a dozen words like 'ought', 'right', 'duty', 'fairness', 'responsibility', 'justice', and the ethical problems that concern us cannot be reduced to debates between the metaphysical propositions of the proponents of natural law, utilitarianism, common sense and so on [11].

Putnam, like Diamond, reflects Murdoch's legacy, which is to be careful what *we* say.

> There are ethical propositions which, while being more than descriptions, are also descriptions. One is then 'entangled' by descriptive words like 'cruel', 'impertinent', 'inconsiderate' [11].

These *entangled* terms, which are 'both evaluative and descriptive' and *ordinary*, are, for Putnam, at the heart of our ethical life: the elucidation of their uses is part of moral knowledge, which is knowledge or ethics without ontology or metaphysics. "J. McDowell and I have both stressed this, and we are both aware of our debt to Iris Murdoch" [11].

We will return below to Murdoch's legacy and her reliance on the *vision* and ordinary *texture* of language. In contrast to metaphysics, the ethical approach should bring us back to the rough ground of ordinary language. There is nothing ontological in this realist approach to ethics: "logic as well as ethics can be found *there*, in what we do, and something like a fantasy prevents us from seeing it" [11]. The elements of an ethical vocabulary only make sense in the context of our uses, given a particular form of life; they may be said to 'come to life' against the background of a specific *praxis*. Context gives words their meaning; a meaning that is never fixed and always particular. "Only in the practice of language can a word have meaning" ([12], p. 344).

Meaning is defined not only by use or context (as many philosophers have recognized) but is inscribed and perceptible only in the dynamic background of language practice, which is modified *by what we do with it*.

'Beautiful' is linked to a particular game. Similarly in ethics: the meaning of the word 'good' is linked to the very act it modifies. We can only establish the meaning of the word 'beauty' by considering how we use it ([13], p. 35).

One might then be tempted to connect ethics with a particularistic ontology, which would put abstract particulars (e.g., from perception) at the center of a value theory or realism of particulars. However, this would be to lose the meaning of the idea of family resemblance, which is precisely the negation of an ontology, including abstract particulars. Wittgenstein criticizes the craving for generality—the tendency to look for something common to all entities that we standardly subsume under a general term. The idea that a general concept is a property common to its particular cases is related to other primitive and overly simple ideas about the structure of language ([14], pp. 57–58). What is needed, as Hilary Putnam has suggested, is an ethics without ontology rather than an ontology of the particular [15].

Realism in ethics rather requires exploration of the way our ethical preoccupations are embedded in our language and our life, in clusters of words that extend beyond our ethical vocabulary itself and sustain complex connections with a variety of institutions and practices. In order to describe ethical understanding, we would have to describe all of this, all these particular uses of words, of which a general definition cannot be given. From an ordinary language's perspective, the elements of moral vocabulary have no meaning except within the context of our customs and form of life. In other words, they *come to life* against the background that "gives our words their meaning".

For Wittgenstein, meaning is not only determined by use or "context" (as many analyses of language have recognized) but is embedded in, and only perceptible against, the background of the practice of language. To redefine ethics by starting off with what is important means paying this "attention to particulars". A whole cluster of terms, a language game of the particular—attention, care, importance, and what matters—is common to ordinary language philosophy and the ethics of care. Attention to detail is the source of a realistic shift of perspective in moral philosophy: from the examination of general concepts and norms of moral choice to the examination of particular visions, of individual "configurations" of thought. The ethics of care merges with this sensitivity to words and the "realistic spirit" by drawing our attention to the place of ordinary words in the weave and details of our lives and our relation to/distance from our words.

But what kind of interest do we have in particular? The philosophical drive for generality is 'contempt for the particular.' By contrast, moral perception is care for the particular. In her important essay, *Vision and choice in morality*, Murdoch, a disciple of Wittgenstein, invokes the importance of attention in morality [16]. According to her interpretation, the first way of expressing care is to pay attention to something, that is, to be attentive. The word attention is a possible translation of the term care, perhaps drawing it a little too much from the perceptual side but highlighting the anticipatory dynamics of this perception. Murdoch also evokes the differences in morality in terms of differences in Gestalt. She wants to avoid the classical idea of perceiving an object via a concept:

> Moral differences here are less like differences in choice, and more like differences in *vision*. In other words, a moral concept is less like a movable and expandable ring placed over a certain domain of facts, and more like a *Gestalt* difference. We differ, not only because we select different objects from the same world, but because we *see* different worlds. (...)
>
> Here the communication of a new moral concept cannot necessarily be accomplished by the specification of a factual criterion open to any observer ('Approve *this* field!') but involves the communication of a completely new vision [16].

Here again, Murdoch operates a critique of the *general* in ethics. There are no univocal moral concepts that can only be applied to reality to delimit objects. Nevertheless, our concepts depend on it through their application relative to a domain of interest, a given narrative or description, and our personal interest and desire to explore these in terms of what is important to us. In the idea of importance, we discover the means for another formulation of the concept of care, as attending to what is important or matters to us: what counts.

4. Importance of Importance

This relationship of care to *what matters* was highlighted by Harry Frankfurt in *The Importance of What We* Care *About* (1988). In an analogous spirit, Cavell discusses cinema and the films that *matter* to us, which are the objects of our attention and care, requiring the education of perception and attentive vision:

> The moral I draw is the following: to answer the question 'what happens to objects when they are filmed and projected?'—as well as 'what happens to particular people, places, subjects and motives, when they are filmed by this or that filmmaker?'—There is only one source of data, namely the appearance and meaning of these objects, these people, which we will in fact find in the sequence of films, or passages of films, that matter to us ([17], p. 79).

The importance of cinema lies in the way it brings out visually what matters, in order, as Cavell puts it, "to magnify the feeling and meaning of a moment". However:

> ... it is also his task to go against this tendency and, instead, to recognise this tragic reality of human life: the importance of its moments is not usually given to us with the moments while we are living them, so that it can take a lifetime's work to determine the important crossroads of a life ([17], p. 79).

It is possible to understand the concept of care by adverting to the specific attention to the *unseen* importance of things and moments required by the appreciation of filmic art. 'The inherent concealment of importance' is part of what cinema also teaches us about our ordinary life. Redefining morality on the basis of importance, and its link to the structural vulnerability of experience produces the necessary elements to constitute the ethics of care. The notion of care is inseparable from a whole cluster of cognate terms, which constitute a language game of the particular. Among these are *attention, concern, importance, significance,* and *mattering*. Our capacity for care becomes, according to Murdoch, "a detached, non-sentimental, non-selfish, objective version of care". The emergent attention is the result of the development of a perceptual capacity. It is rooted in the ability to see the detachment of detail, of expressive gesture, against a background, without undue ontological reification. We can now see more clearly the contribution made by the ethics of care to transforming ethics into attention to the human form of life.

This particularism of attention to detail has also been propounded by Diamond, notably in her important essay *Getting a Rough Idea of What Moral Philosophy Is*, which closes *The Realist Spirit* ([1], pp. 495–515). According to her construal, moral philosophy needs to shift its focus from examining general concepts to examining particular visions, the 'configurations' of thought of individuals. On this point, Murdoch is radical:

> We consider something more elusive which may be called their total view of life, as manifested in the way they speak or remain silent, their choice of words, their ways of appreciating others, their conception of their own lives, what they find attractive or praiseworthy, what they find amusing: in short, the configurations of their thinking which are continually manifested in their reactions and conversations. These things, which may be shown openly and intelligibly or elaborated intimately and guessed at, constitute what may be called a man's texture of being, or the nature of his personal vision ([18], p. 49).

It is indeed in the use of language, in the choice of words, style of conversation, etc., that the moral vision of a person is openly shown or intimately elaborated, which for Murdoch is not so much a theoretical point of view as a *texture of being*—which is still a Gestalt term since texture can appear in various modalities, visual, aural and tactile. This texture has nothing to do with moral choices or ethical argumentation. Rather, it once again pertains to 'what matters' and what makes and expresses the differences between individuals.

> We cannot see the moral interest of literature unless we recognise gestures, manners, habits, turns of speech, turns of thought, styles of face, as morally expressive—of an individual or a people. The intelligent description of these things is part of the sharp, intelligent description of life, of what *matters*, what makes a difference, in human lives ([1], p. 507).

It is these differences that are to be the object of 'the intelligent, sharp description of life'. The idea of a human life alludes to the Wittgensteinian form of life, which similarly defines a texture. Texture, then, designates an unstable reality, which cannot be pinned down by concepts or by particular determined objects, but is nevertheless accessible through the recognition and identification of gestures, manners, styles, etc. The form of life is, from an ethical point of view, defined by perception. In literature, the paradigmatic exploration of attention to moral textures or patterns belongs to Henry James's novels, described by Diamond and Nussbaum in their essays on James's magnificent oeuvre. Jamesian motifs are, of course, perceived as 'morally expressive', as no doubt they were intended to be. What is perceived, moreover, is not an objective realm of moral values or moral concepts but the moral expression itself, which is only possible and indeed apprehensible in relation to the life forms depicted in the background of the novel. Literature is, arguably, the privileged place for moral perception, which at its culmination in the later Jamesian novel, achieves its aims by creating a background that makes the significant differences between life forms appear in bold outline. We return to the use of *Gestalt* later on, in reflecting on the direct perception of meaning—yet with a lighter emphasis on the constitution of an object as on the perception of its 'possibilities', which we are invited to explore. To perceive is always to give oneself with the perceived object an immediate opening onto an anticipatory perspective whose guidelines call for acts of exploration. Attention to ordinary expression and human voice and texture leads to re-considering the question of women's expression, which has been stifled or neglected by philosophy. Once again, ordinary language is not to be envisioned as having only a descriptive or even agentive function but as a perceptual instrument that allows for subtlety and adjustment in perceptions and actions.

5. Moral Competence and Education

The definition of ethical competence in terms of refined and active perception (versus the ability to judge, argue and choose) is taken up by Nussbaum [19]. For Nussbaum, morality is substantively a matter of perception and attention and not of moral argument. One immediate objection to Nussbaum is that her arguments reinstate a schematic and dubious opposition between feeling and reason. Yet Nussbaum's discussion is relevant in terms of the refocusing of the ethical question on a form of moral psychology rooted in a fine and intelligently educated perception.

According to Nussbaum, moral competence is not only a matter of knowledge or reasoning but also a matter of learning the right expression and educating the sensibility. A truly artistically successful author is capable of educating the reader's sensibility by rendering particular characters or situations perceptible to the reader through the apt choice of literary frame. A literary education of sensibility is one that is capable of generating meanings. See, for example, Diamond's chosen illustration of the life of Hobart Wilson in her 'Differences and Distances in Morals'; or any of James's characters, who, in the course of his endlessly subtle narratives, never fail to teach the reader the correct and clear apprehension of things. In his preface to *What Maisie Knew*, James notes: "The effort to see really and paint really is no small thing in the face of the *constant* force that works to confuse everything" ([20], pp. 165–166). This novel, Diamond notes, is entirely a critique of perception through the description of "a social world where the perception of life is characterised by the inability to see or gauge Maisie's alertness" ([20], p. 418).

For these reasons, the idea of description or vision (qua the orthodox or objectivist model of perception) no longer suffices to account for moral vision: it consists in seeing not objects or situations but the possibilities and meanings that emerge in things; in anticipating, in *improvising* (says Diamond) at every moment in perception. Perception is then active, not in the Kantian sense of being conceptualized, but because it involves a constantly changing perspective. We thus rediscover the alternation of 'duck' and 'rabbit' that Gilligan adverts to in the wake of Wittgenstein. We think, moreover, of Nussbaum and Diamond's analysis of Henry James: the novel teaches us to look at moral life as 'the scene of adventure and improvisation', which transforms the idea we have of moral *agency* and makes visible to us the values that reside in moral improvisation.

There are thus constraints on perception, not because it is voluntary, but because it is necessary to see the emergence of dynamics and apprehend the possibilities inherent in things. As Diamond observes, "[s]eeing the possibilities in things is a matter of a transformation in one's perception of them" ([20], p. 418). Wittgenstein [2] (§90) similarly notes that we are dealing not with phenomena but with the *"possibilities* of phenomena".

Learning a language is learning to perceive the possibilities of phenomena, or the possibilities in things, which form the background to moral expression. This is an essential point, which comes out clearly in *The Claim of Reason*:

> In 'learning language' you learn not only what the names of things are, but what a name is; not only what the form of expression suitable for the expression of a desire is, but what it is to express a desire; not only what the word for 'father' is, but what a father is; not only the word 'love', but what love is ([21], p. 271).

Thus, according to Cavell, the learning of morality is inseparable from the learning of language and of the accompanying form of life. When construed on these lines, care is not a subordinate or marginal element of ethics but lies at its very root. Integration into a form of life is in itself important to us, and it consists in learning what is important in terms of *significance* as well as meaning [22,23]:

> Rather than saying that we *tell* beginners what words mean, or teach them what objects are, I would say: we *introduce* them to the relevant forms of life contained in language and gathered around the objects and persons of the world we live in [22,23].

Learning a language thus defines ethics both as attention to reality and to others—in short, to a form of life. Language learning is the grasping of an expressive structure, specifying adequate modes of expression as well as the *meaning* of words. It is an initiation into a form of life that undertakes the training of the quality of sensitivity (and affiliated faculties) through exemplarity. Morality, then, also concerns our capacity to interpret moral expression—not only the capacity to elicit and form a moral judgment and moral choice but their various moral readings. However, this expressive capacity is not purely instinctive or affective. It is also conceptual and linguistic—it is our ability to make good use of words and to use them in new contexts, enabling us to respond or react correctly to the moral aspects we encounter. Diamond adverts to Murdoch's construal of moral thinking to argue that despite its renunciation of non-cognitivism, contemporary moral philosophy is still insouciant to language, as well as blind to moral expression:

> We obsess again and again over 'evaluations', 'judgements', explicit moral reasoning leading to the conclusion that something is worthwhile, or is a duty, or is wrong, or should be done; our idea of what is at stake in moral thinking is again and again 'it is wrong to do x' versus 'it is permissible to do x;' the abortion debate is our paradigm of moral debate. 'Distrust of language' has become the inability to see all that is involved in using it well, responding to it well, tuning into it well; the inability, then, to see the kind of failure that can be involved in using it badly. How do our words, our thoughts, our descriptions, our philosophical styles let us down? How do they, used to their fullest extent, enlighten us? [Cora Diamond, *The Realistic Spirit*] (pp. 379–380).

The capacity for moral expression is rooted in a plastic form of life, as it is vulnerable to good and bad uses of language. It is the form of life in the natural as well as the social sense that determines the ethical structure of expression, which in turn reworks it and gives it form [24]. Our relationship with others, reflecting the kind of interest and concern we invest in them and their importance to us, exist only in the possibility of the unveiling of oneself in expression, be it successful or failed, voluntary or involuntary.

> In order to recognise each other's readiness to communicate, which is presupposed in all our expressive activities, we must be able to 'read' each other. Our desires must be manifest to others. This is the natural level of expression, on which true expression is based. Mimicry and style are based on this (...). But there would be nothing to rely on if our desires were not embodied in the public space, in what we do and try to do, in the natural background of self-disclosure, which human expression works endlessly [25].

What is described in a skeptical mode by Cavell, alluding to the difficulty of self-expression and of recognizing and reading the self-expressions of others, is captured in the hermeneutic mode by Charles Taylor. Both Cavell's and Taylor's accounts lead to a moral questioning of mutual expression, the experience of meaning, the constitution of style, and of self and mutual education through learning to pay attention to the range of human expressions: "Human expressions, the human figure must, in order to be grasped, be *read*" ([21], p. 508). In other words, the reading of human expression, which makes it possible to *respond*, is a product of attention and care. It is the result of learning to be sensitive. We find in these remarks the Cavellian theme of education throughout all the stages of adult life. By recognizing that education does not cease upon reaching adulthood, we understand that education is not only a matter of the acquisition of factual knowledge but of further refinements and attunements, a life-long process rather than an early stage in development. This, as it happens, is also the point of Wittgenstein's insistence, from early on in the *Investigations*, on the idea of learning a language. The latter process consists in grasping not meanings but a set of practices that are not 'founded' in a language or causally linked to a social or natural background but learned at the same time as the language itself and which constitute the shifting texture of our life.[12] The relation to others, the type of interest and care we have for them, the importance we give them, take on their meaning within the context of a possible expression and/or unveiling of oneself (Laugier, [26]).

As Lovibond has shown, moral education consists of acquiring mastery of the contexts, connections, and backgrounds of moral actions so as to perceive moral reality and expression directly [27]. Lovibond's realist approach is in line with McDowell's emphasis on *Bildung* and second nature: a specific linguistic competence is developed in the field of morality, as the acquisition through moral education of particular sensitivity to appropriate ethical reasons [28]. Then, says Lovibond, "sensitivity to the force of ethical reasons becomes a component of our second nature" ([27], p. 61). In short, we *learn* to see in ethics.

However, beyond these realistic approaches and the alternative model of virtue ethics, it must be understood that learning moral language is also based on a certain authority and a form of blindness, of trust. In Cavell's reading of Wittgenstein, the question of education is permeated by skepticism: learning does not guarantee the validity of what one does since only the approval of one's moral 'masters' can vouchsafe validity. Nothing, therefore, grounds our practice of language except that practice itself—"the whirl of organism which Wittgenstein calls *forms of life*," noted Cavell in *Must We Mean What We Say?*

> We learn and teach words in certain contexts, and we are then expected to be able to project them into other contexts. (...). Human speech and activity, sanity and community, are based on nothing more than this, but also on nothing less. It is a vision as simple as it is difficult and as difficult as it is terrifying ([29], p. 52).

The vision is 'terrifying' because it posits that learning is always infinitely extensible and that once we have learned a word by exposure to a few typical contexts of use, we are expected to project it into new contexts and so to improvise constantly. The stakes involved in the expectation of endlessly fluid improvisation are moral stakes, relating to the 'sanity' or the possibility of sharing and learning a form of life. This goes further than the reference to Aristotelian moral education found in virtue ethics: to learn a word is to learn, and to imagine a form of life ([21], p. 125) [30]. An ethics of the ordinary life, which would simply refer to the authority of 'our practices' against theory, would be hopeless. Ethics does not refer to a description of our practices: "Our practices are exploratory, and it is really only through such exploration that we come to a full view of what we ourselves thought, or meant" ([1], p. 39). We are able to understand what ethics is not, namely, a set of principles or rules, or general concepts. However, ethics cannot be purely descriptive, insofar as our ethical concepts also work on our practices, our form of life; concepts are also a form our life takes.

> Considering use can help us see that ethics is not what we think it should be. But our idea of what it should be has necessarily *shaped* what it is, as well as what we do; and considering use, as such, is not enough ([1], p. 39).

Such a recourse to practice is once again borrowed from Wittgenstein, and in particular from his approach to the concept of a 'rule', conceived not as a determinant of practices but as *visible* against a background of human practices. Normativity is woven into the texture of life:

> We are not just trained to do '446, 448, 450' and so on; we are brought into a life in which we depend on people following rules of all kinds, and they depend on us: the rules, the agreement in how to follow them, the confidence in that agreement, the ways of criticising or correcting those who do not follow them properly—all these are woven into the texture of life ([10], pp. 27–28).

Instead of the perceptual, and perhaps static, theme of background, we might prefer the themes of texture and pattern. Wittgenstein, for one, speaks of a "pattern in the tapestry of life" and of a "vital swarming"; or, as in *Zettel*, of *place* and *connections*: "Pain occupies *such and such a* place in our life, it has *such and such* connections" [31] (§§532–533).

As Diamond observes, connections 'in our lives' are not hidden but have their being right before our eyes. Here she alludes to the well-known simile of the 'figure in the carpet' in James's classic short story (The Figure in the Carpet, by Henry James). We perceive through concepts, including moral ones, because our concepts 'grasp' (their referents) in the unfolding of a texture of life, which is dynamic, and where patterns recur and emerge.

> If life were a tapestry, this or that pattern (pretending, for example) would not always be complete and would vary in many ways. But we, in our conceptual world, always see the same thing repeated with variations. This is how our concepts *grasp* (*auffassen*) [32] (§672).

6. Background and Life Form

The background of the life form is neither causal nor fixed like a set, but living and mobile. Again, we can appeal to *life* forms instead of *forms of life*. The distinction marked is not a definitive or stable form, but the forms that our life *takes* under the attentive gaze —the 'whirl' of our life lived within a language, of our 'visions', as opposed to a stable body of meanings or social rules.

The term background (*Hintergrund*) appears in Wittgenstein to designate a background of description, which brings out the nature of actions. Pace Searle, it is not intended to *explain* anything. The background cannot have a causal role, for it is language itself in its instability and sensitivity to practice:

> We judge an action by its background in human life (...) The background is the train of life; And our concept refers to something in this train [33] (§624–625).

How can the human way of acting be described? Only by showing how the actions of the diversity of human beings blend together in a swarm. It is not what an individual does, but the whole swarming whole (Gewimmel) that forms the background against which we see the action. [33] (§629); [31] (§567).

We see the action, but we are *caught* in the midst of a 'swarming life-form' on which it stands out and becomes sensible and thus important. It is not at all the same to say that the application of the rule is causally *determined* by a background, and to say that it is to be *described in* the background of human actions and connections. This is the difference between a gestalt and descriptive conception of ethics and a 'conformist' conception that attempts to provide justification by reference to prior agreement bestowed by the community. The background does not provide or determine ethical meaning since there is none but allows for a clearer view of what is important and meaningful to us in the important moment: the connections in the texture of our lives. Wittgenstein mentions, in *Culture and Value*, "the background against which what I can express receives meaning" ([34], p. 16). The 'accepted' or given background does not determine our actions (thus, no causality is involved) but allows us to see them clearly.

If we define ethics by such an immanent caring description, it directs our attention to the moral capacities or competencies of ordinary people. Attention to the everyday, what Cavell calls the ordinary other, is the first definition of caring. The celebrated definition of care by Joan Tronto and Berenice Fisher has to be taken seriously as a realistic claim:

> In the most general sense, care is a species of activity that includes everything that we do to maintain, continue, and repair our world so that we can live in it as well as possible. That world includes our bodies, our selves, our environment, all of which we seek to interweave in a complex, life sustaining web ([9], p. 40).

Reflection on care can be construed as a consequence of the turn in moral thought illustrated by the work of Cora Diamond: against what Wittgenstein in the *Blue Book* called the "craving for generality," it is the attempt to valorize, within morality, attention to the particular(s), to the ordinary details of human life, the aspects of life neglected by philosophy and by us. This descriptive aim transforms morality: care, like OLP, brings our attention back to the rough ground of the ordinary, to the level of everyday life.

As Bernard Williams reminds us, "ethical theories are abstract schemes that are supposed to guide everyone's judgement on this or that particular problem." Williams' formulation proves to be a source of linked difficulties—those of moving from the general to the particular, from the rule to its application, from theory to experience. Beyond these epistemological difficulties, his description raises further pertinent questions: why focus ethical reflection on the question of principles, foundation, and justification? Why should it follow the legislative or scientific model? Why give rules instead of simply describing what we do? These are the difficult questions that the ethics of care must face, insofar as methodologically, it goes against the grain of contemporary moral theories.

The mythology of 'moral theory' lies in the idea of elaborating a number of principles that can produce a 'morally correct answer' to most moral problems *in all circumstances*. The anti-theoretical, 'unorthodox' view, on the other hand, rejects the possibility of substantive and general moral principles or metaethical theories about the nature of moral or normative statements, from which one can develop modes of justification and reasoning that would hold for all situations. Most unorthodox moral philosophers (e.g., Anscombe, Baier, Diamond, Lovibond, Williams, McDowell) are influenced by Wittgenstein's thinking. Certainly, many will agree that there is a duty, for example, to care for one's family and friends—but one does not usually want to be loved out of duty, and this very concern not to be loved out of duty would be a more interesting topic for morality than obligation itself (it is, for literature or film). Similarly, as Diamond notes, there may be something mean-spirited and 'ungenerous' about a perfectly rigorous person who is consumed by the idea of doing what they consider to be their duty [35]. The *unlikability* in the strong sense that taints the character of the dutiful ethicist is something that should be part of moral reflection rather than consigned to the litter of ethically marginal questions. Baier suggests that one should be interested in a virtue such as *gentleness*, which can only be treated in both descriptive and normative terms, and "resists analysis in terms of rules" ([36], p. 219), being an appropriate response to the other *according to the circumstances*: it requires an experimental attitude, sensitivity to a situation and the ability to improvise, to 'move on' from certain reactions. Baier frequently draws on Hume to define moral attitudes such as expectation or simply waiting to see what happens rather than applying principles. Without these expectation-based attitudes, moral reflection runs the risk of becoming locked into the 'side view', of losing sight of *what matters* in morality, what it is we care for and about.

Baier criticizes, as Murdoch, Diamond, and Anscombe did, the idea that moral philosophy is reducible to questions of obligation and choice—as if a moral problem, by being formulated in these terms, becomes thus treatable [37]. Baier takes up the irony in Hacking, directed at the obsession of moral philosophy with the game-theory model [38]. To wit, everyone will have noticed the obligatory chapter on the 'prisoner's dilemma' in any serious book on moral philosophy. We may recall the proposition in the *Investigations* where Wittgenstein defines agreement *in* form of life:

> It is what human beings *say* that is true and false; and they agree in the language they use. It is not agreement in opinions but in the form of life [2] (§241).

The model of agreement for Wittgenstein is linguistic agreement: we agree *in* language. This allows us to understand the nature of agreement. We may take it that our language usage and practices are given as a set of rules to which we have no choice but to adhere. However, another of Wittgenstein's discoveries is that usage is not enough. My agreement with or belonging to *this or that* form of life, whether social or moral, is not given. The background is not a priori, but it is modifiable through the practice itself. The acceptance of the form of life as 'a given for us'—which Wittgenstein advocates—is acceptance of a natural given ("the fact of being a man, therefore with that (extent or scale of) capacity for work, pleasure, endurance, seduction" ([24], pp. 48–49). However, the form of this acceptance—the 'extent and scale' of our agreement—is not knowable a priori 'any more than the extent or scale of a word can be known *a priori*', because the use of moral language is improvised. Thus, one does not agree to everything in advance. The moral burden is at all times in "*what* we *should say when*" [39]. The fact that moral language is given to one does not imply that one knows a priori how one is going to get along, to agree *in* this language with one's fellow language users, to find the right expressions to respond, etc. What constitutes language agreement and moral agreement is the ever-open possibility of rupture, the threat of skepticism, or the loss of moral voice [40] (chp. 5).

A form of life can be grasped only by attention to textures or moral patterns, perceived as "morally expressive" in/on the background provided by a form of life. Our capacity for moral expression is rooted in a mutable form of life, vulnerable to our better and worse uses of language. The type of interest, the care that we have for others, and the importance that we give them, do not exist except in the possibility of the display or revelation of the self in its moral expression. The idea of an ethics formulated *in a different voice* and expressed in women's voices is inseparably an ordinary conception of ethics, an expressivist conception of ethics, and a realistic conception of ethics. This ethics rather starts from experiences of everyday life and the moral problems of real people in their ordinary lives.

7. Losing Concepts

According to Diamond, many of the statements found in contemporary moral philosophy are, in Diamond's own words, "stupid or *insensitive* or delusional". She gives, as an example, a passage in which Peter Singer argues in favor of the defense of animals:

> What I mean by 'stupid or insensitive or delusional' can be made clear by a single word, the word 'even' in the quotation: 'We have seen that the experimenter reveals a bias in favour of his own species when he experiments on a non-human in a case where he would not consider it justified to use a human being, even a retarded human being' ([1], p. 33).

What is wrong with such an argument is not the argument itself but the use of the word 'even.' What is wrong is the absence of care. When Diamond says that moral philosophy has become mostly blind and insensitive, she means it has become insensitive to the human specificity of moral questioning and to ordinary moral life.

What matters in moral perception is not agreement and harmony but the perception (sometimes violent) of contrasts, distances, differences, and their expression; that moment when, as Diamond says, there is a 'loss of concepts', when it no longer works. Cavell describes this difficulty in terms of skepticism, as a sensation and temptation of inexpressiveness, as our inability to go beyond our natural reactions to know the other—to go beyond the limits of my understanding and concepts, but also of my *experience*.

> Our ability to communicate with him depends on his 'natural understanding', his 'natural reaction' to our instructions and gestures. It depends, therefore, on our mutual agreement in judgements. This agreement takes us remarkably far along the path of mutual understanding, but it has its limits; limits which, one might say, are not only those of knowledge, but those of experience ([21], pp. 184–185).

What is important in the ethical situation is not just the agreement but the disagreement that the sensitivity to words creates: the exposure of the *loss of our concepts* and the difficulty of applying them in new contexts. Diamond takes the case of animal experimentation, showing that certain forms of argument are unbearable, and create a distance and perplexity fundamental to a definition of ethics:

> Suppose someone were to say in a discussion or in an experiment on animals that one of the reasons why it would be wrong to experiment on 'newborns', to put them in cages, to subject them to chemicals or electric shocks or cancer or extreme fright or anguish or to kill them—a reason that is not applicable to animals—is that it would deprive society of the valuable contributions they could make as adults. This argument would obviously not apply to animals because they cannot make the same kind of contribution (...).

My distance from someone like that is not a matter of refusing what they think they can support. It is rather that I would say to myself: 'Who is he, and how can he think that this is what should be claimed in this discussion? What kind of life is he living, in what life can this discussion take place?' [41]

The important point for Diamond is that there is no opposition between sensibility and understanding, but that sensibility is a form of conceptual life. This explains the 'sensitive' reactions we have to conceptual matters, such as the kind illustrated by Diamond above. There is no need to separate argument and feeling in ethics. Rather, Wittgenstein reveals the properly sensitive character of concepts and the perceptive character of conceptual activity that are at work, which allow for the clear apprehension of conceptual contrasts and divergences. For example, it may be possible to know without further ado and without being able to produce a counterargument that what someone is arguing is 'solemnly comical nonsense', or morally repugnant, utterly stupid and delusional, etc. Ultimately, to give the concept of care its due place, we must place it center-stage in the framework of ethics. That is, we must acknowledge that morality as a whole must become sensitive—a 'sensitivity that would envelop the totality of the mind'.

The question is that of the expression of experience: when and how to trust one's experience, to find the validity of the particular. It is the question of finding a subjective expression and finding one's voice. The history of feminism begins precisely with an experience of non-expression, which the theories of care account for in their own full-fledged way, in their ambition to highlight an ignored, unexpressed dimension of experience. This experience, described by Cavell in the film genre he calls 'melodrama of the Unknown Woman' is one of radical alienation, of the impossibility of expressing this experience in language—what is often called today 'gaslighting.' Both movies we are alluding to here reveal the experience of a woman's inexpressiveness, silencing—losing not only self-confidence but language and perception. This is the problem that Gilligan's ethics of care confronts in a theoretical way.

John Stuart Mill was concerned with the problem of lacking the right concepts and theoretical framework, where one has no voice to make oneself heard because one has lost contact with one's own experience, with one's life. Women's inexpressiveness is the stylization of human inexpressiveness.

> Thus the mind itself is bent under the yoke: even in what people do for their pleasure, conformity is the first thing they consider … so much so that their human capacities are atrophied and lifeless; they become incapable of the slightest lively desire or spontaneous pleasure, and they generally lack opinions or feelings of their own, or truly their own. Is this or is it not the desirable condition of human nature? [42] (III, §6).

This is a description that captures all those situations that involve a loss of experience, language, and concepts, and that can motivate a desire to come out of this situation of loss of voice, to take back possession of ordinary language, and to find a world that would be the adequate context for it. Reconnecting with experience and finding a voice for its expression is perhaps the primary aim of ethics. Care, understood as attention and perception, is thus to be differentiated from a sort of suffocation of the self by affect or devotion. It confronts us with our own inabilities and inattentions, but above all, it shows us how these inattentions are then translated into theories and valuations.

It remains to articulate this subjective expression with the attention to the particular that is also at the heart of care, and thus to define a *knowledge through* care. The moral knowledge, for example, that literature or film brings to light through the education of sensibility (i.e., the training of the virtue of sensitiveness) is not necessarily fully translatable into rational or moral argumentation, but it *is* nevertheless knowledge. This idea is playfully at work in Nussbaum's ambiguous book title, *Love's Knowledge*. The syntactical play on meanings signals not the knowledge of a general abstract object 'love', but the particular knowledge that the sharpened perception of love grants us through the experience of love itself. As her title wryly suggests, there is no contradiction between sensibility and knowledge, care and rationality.

Hence Diamond's redefinition or redescription of morality from literature. "I have tried", she says, "to describe certain features of what the moral life *looks like*, without saying anything at all about what it *should* look like." This phenomenal description of the moral life allows for a transformation of the field of ethics, the refocusing on sensibility, but also a disappearance of ethics as a specific field:

> Just as logic is not, for Wittgenstein, a particular subject, with its own body of truths, but permeates all thought, ethics has no particular subject; an ethical *spirit*, an attitude towards the world and life, can permeate any thought or discourse ([35], p. 153).

In short, ethics is an attention to others and to the way they are caught along with us in connections. The ethics of care is, in this sense, inspired by Hume, Mill, and Wittgenstein.

Cavell and Diamond opposed, as did Murdoch, the non-cognitivist meta-ethics, which analyses moral statements by discerning an emotional or affective component and a factual component (erroneously relying on Wittgenstein and his rejection of the ethical propositions in the *Tractatus*). The problem, as Putnam and Cavell have amply noted, lies in the claim to deliver an *analysis* of moral statements, amounting to a *theory* of the fixed meaning of statements. If we want to analyze these statements, we will obtain a statement of a fact together with an expression of emotion (such as an exclamation or expression of appreciation or disgust). The problem with emotivism in meta-ethics is, therefore, semantic. It is as if a moral statement could be reconstructed as an additive equation relating a statement *plus* a feeling. It is as if the expression was added to the statement and was not the statement itself. This is an untenable philosophy of language, which was questioned by Wittgenstein himself [43]. The meta-ethical theorizing of the 1930s invented the blindness that Diamond wants to criticize in *The Realistic Spirit*:

It is striking that, although this approach in moral philosophy has virtually disappeared, what Murdoch meant by 'distrust of language' is as relevant as ever; (it) has become the inability to see all that is involved in making good use of it, responding well to it, tuning in to it; the inability, therefore, to see the kind of failure that may be involved in using it badly ([1], p. 51).

8. Adventure of Perception and Agency of Care

The philosophical lines explored so far make it possible to understand a fundamental requirement of the ethics of care. Through a 'loving and attentive' and thus *caring* reading, we perceive moral situations differently and actively. This changes our perception of the responsibility of the moral agent and of agency itself. The attention to others, both enjoined and creatively explored by literature and the arts, does not give us new certainties or the literary or artistic equivalent of theories. Rather, it confronts us with uncertainty and skepticism. By focusing on a narrow conception of ethics and perception, one risks *missing the adventure*, in Diamond's phrasing. That is, one risks missing a dimension of morality, specifically the visible *aspect of* moral thinking, or "what moral life looks like" ([1], p. 36). Moreover, it is owing to a lack of care that we manage to miss this all-important aspect of ethics.

Conceptual adventure is hence a component of moral perception. There is adventure in any situation that mixes uncertainty, instability, and 'the sharp sense of life.' Diamond and Nussbaum refer to a passage from James that beautifully makes explicit this adventurous *form* that moral life takes:

> A human, personal 'adventure' is not an a priori, positive, absolute and inextinguishable thing, but just a matter of relationship and appreciation—in fact, it is a name we give, appropriately, to any passage, any situation that has added the sharp taste of uncertainty to a sharp sense of life. Hence the thing is, quite admirably, a matter of interpretation and of particular conditions; and without a perception of these, the most prodigious adventures may vulgarly count for nothing ([20], p. 307).

Famous passages in James' novel *The Ambassadors* highlight this adventure of perception. The novel's hero, the sensitive, aging, sheltered Lambert Strether, comes to acquire a new moral attitude and a "new standard of perception" in the "great swarm" of Parisian life, which turns out to be difficult, uncertain, and dangerous:

> *I see it* now, I haven't seen it enough before and now I'm old! Too old for what I see. Oh, but at least I *do* see ([44], p. 615).

These troubling moments in the novel define caring as seeing and, conversely, attentive and anticipatory perception as caring. Caring is activity, mobility, and improvisation.

> What happens to her becomes an adventure, becomes interesting, exciting, by the nature of the attention she gives it, by the intensity of her awareness, by her imaginative response. (...) The inattentive reader thus misses out doubly: he misses out on the adventure of the characters (for him, 'they count for nothing'), and he misses out on his own adventure as a reader [1].

Thus, we can see the moral life as an adventure that is both conceptual (one extends one's concepts) and sensitive (one exposes oneself). Put another way, it is both passive (one allows oneself to be transformed, to be touched) and agentive (one seeks 'an active sense of life'). There is no need to separate conceptual life and affection, just as there is no need to separate, in moral experience, thought (spontaneity) and receptivity (vulnerability to reality and to others). James adds that it is necessary that nothing escapes the attention: "*Try to be one of the people on whom nothing is lost.*" With this magisterial Jamesian insight, we have reached our final construal of the ethics of care.

In our times, it is arguably in film and television productions that the Wittgensteinian attention to detail and its proximity to care is most strongly evinced. A number of examples drawn from recent cinema describe, through the description and fine narration of caring and what the agential dimension of care is within the great diversity of forms of care. It is as if cinema, having exhausted the representations and conversations of *romance* (emblematized in the comedy of remarriage and the melodrama genres of the Golden age of Hollywood), now describes a wider variety of forms and objects of affection. We might think of the accentuation of care in disaster or science-fiction films, whose plots are often centered on the preservation or survival of a family structure (*The Day After Tomorrow*, R. Emmerich, 2004 [45]; *War of the Worlds*, S. Spielberg, 2005 [46]; *Don't Look up*, 2021 [47]). Recent TV series focus on care (*The Leftovers* [48], *This is Us* [49]) and some even present the concrete work of care: *Unbelievable* (2019) [50], *Maid* (2021 [51] Film and TV have the capacity to highlight the necessity and importance, for all humans, of this dimension of our lives [52]) —and to shape both perception and morality altogether.

Funding: This publication has received funding from the European Research Council (ERC) under the European Union's Horizon 2020 research and innovation programme (grant agreement N° 834759).

Conflicts of Interest: The author declares no conflict of interest.

References

1. Diamond, C. *The Realistic Spirit: Wittgenstein, Philosophy, and the Mind*; Bradford Books; MIT Press: Cambridge, MA, USA, 1991.
2. Wittgenstein, L. *Philosophical Investigations*; Anscombe, G.E.M., Translator; Prentice Hall: Englewood Cliffs, NJ, USA, 2000.
3. Gilligan, C. Moral Orientation and Development. In *Justice and Care*; Held, V., Ed.; Westview Press: Boulder, CO, USA, 1987.
4. Diamond, C. Moral Differences and Distances. In *Commonality and Particularity in Ethics*; Alanen, L., Ed.; Macmillan: Basingstoke, UK, 1997; pp. 197–234.
5. Wittgenstein, L. *Cambridge Lectures, 1930–1933*; Oxford University Press: Oxford, UK, 1954.
6. Wittgenstein, L. *Tractatus Logico-Philosophicus*; Ogden, C.K.; Ramsey, F.P., Translators; Routledge & Kegan Paul: London, UK, 1922.
7. Diamond, C. *Wittgenstein, Dictionnaire de Philosophie Morale*; Presses Universitaires de France: Paris, France, 1996.
8. McDowell, J. Non-Cognitivism and Rule-Following. In *Wittgenstein: To follow a Rule*; Holzman, S., Leich, C., Eds.; Routledge and Kegan: London, UK, 1981.
9. Diamond, C. Rules: Looking in the right place. In *Attention to Particulars*; Phillips, D.Z., Winch, P., Eds.; St Martin's Press: New York, NY, USA, 1989.
10. Laugier, S. *Politics of the Ordinary*; Care, Ethics, and Forms of Life; Peeters: Leuven, Belgium, 2020.
11. Putnam, H. A conversation with Jacques Bouveresse. *Monist* **2021**, *6*, 25–70.
12. Wittgenstein, L. *Bemerkungen über die Grundlagen der Mathematik*; Anscombe, G.E.M., von Wright, G., Rhees, R., Eds.; Suhrkamp: Frankfurt, Germany, 1954.
13. Wittgenstein, L. *Cambridge Lectures, 1932–1935*; Ambrose, A., Ed.; Oxford University Press: Oxford, UK, 1979.
14. Wittgenstein, L. *The Blue and Brown Books*; Rhees, R., Ed.; Blackwell: Oxford, UK, 1958.
15. Putnam, H. *Ethics Without Ontology*; Harvard University Press: Cambridge, MA, USA, 2004.
16. Murdoch, I. Vision and Choice in Morality. In *Existentialists and Mystics: Writings on Philosophy and Literature*; Murdoch, I., Conradi, P.J., Eds.; Chatto and Windus: London, UK, 1997.
17. Cavell, S. The Thought of Movies. In *Themes Out of School*; North Point Press: San Francisco, CA, USA, 1984.
18. Laugier, S. *Why We Need Ordinary Language Philosophy*; University of Chicago Press: Chicago, IL, USA, 2013.
19. Nussbaum, M. *Love's Knowledge, Essays on Philosophy and Literature*; Oxford University Press: Oxford, UK, 1990.
20. James, H. *What Maisie Knew*; Heinemann: London, UK, 1897.
21. Cavell, S. *The Claim of Reason*; Oxford University Press: Oxford, UK, 1979.
22. Ogden, C.K.; Richards, I.A. *The Meaning of Meaning*; Kegan Paul, Trench, Trubner and Co.: London, UK, 1923.
23. Cavell, S. *This New Yet Unapproachable America*; North Point Press: San Francisco, CA, USA, 1989.
24. Taylor, Charles, Action as Expression. In *Intention and Intentionality*; Diamond, C.; Teichman, J. (Eds.) The Harvester Press: Brighton, UK, 1979.
25. Lovibond, S. *Realism and Imagination in Ethics*; University of Minnesota Press: Minneapolis, MN, USA, 1983.
26. Lovibond, S. *Ethical Formation*; Harvard University Press: Cambridge, MA, USA, 2003.
27. Laugier, S. The Ethics of Care as a Politics of the Ordinary. *New Lit. Hist.* **2015**, *46*, 217–240. [CrossRef]
28. McDowell, J. *Mind and World*; Harvard University Press: Cambridge, MA, USA, 1994.
29. Cavell, S. *Must We Mean What We Say?* Cambridge University Press: Cambridge, UK, 1969.
30. Ferrarese, E.; Laugier, S. (Eds.) *Formes de Vie*; CNRS Editions: Paris, France, 2018.
31. Wittgenstein, L. *Zettel*; Anscombe, E., Von Wright, G., Eds.; Oxford University Press: Oxford, UK, 1967.
32. Wittgenstein, L. *Bemerkungen über die Philosophie der Psychologie Vol I*; Anscombe, G.E.M., von Wright, G., Eds.; Blackwell: Oxford, UK, 1980.

33. Wittgenstein, L. *Vermischte Bemerkungen (Culture and Value)*; Von Wright, G., Ed.; The University of Chicago Press: Chicago, IL, USA, 1980.
34. Wittgenstein, L. *Bemerkungen über die Philosophie der Psychologie Vol II*; Anscombe, G.E.M., von Wright, G., Eds.; Blackwell: Oxford, UK, 1980.
35. Williams, B. De la nécessité d'être sceptique. *Magazine Littéraire 361, Les Nouvelles Morales*, January 1998; p. 54.
36. Diamond, C. Ethics, Imagination and the Method of the Tractatus. In *The New Wittgenstein*; Crary, A., Read, R., Eds.; Routledge: London, UK, 2000; pp. 149–173.
37. Baier, A. *Postures of the Mind*; University of Minnesota Press: Minneapolis, MN, USA, 1985.
38. Baier, A. What do Women Want in a Moral Theory? In *Moral Prejudices*; Harvard University Press: Cambridge MA, USA, 1995.
39. Hacking, I. Winner Take Less. In *New York Review of Books*; Rea S. Hederman: New York, NY, USA, 1984; Volume 31.
40. Austin, J.L. *Philosophical Papers*; Oxford University Press: Oxford, UK, 1962.
41. Diamond, C. Losing your concepts. *Ethics* **1998**, *98*, 255–277. [CrossRef]
42. Mill, J.S. *On Liberty*; Longman, Roberts, & Green Co.: London, UK, 1869.
43. Laugier, S. Varieties of Invisibility of Carework. *Philos. Top.* **2021**, *49*, 61–79. Available online: https://muse.jhu.edu/issue/47523 (accessed on 25 July 2022).
44. James, H. The Art of Fiction. In *Theory of Fiction*; James, E., Lincoln, M., Jr., Eds.; University of Nebraska Press: Lincoln, NE, USA, 1972.
45. Emmerich, R. *The Day after Tomorrow*; 20th Century Fox: Los Angeles, CA, USA, 2004.
46. Spielberg, S. *War of the Worlds*; Paramount Pictures: Los Angeles, CA, USA, 2005.
47. McKay, A. *Don't Look Up*; Netflix: Los Gatos, CA, USA, 2021.
48. *The Leftovers*; HBO: New York, NY, USA, 2014–2017.
49. *This Is Us*; ABC: Tokyo, Japan, 2016–2022.
50. *Unbelievable*; Netflix: Los Gatos, CA, USA, 2019.
51. *Maid*; Netflix: Los Gatos, CA, USA, 2021.
52. Laugier, S. Nos Vies en Séries. In *Éthique et Politique d'une Culture Populaire*; Flammarion; Climats: Paris, France, 2019.

MDPI

St. Alban-Anlage 66

4052 Basel

Switzerland

Tel. +41 61 683 77 34

Fax +41 61 302 89 18

www.mdpi.com

Philosophies Editorial Office

E-mail: philosophies@mdpi.com

www.mdpi.com/journal/philosophies